A User's Guide to the Crisis of Civilization

A USER'S GUIDE TO THE CRISIS OF CIVILIZATION

And How to Save It

Nafeez Mosaddeq Ahmed

First published 2010 by Pluto Press
345 Archway Road, London N6 5AA

www.plutobooks.com

Distributed in the United States of America exclusively by
Palgrave Macmillan, a division of St. Martin's Press LLC,
175 Fifth Avenue, New York, NY 10010

British Library Cataloguing in Publication Data
A catalogue record for this book is available from the British Library

ISBN 978 0 7453 3054 9 Hardback
ISBN 978 0 7453 3053 2 Paperback

Library of Congress Cataloging in Publication Data applied for

This book is printed on paper suitable for recycling and made from fully managed and sustained forest sources. Logging, pulping and manufacturing processes are expected to conform to the environmental standards of the country of origin.

10 9 8 7 6 5 4 3 2

Designed and produced for Pluto Press by
Chase Publishing Services Ltd, 33 Livonia Road, Sidmouth, EX10 9JB, England
Typeset from disk by Stanford DTP Services, Northampton, England
Simultaneously printed digitally by CPI Antony Rowe, Chippenham, UK and
Edwards Bros in the United States of America

Contents

Preface and Acknowledgements

Fragments of the ideas that led me to write this book began surfacing around 2003, but I only began working on the project concertedly around 2006. Since then, it has been a case of trying desperately to keep up with the pace of events. The project is certainly ambitious – aiming to demonstrate the interconnection of global crises of ecology, energy, economy and extremism; forecast their probable converging trajectories; diagnose their systemic causes; and finally recommend a series of key structural 'reforms' that demand urgent attention.

Early on, I had wanted to write a quick, populist book summarizing key issues. But I quickly realized that my own ambitions for the project made this virtually impossible – as my research progressed, it became unequivocally clear that there was considerable confusion throughout the literature (academic and non-academic) about the subjects I was exploring. While there have been huge quantities of books published separately on different global crises, almost none attempted to understand them collectively and systemically; and those few that tried suffered from considerable deficiencies.

It became clear that a much more rigorous approach was essential. But I was keenly aware that a tome read by only a few scholars would contribute little good to the task of tackling the global crises that this book is concerned with. The book is definitely still a 'tome', but I've ensured that it remains accessible to a general audience. While not dumbing it down, I've taken pains to avoid technical jargon unless it is absolutely necessary, introduce readers to the nuances involved in interpreting the available empirical data on global crises, and walk them through the theoretical issues that arise when analyzing these crises' systemic causes and potential solutions.

The source material for the book cuts across a wide variety of disciplines in both 'harder' and 'softer' sciences, including but not limited to Climate Sciences (such as Solar Physics, Atmospheric & Earth Sciences, Oceanography, Geography); Geology; Human Ecology; Development, Monetary and Financial Economics; Security Studies; Systems Theory; International Relations Theory; and Social Theory. This is all tied together by my own distinctive theoretical approach to the study of International Relations, on the linkage between political violence and social crisis in the context of imperial social systems. I am most fortunate to have benefited immeasurably from the guidance and criticisms of an interdisciplinary panel of twelve academic peer-reviewers who provided comments on the parts of the book relevant to their specific area expertise. These are (in alphabetical order): Robert D. Crane, former Deputy Director (for Planning) of the US National Security Council; former foreign policy advisor to President Richard Nixon; co-founder of Center for Strategic and International Studies; David Cromwell, National Oceanography Centre, University of Southampton; Alan Dupont, Michael Hintze Chair of

International Security and Director, Centre for International Security Studies, University of Sydney; Daniele Ganser, Lecturer, Department of History, Basel University; former director of Secret Warfare Project and Senior Researcher, Swiss Federal Institute of Technology; Richard Heinberg, Senior Fellow-in-Residence, Post-Carbon Institute; Jon Hughes, former Deputy Editor, *The Ecologist*; Richard Levins, John Rock Professor of Population Sciences, Department of Global Health and Population, School of Public Health, Harvard University; Simon Mouatt, Senior Lecturer in Economics & Business Modelling, Southampton Business School, Southampton Solent University; Ola Tunander, Research Professor, International Peace Research Institute, Olso; Consultant for Norwegian Ministry of Foreign Affairs, Norwegian Ministry of Defence; Special Expert to Swedish government inquiry into Western covert operations; Jeff Vail, former US Department of the Interior energy infrastructure counterterrorism analyst; former US Air Force intelligence officer; Contributing Editor, *The Oil Drum*, journal of the Institute for the Study of Energy and Our Future; David Wasdell, Director of the Apollo-Gaia Project; lead scientist on feedback dynamics of complex systems for the Global System Dynamics and Policies Project of the European Commission; Paul Zarembka, Professor of Economics, State University of New York.

I cannot thank the above individuals enough for so generously dedicating their valuable time, despite their own overwhelming work schedules, to contributing their expertise on different areas of the book relevant to their knowledge base, and/or on its overall argument. In particular, I would like to thank Richard Levins and Jeff Vail for perusing almost the entirety of the draft manuscript, and providing detailed comments on most areas of the book. Their wide-ranging feedback allowed me to greatly increase the integrity of the argument. Obviously, the aforementioned bear no responsibility for any mistakes, errors or deficiencies that might remain, which are mine alone.

Others to whom thanks is due include my dedicated team at the Institute for Policy Research and Development: Farid Froghi, Toufic Machnouk, Wasim Ukra, Pedram Emrouznejad, and Aisha Dennis. Thanks also to Roger van Zwanenberg and the team at Pluto Press for their enthusiasm and professionalism in bringing this project to life. Last but not least, I must thank my wife, Akeela, for her years of endurance and support; my daughters Amina and Zainab for their endless enthusiasm; my in-laws for their patience and advice; my sisters for their unspoken patronage; and my parents for their hard work and prayers – whatever I have achieved rests on their shoulders.

Introduction

BETWEEN DANGER AND HOPE

Only 500 generations ago, hunter-gatherers began cultivating crops and forming their tiny communities into social hierarchies. Around 15 to 20 generations ago, industrial capitalism erupted on a global scale. In the last generation, the entire human species, along with virtually all other species and indeed the entire planet, have been thrown into a series of crises, which many believe threaten to converge in global catastrophe: global warming spiralling out of control; oil prices fluctuating wildly; food riots breaking out in the South; banks collapsing worldwide; the spectre of terror bombings in major cities; and the promise of 'endless war' to fight 'violent extremists' at home and abroad.

These crises continue to provoke anxiety, confusion, and questions, clear answers to which are increasingly difficult to come by – or at least that's the impression one can get from the headlines. Is climate change real, and if so, is it really caused primarily by human activities? Are we running out of oil, or is the problem simply a question of supply and demand? Are high food prices due to overpopulation and insufficient food to go round, or to the way food is distributed around the world? What does the global financial crisis mean for neoliberal capitalism – is there a viable alternative? And what, in this context, is the root cause of the rise of violent extremism and international terrorism? Is the 'War on Terror' – military interventions abroad and curtailment of civil liberties at home – the best solution?

Not only are these seemingly different crises worsening at the same time, there is also a growing sense of bewilderment about their causes and consequences. How then will these crises affect the stability of the international system? And indeed, what role has the international system played, if any, in exacerbating these crises? The argument of this study is that these crises cannot be easily abstracted from the very structure and nature of the civilization which has incubated them. The world is, therefore, in a period of momentous transition, fraught with unprecedented danger, yet simultaneously holding the prospect of unprecedented opportunity.

In terms of danger, we face the unfolding of interlocking ecological, energy and economic crises that threaten to unravel the social and political fabric of human communities everywhere. These crises are so unprecedented in their potential magnitude that without urgent preventive, mitigating and transformative action, they may threaten the survival of the human species. These threats have now become *global crises*, in that 1) their impact is now felt and experienced across the world even if their origins may have been local or regional, and 2) their impacts have reached critical points in terms of their destruction of human and natural life, potentially passing a point of no return.

But there is hope. These crises speak to the fact not merely that 'another world is possible', but that it is indeed inevitable. They demonstrate that modern industrial civilization as we know it *cannot survive in its current form* beyond the twenty-first century.[1] This means that we are in an age of transition to the dawn of a new era: the post-carbon age. This transition clearly has major ramifications for our conceptions of security, risk, and prosperity. While this is obvious cause for alarm, it is also cause for great optimism. Indeed, the most important implication concerns the question of what comes after. If global crises are symptoms of a period of civilizational transition, then the onus is on us to work actively toward developing a system that is more just, equitable and free than is currently even conceivable.

What this new post-carbon world will look like is entirely up to us. And the only way to ensure that it will be a world that fulfils humanity's highest aspirations is to revitalize our conceptions of security, risk and prosperity by accounting for the dangers and opportunities of transition. This can only be done, however, by fully understanding the root causes – *the systemic causes* – of the crises of industrial civilization. Doing so would allow us also to develop a clearer idea of how these crises are likely to play out over the coming decades: 1) how they may shape and disrupt the international system; 2) how the international system may respond to and adapt to the new conditions these crises create; and ultimately, 3) how the international system *should* change if we are to prepare more effectively and pragmatically for the post-carbon age.

This study thus aims to map out the systemic failures at the root of global crises, and forecast their probable trajectories not in isolation, but in their mutual interaction. This provides the basis for highlighting a set of core systemic solutions that we – policy-makers, scholars, activists and citizens alike – will need to explore during this uncertain period of flux.

THE INVISIBLE CRISIS

Global crises now encompass almost every sphere of human activity – social, political, economic, cultural, ethical and psychological. As these crises escalate, they are increasingly likely to aggravate one another over the coming years and decades. Expert projections suggest that the following problems – climate change, hydrocarbon energy depletion, water and resource scarcity relative to exponential population growth, declining food production, inter- and intrastate conflict, escalating impoverishment and inequalities, growing instabilities in the global economy, social malaise and declines in well-being, the legitimization of far-right politics, and normalization of political violence – do not only follow their own individual developmental trajectories, but are inherently interconnected. This means they feed back into each other in ways that are little understood and increasingly difficult to predict on the basis of conventional modelling techniques in the social and physical sciences. So there is an urgent need not simply to view global crises as individual stressors on the global system with their own dynamics, but as *interdependent* stressors each of whose dynamics will continuously and mutually influence the others.

Global crises threaten the long-term viability of industrial civilization. As the United Nations Intergovernmental Panel on Climate Change (IPCC) reported in 2007, *at current rates of global warming the earth will be uninhabitable by the end of the twenty-first century*. Yet the world faces much more than the danger of climate change. We also face an energy crisis that manifested in soaring fuel prices up to 2008, and will culminate in an oil supply crunch; a global food crisis the results of which have been increased hunger, malnutrition and impoverishment around the world, and persistently high prices; a global financial crisis in the form of the 'Credit Crunch' which some see as the precursor to a new Great Depression; an epidemic of socio-political unrest due to faith-based terrorism; intensifying intra- and international conflicts, prolonged costly military interventions and occupations in the Middle East and Central Asia; the breakdown of community cohesion, suppression of civil liberties, the rise of racial discrimination and violent crime; increased prevalence of mental illness; greater threat of virulent new strains of disease, and even the prospect of regional and global pandemics. Although security experts increasingly recognize that these diverse global crises urgently need to be factored into conventional approaches to security, social and investment policies, they continue to seriously underestimate the cumulative effects that major global trends will have over the coming years.

An obvious question that is rarely asked, for example, is why are all these crises – in the economy, in the environment, in energy, in extremism, in demography (dysfunctional population distribution and an ageing population), and in intellectual terms (reflecting the enormous gap between our scientific knowledge and the crises faced by our species) – happening at the same time? This book explores the hypothesis that these seemingly separate crises are in fact manifestations of the dysfunctional global, political, economic, ideological, and ethical system that characterizes industrial civilization *in toto*. To truly understand the unprecedented new security landscape represented by global crises requires study of the interconnected dynamic of the interlocking web of factors from which they emerge. In the era of 'globalization', it no longer makes much sense to speak of 'outsides' and 'insides'. We have to understand local events in relation to global processes – global processes that are dominated by a core network of states, corporations, and institutions. The real threat to civilization is not from outside. It is from itself.

THE FAILURE OF CIVIL SOCIETY

Part of the motivation for undertaking this study is the paucity of interdisciplinary analyses of global crises as systemic and mutually interdependent. This lacuna has undermined the ability of mainstream policy institutions to accurately predict the trajectory of these crises, and hindered the production of viable and far-reaching solutions. Overall, mainstream institutions are reluctant to recognize that the structure of contemporary civilization *is part of the problem*. The result is a security framework obsessed with *responding* to emergencies and *anticipating* contingencies, rather than advocating deep systemic reforms essential to *prevent* such emergencies and contingencies in the first place.

Since the 2008 global banking crisis, all eyes have been on the financial and economic system. Neoliberal economists had not only failed to predict the crisis, but the forms of government intervention deployed to alleviate it violated the sacred cow of the 'free market' as advocated by neoliberalism – the ideology that eschews state intervention in the economy. Yet this failure has not resulted in official recognition that neoliberal economic theory is based on faulty assumptions. The irony is that although neoliberal economics is supposed to be a science, it could not foresee the global financial crisis. This indicates the invalidity and ultimately unscientific character of conventional economic models, underscoring their incapacity to understand the global economy. Worse, precisely because of the epistemological constraints of neoliberal economic theory, the financial and economic crisis is largely not understood in the context of its relationship to other accelerating global crises such as hydrocarbon resource depletion and environmental degradation. On the contrary, neoliberal capitalism is 'naturalized' – seen as the inevitable expression of chronically self-interested human economic behaviour – and economics portrayed as a 'science' which simply captures the laws and dynamics of this behaviour.

Omitted from this view is the possibility that in fact the behaviour of consumers and producers in the world economy is not simply 'natural', but related to the historically specific socio-political constitution and structure of the international system, and that *this constitution and structure are deeply out of sync with the natural order*. Of course, acknowledging this possibility undermines the general sense that modern industrial civilization is on an inexorable path of progress, development and advancement. The idea that there is a distinctly destructive dynamic to our civilization is disconcerting. Yet the Crisis of Civilization does not necessarily mean the end of 'progress' as such; although it may well signal the end of a particular type of progress. Rather, it brings to attention the fact that our preoccupation with how 'progressive' our civilization is has cultivated a kind of civilizational apathy – a belief that we can't make our way of life any better than it already is, and a blind faith that technology will solve all of our problems.

THE STRUGGLE FOR CIVILIZATION

Since 9/11, the idea of civilization has featured most prominently in relation to the 'War on Terror', often implicitly constructed as the titanic clash of two opposing civilizations, 'Islam' and 'the West'. The 'War on Terror', in this context, is not just about defending our security, but about preserving our values and way of life. On the fifth anniversary of 9/11, then US President George W Bush said: 'This struggle has been called a clash of civilizations. In truth it is a struggle for civilization.'[2] British Prime Minister Tony Blair echoed this declaration in early 2007, when he described the 'War on Terror' as 'a clash not between civilizations', but rather 'about civilization'. It is a continuation of 'the age-old battle between progress and reaction, between those who embrace the modern world and those who reject its existence'.[3] This focus on the centrality of modern Western civilization can be traced to the work of two key policy thinkers.

The most relevant is Harvard professor Samuel Huntington – a former US government adviser – who in the late 1990s argued that international relations was best understood in terms of contestations between different civilizations, rather than between states. Huntington defined civilizations according to dominant religio-cultural ideas and values which he saw as geographically confined to specific continents: 'Nation states will remain the most powerful actors in world affairs, but the principal conflicts of global politics will occur between nations and groups of different civilizations. The clash of civilizations will dominate global politics.' In particular, Huntington predicted the likelihood of a civilizational clash between Islam and the West, and was apparently vindicated by the events of 9/11. But Huntington also emphasized that the global predominance of Western civilization was not a consequence of inherent ideological or moral superiority, but rather of 'superiority in applying organized violence'. He thus expressed scepticism about the universal acceptance of Western ideals and values.[4]

Anglo-American policymakers were also inspired by the work of Francis Fukuyama, former director of policy planning in the State Department, who in 1989 predicted 'the end of history' due to the fall of communism. According to Fukuyama, by removing the last major ideological rival on the political scene, the Soviet collapse had proven the inherent superiority of Western democratic capitalism over all historic alternatives. The Western system would thus spread throughout the world, marking 'an unabashed victory of political and economic liberalism' and thus the 'end point of mankind's ideological evolution'. The 'end of history' thesis suggested that since there were no viable alternatives to Western democratic capitalism, we had effectively reached the peak of social and political progress. Nearly twenty years later, as a longtime professor at Johns Hopkins University, Fukuyama conceded that he no longer saw the emergence of a democratic capitalist world order as inevitable. But he insisted on the intrinsic superiority of liberal capitalist democracy as the optimum mode of human social organization, to which there is no viable ideological rival. Communism is dead, and radical Islamism is primarily a byproduct of the alienation generated 'when traditional cultural identities are disrupted by modernization and a pluralistic democratic order that creates a disjuncture between one's inner self and external social practice'.[5]

Anglo-American policymakers have generally fused the streams of thinking associated with Huntington and Fukuyama in formulating their approaches to the 'War on Terror', but they have failed to ask the right questions – unsettling questions that strike at the heart of their underlying understanding of modern civilization.[6] *Should* we be satisfied with the current political and economic order *as it is*? Is democracy in its current form *good enough* to ensure popular political participation? Is capitalism the *best* we can do to sustain high levels of material well-being? Are global crises a result of the relative *absence* of democracy and capitalism from certain parts of the world, and thus do we need *more* democratic capitalism to solve our growing social problems?

In the era of 'globalization', a single transnational economic order has left almost no corner of the earth unpenetrated. Local events cannot be understood in isolation from global processes, which are dominated by a core group of states, corporations, and institutions based largely in the West, or, perhaps more precisely, the North. Thus, to speak of mutually exclusive civilizations within this global system, contending for politico-economic supremacy, is vastly oversimplifying. It avoids the key question of how these contending forces should be understood in the context of the overarching *global system* that incubates them both.

Bringing in this wider *global systemic* context immediately shifts our focal point to the intensifying and converging trajectories of global ecological, energy and economic crises. This context also allows us to raise hitherto inconceivable questions, not merely about the global systemic *origins* of these crises, but also about their probable global systemic *impacts* over time, and the question of what global systemic *solutions* can be implemented to resolve them – and thus about the integrity, viability and continuity of our civilization.

THE CORE ARGUMENT

This book provides an integrated, interdisciplinary reassessment of our current global predicament. It is an empirically driven analysis of global crises, developing a body of data from which a reinvigorated human-centred global vision for security through civilizational renewal can be developed, and through which can be revealed the myriad points of interconnection, so often missed by conventional security experts, between different global crises. It proceeds by reviewing the complex systemic interrelationships between global crises, explaining their shared trajectories, and developing a single qualitative map by which to chart their mutual convergence over the coming decades. It makes the following key sub-arguments:

1. Global crises are not aberrations from an optimized global system which require only minor adjustments to policy; they are *integral* to the ideology, structure and logic of the global political economy.
2. Therefore, global crises cannot be solved solely by such minor or even major policy reforms – but only by drastic reconfiguration of the *system itself*. Failure to achieve this will mean we are unable to curtail the escalation of crises.
3. Conventional expert projections on the impact of global crises on the political, economic, and ecological continuity of our civilization are flawed due to their view of these crises as separate, distinctive processes. They must be understood holistically, intertwined in their causes and hence interrelated in their dynamics.

Against this background, an urgent scenario emerges for global business-as-usual – one in which multiple, global crises systematically *converge* via interdependent positive feedback mechanisms, greatly accelerating the pace of global systemic disruption, as well as exacerbating and magnifying the impact of specific crises throughout the system. This leads to the conceptualization of global crises as symptoms of *systemic transition* representing both danger and opportunity: either the danger of collapse and the emergence of regressive social forms; or the unprecedented opportunity to grasp the reins of history, mobilizing the best of human culture, values and ideas to move toward an enlightened civilizational revival. Indeed, the notion that the very danger of societal collapse means that we are potentially on the threshold of a positive renewal of society is increasingly supported by experts who have studied the rise and fall of human civilizations throughout history.

Thus, we should not look upon these crises with paralyzing horror, but rather with the recognition that they signify the *inevitability* of civilizational transition. Indeed, the global crises examined here demonstrate the inevitability of two world events before the end of this century: 1) the end of industrial civilization as we know it; and 2) the coming of a post-carbon society. Arriving after the industrial era, post-carbon civilization will be capable of harnessing sciences, technologies, values and cultures made possible only because of the advances of that previous period. For the first time, all of world history and culture is at our fingertips. The resources and lessons of the long history of human civilization are at our disposal. Thus, post-carbon civilization signifies not a step backwards, but a step forwards, using industrialization as a stepping stone to achieve the goal of creating truly sustainable, harmonious and prosperous societies that previously were considered beyond reach.

Acceptance of the inevitability of these world-historical events allows us to recognize the need to prepare ourselves and our societies – philosophically, culturally, ethically, politically, economically, and technologically – for what can be understood as the coming *Post-Carbon Revolution*. This study thus aims to interrogate the features of industrial civilization bound up with these global crises, in order to identify the key issues that we will need to explore in preparing for post-carbon civilizational revival.

Theoretical and Methodological Approach

This argument is held together by a core theoretical argument and a specific empirical methodology. Mobilizing a Political Marxist framework for the analysis of imperial social systems,[7] the book advances the following overarching theoretical case: global crises are generated directly by the operation and structure of the global system. The social form of this global system is neoliberal capitalism, currently centred around Anglo-American power (premised on monopoly ownership of the world's productive resources by core Northern states, primarily the US, Britain, Western Europe, Japan, China, Russia, Australia). This social form is currently co-extensive with a political structure (the sovereign states-system and US-dominated multilateral global governance institutions) by which it is regulated, made consensual through an implicit ideology/worldview (a form of ultra-materialism premised on dividing and reducing nature and life to material quantities and values) and a corresponding value system (which places excessive 'value' on unlimited material consumption and production and thus rationalizes capitalism's drive for endless growth). Global crises are, therefore, integral manifestations of the very nature of the political, economic, ideological and ethical structures of contemporary industrial civilization. They are, in other words, different faces of a single Crisis of Civilization. Hence, they cannot be averted until it is recognized that these very structures themselves need to be transformed. Failure to do so will mean that efforts to address global crises will continue to be futile over the long term.

The bulk of the book interrogates this *problematique* through a primarily empirical analysis, which provides the data through which the theoretical diagnosis of our global predicament is developed. This analysis focuses on six global crises (identified below), each of which has a whole chapter devoted to it. These chapters proceed by reviewing the data and analysis provided by leading experts in the relevant fields, and by identifying their key issues of disagreement, with a view to navigate the arguments and counter-arguments in order to discern a possible resolution. In most cases, one finds that there is either an overwhelming or an emerging consensus among experts – this consensus, however, is not accepted at face value, but tested against the strength of counter-arguments. Further, a key objective is to not simply look at these crises in isolation, but in their mutual interdependence. This is achieved by way of a simple empirical procedure: while most of each chapter is devoted to exploring the relevant data and resolving the internal debates within the field, toward the end of the chapter the discussion turns directly to consideration of the *causal* and *consequential* connections between the crisis under focus and the five other global crises discussed in more detail in their own chapters.

STRUCTURE OF ARGUMENT

This book identifies and reviews trends in and across six specific global crises. It begins with a discussion of: 1) *climate change*; 2) *energy scarcity*; 3) *food insecurity*; and 4) *economic instability*. Against this background, it critically examines 5) the political economy of *international terrorism* and its direct relationship to global crises. The book then assesses the character and efficacy of the state-security response in terms of 6) the tendency toward *militarization* in the domestic and foreign policies of Western societies. Of course, these are by no means the only crises we face, but their sheer number and magnitude necessitate the focus on those which appear to be most fundamental in terms of causation. Two others that this book is unable to explore in detail, which are briefly dealt with where relevant, are worth highlighting here: *demography*, not simply in terms of population growth, but in terms of its uneven character in the form of massive centralization of populations in urban regions, over-exploitation of natural resources and mass displacement in the context of concomitant environmental catastrophes and social conflicts, a 'youth bulge' linked to chronic unemployment and poverty in regions of scarce resources, combined with an unsustainable expanding elderly population in the North relative to too few economically active young people; and *epidemiology*, in terms of the emergence of new and increasingly virulent diseases – such as avian flu and swine flu – with increasingly deadly consequences, facilitated by the conditions of industrial society such as agricultural techniques and long-distance transport. There are other crises still, such as regional and global water shortages, but arguably the six global crises emphasized in this book are largely causally prior to these secondary crises, which can be understood in many ways as symptomatic, themselves interdependent offshoots of global systemic dysfunction.

On this basis, the book provides a sustained theoretical critique of the structure of neoliberal capitalism. Finally, the conclusions outline the contours of the sort of wide-ranging social-structural transformations that will be necessary to sustain a viable form of civilization capable of surviving the impact of these crises and functioning in harmony with the natural world.

Climate Catastrophe

This chapter begins by reviewing the public media debate over whether climate change is actually occurring, and if so, whether it is anthropogenic (caused by human activities) or natural. Drawing on relevant scientific literature and discussions, I explore in particular 'sceptic' arguments trying to explain away climate change in the context of solar activity, natural climate fluctuations, and/or global cooling. I find not only that the 'sceptic' arguments are overwhelmingly inaccurate, but that there is a growing body of scientific evidence indicating that conventional assumptions about the pace and impact of global warming are likely to be very conservative.

At current rates of increase of fossil fuel emissions, global warming is well on the way to surpassing as much as 4° Celsius (C), by mid century. The import is that we are entering a new era of accelerated warming, potentially triggering rapid and abrupt climate change that could lead to irreversible conditions on earth which, at worst, could precipitate runaway increases in temperature making the planet extremely hostile for all species of life. New evidence shows that previous assessments for the atmospheric concentration of greenhouse gases underestimated the extent of the problem. Recent studies confirm not only that the safe level of atmospheric concentration of carbon dioxide (CO_2) should be revised downwards to below 350 parts per million (ppm), but that the current level of concentration is already over 386 ppm, and that the current level of equivalent concentration of CO_2 (which includes other dangerous greenhouse gases like methane) is 445 ppm. This level guarantees a rise in global average temperature of at least 2°C, beyond which the world enters an era of potentially sudden and dangerous climate scenarios. Scientific assessments increasingly confirm that the window of opportunity to prevent dangerous climate change will be closed by 2012 at latest. Beyond that point, it is likely that there will be far-reaching consequences that cannot be accurately forecasted. However, at current rates of increase of fossil fuel emissions, a global mean temperature rise of *at least* 4°C this century is guaranteed, and a rise of 6°C is highly probable, without mitigating action. The chapter then explores the relationship of climate change to hydocarbon over-dependence, as well as the consequences of climate change in terms of geopolitical instability, food production, escalating costs to national economies, the magnification of security threats, and the tendency of states to respond to the issue not by addressing root causes, but by consolidating state power. This tendency is further identified as being related directly to the structure of neoliberal capitalism, with its inability to recognize long-term human costs as opposed to short-term profits, and its reliance on the coercive powers of the state to mobilize against resistance to neoliberal policies.

Energy Scarcity

Despite the grave implications of anthropogenic climate change and the implied necessity for industrial civilization to transfer to more efficient, sustainable, and clean forms of renewable energy, industrial civilization remains firmly on a course of escalating exploitation of fossil fuels. This chapter aims to tackle the hypothesis that we face an increasing scarcity of cheap energy in the form of hydrocarbon resources easily accessible through known technologies, in particular the concept that we are near (or have passed) the peak of world oil production. I critically review the official position of most governments and the energy industry, which is that the world retains abundant hydrocarbon resources to sustain continued economic growth for the foreseeable future. I find that this position is contradicted by credible independent studies showing that world oil production most probably peaked in 2006. This provides a primary geophysical explanation for oil price hikes, whose height in early 2008 was also due to the role of unrestricted financial speculation. Producing oil from unconventional sources, such as tar sand and oil shale, is not only more destructive environmentally than conventional oil, but is also incapable of contributing meaningfully to world demand. World production of natural gas and coal, along with exploitation of world uranium reserves, will all peak and decline before mid century, even as mushrooming populations and swelling industrial economies generate increasing demand. I further explore the interlinkage between peak oil, world economic recession, and fluctuating oil prices.

In this respect, the chapter further interrogates how excessive hydrocarbon exploitation driven by the profit imperatives of neoliberal capitalism is undermining the global system's own resource base, yet systematically sidelining viable efforts to pursue more decentralized renewable energies, primarily because the very structure of power in the global system is inextricably conjoined with military-backed geopolitical domination of territorially demarcated hydrocarbon energy deposits in strategic regions like the Middle East and Central Asia. Yet the long-term outlook is that growing hydrocarbon energy scarcities could lead to increasingly volatile, and eventually permanently high, oil prices, with devastating consequences for the global economy. This in turn would create conditions conducive to 'energy wars' - violent geopolitical competitions for the domination of strategic hydrocarbon resources.

Food Insecurity

This chapter attempts to explore the origins of the global food crisis in the context of the end of the age of cheap oil, and our entry into a new age of scarce hydrocarbon energy resources, which I argue had an immediate impact on world food prices. These price hikes were not only due to rises in the cost of agricultural production and long-distance transport, but also due to the ecological limits of transnational agribusiness and industrial agricultural methods. Despite most of the earth's fertile land being devoted to agriculture, year by year world grain shortfalls show that global agribusiness is reaching the limits of its productivity, exhausting the soil, and encroaching unsustainably on local environmental resources. Climate change is exacerbating these conditions by already leading to droughts in many areas, provoking unprecedented crop failures; and US investment in biofuels as an alternative energy source is worsening the situation by reallocating agricultural land from production for food to production for energy. The unequal structure of global food distribution means the immediate losers from this process are the less developed countries of the South, leading to civil unrest and rioting. Without a change of course, it is only a matter of time before this experience extends to the countries of the North, and in some cases it arguably already has.

While revealing the unequal structure of global corporate agribusiness and its complicity in generating mass hunger and food dependency for large swathes of the South, the chapter moves on to examine how these circumstances are being continuously degraded and exacerbated not only by the debilitating impact of industrial farming techniques which ravage the environment and denude the soil, but also by the impact of climate change and energy depletion. While climate change is changing weather patterns, dramatically leading to droughts and thus large-scale harvest failures, oil depletion is set to permanently undermine industrial forms of agriculture well within the next decade, with dramatic economic and social costs. Mass migrations to escape drought and starvation may result across areas of the South in particular, magnifying geopolitical insecurities and exacerbating Northern attempts to 'wall off' asylum seekers and foreigners.

Economic Instability

High oil prices in the context of peak oil not only triggered food price hikes, but were also a major spark for the onset of global economic recession in general, of which food insecurity is merely one dimension. This took place in a global political economy whose structure is built not only on the systematic generation of massive global inequality through the exacerbation of Southern impoverishment and Northern overconsumption, but also on the creation of profit through the systematization of debt. Part of this process in the US involved unsustainable debt-driven booms in consumer and real estate markets due to predatory lending practices by banks and financial services firms. Such practices were not merely erroneous or the outcome of sheer greed, but were actually rational responses to systemic pressures generated by the structure of neoliberal capitalism (discussed in chapter seven).

Throughout the period from 2000 to 2008, leading economists and financial institutions issued warnings of an impending global financial crisis that would begin with the collapse of housing markets. Governments not only ignored these warnings, they encouraged the predatory and risk-accumulating strategies of investors and speculators. While the credit bubble was unsustainable and would have had to burst at some point, the trigger came in the form of inflationary pressures induced by the excessive post-peak oil prices, raising the cost of living for the previously stable middle classes to exorbitant levels that made it impossible to meet mortgage obligations. The spate of defaults that became the subprime mortgage crisis triggered the unravelling of the bubble of bad debt which had made possible the preceding years of growth. Critically, the neoliberal 'Washington Consensus' proved not only powerless to prevent the crisis, but was in fact among the key promoters of risk-generation and debt-proliferation that created it. The economic events of 2008 were, however, only the first stages of a deepening recession and eventual depression, involving the dramatic deflation of the real economy and temporarily bringing fuel prices down again. The next decade will be a period of fluctuating prices as the economy contracts, then grows again, before permanently breaching oil capacity limits – an event with grave and irreversible consequences that will signal the danger of permanent economic contraction, amplifying the erosion of social cohesion and the normalization of political violence.

International Terrorism

Against this convergence of global ecological, energy and economic crises, endangering the very continuity of industrial civilization, the threat of 'international terrorism' from predominantly Islamist extremist networks is marginal. However, in terms of expenditure, Western states have elevated terrorism to a major priority compared with other global crises. While hundreds of billions of pounds, if not trillions, are being expended on military expansion in strategic regions, the far more serious threat to security posed by global crises has been virtually ignored. This contradiction is no accident, but rather a fundamental feature of the global political economy.

Like climate change and peak oil, the globalization of international terrorism is a direct consequence of Western states' *over-dependence* on oil – to the extent that reducing our reliance on oil is a structural impossibility unless the global political economy itself is radically transformed. Extensive historical and empirical evidence confirms that al-Qaeda terrorist networks are covertly sponsored by several key Muslim states in the Middle East and Central Asia, such as Saudi Arabia and the Gulf states, Pakistan, Algeria, and Azerbaijan, among others. Yet these regimes, which thus constitute the locus of al-Qaeda's operational capabilities, are financially and militarily sponsored by the West, largely due to their function as major oil exporters. Saudi Arabia and the Gulf states for example, containing the world's largest reserves of oil and gas, are powerful client states of the US, UK, Western Europe and Japan. The United States and Britain have failed to shut down the financial arteries of international terrorism in the Gulf states due to their pivotal geostrategic significance with respect to Western energy security, a relationship that is further entrenched due to the cross-cutting nature of financial investments between the Gulf states and the West.

The other element to this relationship is the direct mobilization of al-Qaeda terrorist networks by Western intelligence services as a mercenary proxy force to secure access to strategic regions. Although it is conventionally believed that Osama bin Laden was supported by the West solely during the Cold War, to repel the Soviet occupation of Afghanistan, in fact Western intelligence services have continued to selectively sponsor al-Qaeda networks from the end of the Cold War until now to control particular strategic energy reserves.

My analysis focuses on the material infrastructure of international terrorism related to Islamist extremism, and asks the question: what are the financial and political conditions that have facilitated the emergence and expansion of a marginal sectarian (and arguably false) interpretation of Islam into a concrete political ideology mobilizing in multiple regions? The answer, partly, is that international terrorism is a consequence of a specific structural feature of the global political economy – its over-dependence on oil and gas – and the consequent financial and geostrategic entanglement of Western interests and investments with those of client regimes in the Middle East and Central Asia. The convergence of global ecological, energy and economic crises will increasingly fuel terrorism, both as a tool of covert action and as a resort of disenfranchized communities, ramping up social anxieties and their accrual into the demonization of 'Others'.

The Militarization Tendency

Western state responses to this convergence of global crises continue to be based on traditional premises of protecting existing structures of political and economic power. Yet this has had a debilitating impact on the formulation of domestic and foreign policies. Abroad, the pattern of the 'War on Terror' has projected Anglo-American power into the world's most strategic energy reserves in the Middle East, Central Asia, and North Africa, which happen to be Muslim-majority regions – a regressive intensification of longstanding policies dating back to the Cold War and earlier. Simultaneously, it has brought pervasive regimes of comprehensive state surveillance into the domestic arena, legitimizing massive discriminatory policing of Muslim communities within the West. Correspondingly, new security frameworks attempting to address the impact of global crises, rather than addressing their underlying systemic causes, in effect append themselves to the concerns of the 'War on Terror' by focusing on these crises merely as 'threat-multipliers' of quite traditional security concerns.

The result is the emergence of an increasingly draconian and interventionist security paradigm concerned overwhelmingly with the task of domestic and foreign population control in the context of uniquely hostile ecological, environmental and socio-political conditions in strategic regions. Lacking a deeper theoretical interdisciplinary basis capable of interrogating the ideology, value system and politico-economic structure of the global system, this enhanced security paradigm effectively elides even a slim possibility of averting global crises. Yet this trend – cutting across both Republican and Democratic political parties in the US, and Labour and Conservative parties in the UK (and their equivalents in Western Europe) – sets off alarm bells. Massive social, political and economic crises in the past have often precipitated the increasing power of right-wing politics, the exclusion and labelling of particular minority groups as responsible for the crisis, followed by the escalation of murderous violence toward them which often becomes genocidal. The danger is that the 'securitization' of global crises along traditional lines may increase the probability of international conflict, and even genocidal violence, as strategies of state consolidation.

Final Analysis – the Global Political Economy

On the basis of these six largely empirical chapters, the final chapter articulates a conceptual framework for diagnosing their findings. This is the most difficult chapter in the book, and non-academic readers may find this section challenging due to its theoretical nature. However, a theoretical detour is essential if one is to understand these crises as factors within a *global system* – where system refers to a network of elements or processes that influence and transform each other through their interrelations, creating an overall dynamic that can only be understood by examining these elements or processes in the context of the complex whole of which they are part.[8] This necessarily requires tightening up our conceptual apparatus and approach. The theoretical framework outlined here is based on an integrated conceptualization of ideological, economic, military and political forms of social power as inherently interdependent and mutually constituted in any given social system, but varying in conjunction with their overarching social form (that is, the forms by which a society exploits natural resources and converts them into products for distribution and consumption), identified in terms of historically specific *social property relations* – a concept designed to convey the fundamental *power relations*, exemplified through *ownership structures*, in which all socio-economic transactions and exchanges are embedded.

This theoretical framework is applied to the contemporary world system, situated as the culmination of a long history of European colonization and imperialism, and since the eighteenth century consisting of a capitalist world system mutually constituted by a set of geopolitical, ideological and ethical structures. This allows me to develop a holistic account of the causes of global crises in the context of the structure of this system, aimed at identifying the crisis tendencies inherent to the structure of neoliberal capitalism. This argument touches on, but goes beyond, the traditional Marxist concern with financial instability and the crisis of over-accumulation, to examine the interrelationship between capitalist 'social property relations', ecological degradation and over-exploitation of natural resources.

It is impossible to understand these crisis tendencies without accounting for the structural contours of capitalism as a fundamental relation of production. Equally, these crisis tendencies are exacerbated in the context of ongoing interstate rivalry over the international regulation of the capitalist system through global governance institutions, and further intensified when such rivalry leads to geopolitical competition for the domination of regions viewed as strategically critical for the functioning of the capitalist world system (due to their geography, resources and population densities, and so on). Such geopolitical manoeuvering is legitimized for domestic populations by the 'War on Terror' – an increasingly polarizing security narrative that conceals the instrumental role of global crises in accelerating geopolitical rivalry, and thus justifies the increasing turn of 'liberal' societies to illiberal, extra-legal mechanisms of control based on military and police power.

What makes opposition to this fast imploding system of self-destruction even more difficult is that the most powerful agents in the system – governments, banks, corporations and so on – both believe and actively seek to convince themselves and everyone else that it is the most progressive and advanced mode of existence possible for humanity, and therefore that its devastating consequences are merely natural, unavoidable but ultimately fixable aberrations from an optimal way of life that need not and must not change. Underpinning this ideology of self-justification is a complex, eclectic and evolving array of reductionist and crude materialist philosophical ideas and moral values about humanity's alleged place in the world, and what kinds of activities are 'worth' doing.

In the Diagnosis, the implications of this theoretical analysis are set out by exploring the inherent structural fragilities in industrial civilization and the form of neoliberal capitalism which has grown around it. The era of seemingly unlimited growth – premised on untrammelled exploitation of the earth's resources for the maximization of an elite minority's profits – will near its end over the coming decades. While the passage to viable post-carbon social forms will be fraught with danger and uncertainty, it is the only path available that can hope to guarantee the continuity of human civilization, well-being and prosperity. The post-carbon age is beginning.

This background allows me in the concluding Prognosis to develop the outlines of a comprehensive vision for a radically different kind of civilization. This vision seeks to identify the kinds of issues, concepts and ultimately *key structural reforms* that we will urgently need to explore and discuss if we are to overcome the Crisis of Civilization, and ease the passage to post-carbon life. To name a few, these must include widening access and control over productive resources; greater community-led governance; mechanisms for more equal wealth distribution; monetary reform based on the abolition of interest; large-scale community-level investment in decentralized renewable energy technologies; smaller, localized organic agricultural enterprises; and a deep-seated, scientifically grounded yet non-reductionist ideological and cultural re-evaluation of the human condition, among many others.

1
Climate Catastrophe

What is climate change? Is it a product of natural cyclical variations in the earth's ecological systems, or is it a consequence of human activities? What are the implications of climate change for the international system? How serious are the ramifications of climate change for the continuity of modern industrial civilization? This chapter begins by confronting the major public media debates regarding the causes of climate change, reviewing the main arguments against the idea that contemporary global warming is due to fossil fuel emissions and is therefore human-induced (anthropogenic). The relevant scientific literature is explored to discern whether we can be sure if, and why, climate change is happening.

I then explore the implications of climate change for national security, finding that a variety of Western security agencies recognize that it will drastically alter the global security landscape for the foreseeable future unless there is significant preventive action. The focus of this analysis is not to list the specific conflicts that might arise (an exercise performed frequently elsewhere),[1] but to assess the overarching ramifications of global warming for the *ability of modern industrial civilization in its current form to survive*. The analysis then extends to a critical examination of the conventional narrative of global warming's progress as described by the United Nations Intergovernmental Panel on Climate Change (IPCC), and as generally endorsed by Western states. I argue that cutting-edge scientific research provides compelling evidence that the current rate of global warming is far greater than the UN models predicted. Integrating the impact of positive feedbacks in the earth's climate systems, the research suggests the probability of a worst-case climate scenario well before the end of the twenty-first century – unless significant preventive and mitigating actions are taken.

But such actions must go far beyond the mere question of reducing emissions. Emissions reductions have largely been addressed as if in a socio-political and economic vacuum, divorced from the real-world systemic changes required to drastically reduce energy consumption in general, and utilize cleaner and more energy-efficient technologies based on renewable fuel sources in particular. Yet this inattention to the global systemic origins of the ecological crisis is part of a long-term trend, evidenced by the fact that policymakers have largely ignored several decades of dire warnings issued by the world's leading climate and environmental scientists. Therefore, for civilization to survive beyond the twenty-first century will require fundamental *global systemic change* at the very heart of modern industrial social relations. Only in the context of such systemic change can the prospect be realized of a post-carbon civilization that is no longer dependent on the unrelenting exploitation of hydrocarbon energies.

A DEBATE RESOLVED? CURRENT CLIMATE CHANGE IS UNEQUIVOCALLY ANTHROPOGENIC

The Scientific Consensus

Climate change generated by human emissions is perhaps the global crisis most prominent in public consciousness – its existence is now readily acknowledged by most governments including the United States, even if reluctantly, and it is generally recognized that urgent steps are required to prevent the prospect of mass extinction. What is missing from the official discourse on climate, however, is not simply an acknowledgement of the real extent and gravity of the civilizational catastrophe it poses, but the corresponding measures required to prevent or avert such catastrophe.

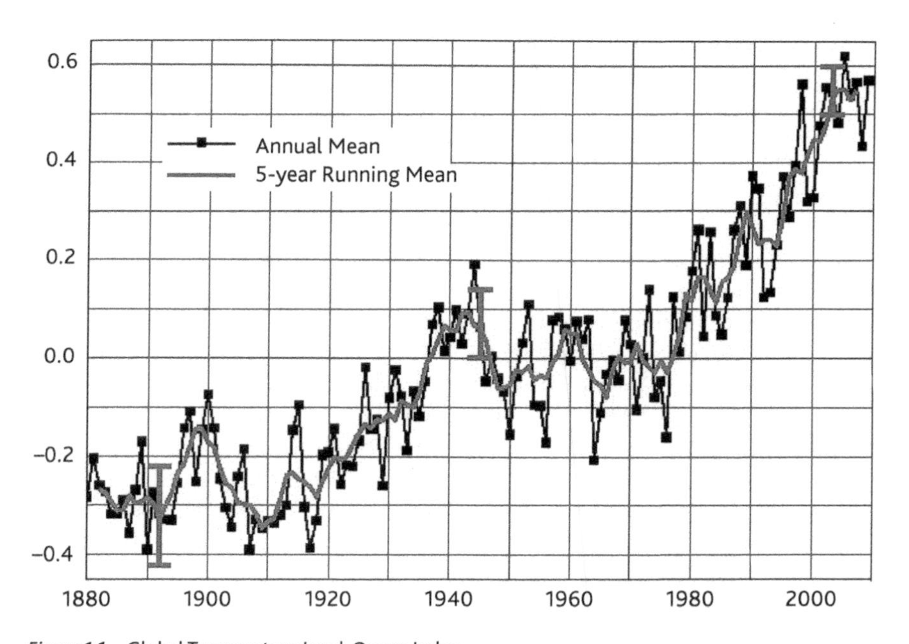

Figure 1.1 Global Temperature Land-Ocean Index.

Source: NASA Goddard Institute for Space Studies (11 January 2008)

Since 1900 there has been an approximately 0.7°C rise in global average temperature (see Figure 1.1). This increase cannot be accounted for by natural variations of solar and volcanic activity, nor by human-induced sulphate emissions, which act to reduce global temperature. It is only by including the impact of human-induced carbon dioxide (CO_2) emissions that climate models are able to accurately simulate the rise in global temperatures over the last century of industrial civilization.

Industrial civilization derives almost all its energy from the burning of fossil fuels, pumping carbon dioxide into the atmosphere – with the exception of approximately 2–3 per cent from renewable and nuclear sources. The emissions of primarily CO_2 – but also nitrous oxide, methane, and chlorofluorocarbons, among other greenhouse gases – from the industries that drive our economies and sustain our infrastructures, are the main engine of global warming in the last few decades. This does not mean that all climate change is solely due to our CO_2 emissions. Scientists acknowledge that there are many other factors involved in climate change, such as solar activity, as well as periodic changes in the earth's orbit. Yet they have overwhelmingly confirmed that these are not the primary factors currently driving global warming.

Global warming sceptics often point to the fact that human-induced CO_2 emissions are tiny compared to natural emissions from the ocean and vegetation. What they forget, however, is that natural emissions are balanced by natural absorptions by ocean and vegetation. This natural balance has become increasingly unstable due to additional CO_2 emissions from industrial activities. In terms of natural emissions, consumption of vegetation by animals and microbes accounts for about 220 gigatonnes (Gt) of CO_2 per year. Respiration by vegetation emits around 220 (Gt). The ocean releases about 330 Gt. This totals about 770 Gt of natural *emissions*. In terms of natural *absorptions*, land plants absorb about 440 Gt of carbon per year, and the ocean absorbs about 330 Gt, again roughly totalling about 770 Gt. This emission-absorption parity (770 Gt released and 770 absorbed) ensures that natural atmospheric CO_2 levels remain in overall balance even as emissions and absorptions fluctuate over time. In comparison, human emissions are only around 26.4 Gt per year. The problem is that this seemingly small addition of CO_2 into the atmosphere by industrialization *cannot be absorbed by the planet*. Only about 40 per cent of it is actually absorbed, largely by oceans, leaving 60 per cent in the atmosphere. Worse still, the oceans are increasingly losing their ability to absorb CO_2, with the Southern Ocean and the North Atlantic both approaching saturation point in 2007. This means that, with time, unprecedented concentrations of CO_2 are accumulating in the earth's atmosphere. Just how unprecedented can be gauged by a simple example – while a natural change of 100 parts per million (ppm) takes between 5000 and 20,000 years, the recent increase of 100 ppm took only 120 years.[2]

The majority of scientific studies show that climate sensitivity to CO_2 emissions is high, or, in other words, that CO_2 emissions induce large increases in global temperature. Despite the media images of a raging debate among climate scientists, the fundamentals are agreed upon – the direct connection between CO_2 and global temperatures has been empirically observed by analysis of ice cores, paleoclimate records, observations of ocean heat uptake, and temperature responses to the solar cycle, among other data. The empirically focused studies, including published research from the 1990s to 2009, show that doubled CO_2 emissions would contribute to warming within the range of at least 1.4 to 4°C.[3] Furthermore, the link between CO_2 and warming is confirmed by fundamental physics, including laboratory analysis on the degree to which CO_2 and other greenhouses gases absorb infrared, as documented by University of Chicago geoscientist Dr. Ray Pierrehumbert in his physics textbook on climate change.[4]

The origins of current climate change are therefore no longer a matter of serious scientific debate. The landmark declaration came in 2007, when the United Nations Intergovernmental Panel on Climate Change (IPCC) published its Fourth Assessment Report, based on a meta-analysis of the scientific literature by 600 scientists from 40 countries, peer-reviewed by 600 more meteorologists. The report confirmed that human-induced global warming is 'unequivocally' happening, and that the probability that climate change is due to human CO_2 emissions is over 90 per cent.[5]

Yet the waters have been increasingly muddied by the perception that there is no real scientific consensus about climate change – either that global warming is not happening, or that if it is, it has little or nothing to do with human activities. In fact, this self-styled 'sceptical' agenda has revolved around a network of ideological and advocacy organizations funded largely by leading players in the fossil fuel industry. Between 1998 and 2005, ExxonMobil has funnelled about $16 million to such groups with the aim of manufacturing uncertainty about even the most indisputable scientific evidence. This has not only generated considerable confusion in the media about climate change, it has also influenced US government policy.[6]

It is therefore important to recognize that claims by sceptics that there is no scientific consensus on climate change are deeply misleading. The scientific consensus can be discerned not only from the IPCC report, but from other meta-analyses of the peer-reviewed literature. In 2004, US geoscientist Naomi Oreskes, Professor of History and Science Studies at the University of California, San Diego, conducted a survey of the 928 peer-reviewed scientific papers on global climate change from 1993 to 2003. She found that 75 per cent explicitly or implicitly accepted the consensus view, while 25 per cent took no position and dealt purely with methods or paleoclimate:

> Remarkably, none of the papers disagreed with the consensus position... Many details about climate interactions are not well understood, and there are ample grounds for continued research to provide a better basis for understanding climate dynamics. The question of what to do about climate change is also still open. But there is a scientific consensus on the reality of anthropogenic climate change. Climate scientists have repeatedly tried to make this clear. It is time for the rest of us to listen.[7]

Efforts to disprove the existence of this scientific consensus have been poor in quality. For instance, although social anthropologist Benny Peiser attempted to refute Oreskes' findings in his own survey of the same peer-reviewed papers, he managed to flag up only 34 studies which he claimed raised doubts about anthropogenic global warming. This is a tiny fraction – only 3.6 per cent – of the scientific papers from this period. But close inspection of the actual abstracts shows not only that the vast majority do not in fact reject the scientific consensus at all, and that those few which can be interpreted as casting some doubt were not actually peer-reviewed.[8] In the end, Peiser himself was forced to retract his criticisms: 'Only few abstracts explicitly reject or doubt the AGW (anthropogenic global warming) consensus which is why I have publicly withdrawn this point of my critique... I do not think anyone is questioning that we are in a period of global warming. Neither do I doubt that the overwhelming majority of climatologists is agreed that the current warming period is mostly due to human impact.'[9] Indeed, when pressed to clarify which specific papers he thought expressed doubt about anthropogenic climate change, he was able only to identify one – which was not peer-reviewed.[10]

Outside the realm of scientific research, there have been several efforts by vested political interests to demonstrate not only that there is a lack of scientific consensus about anthropogenic climate change, but further that an alternative scientific consensus undermines it. In December 2007, Senator James Inhofe, the ranking minority member of the US Senate Committee on Environment and Public Works, released a list of over 400 'prominent scientists' including 'current and former participants in the UN IPCC' who allegedly 'disputed man-made global warming claims' that year.[11] His list was widely publicized by the media. Yet by the time of writing Senator Inhofe has received at least a million dollars in campaign contributions from individuals and companies linked to the US oil and gas industry.[12] Detailed analysis of Inhofe's list of scientists and their actual research on climate change reveals other awkward facts: 1) 84 individuals listed had either taken money from, or were connected to, fossil fuel industries or think tanks founded by them; 2) 44 are television weathermen; 3) 20 are economists; 4) 70 simply have no expertise or qualifications in climate science; 5) increasing numbers of scientists cited as 'man-made climate sceptics' in the Senate report have since been found to support anthropogenic climate change and despite repeated efforts to dissociate themselves from the report, continue to remain on the list.[13]

Examples of flagrant misrepresentation in the report are rife. On the list, for instance, is 'prominent scientist' Ray Kurzweil – not a scientist but an inventor. Worse, Kurzweil is not even a global warming sceptic. Rather, he argued that Al Gore's arguments about climate change were 'ludicrous' for failing to account for the potential of new technologies: 'nanotechnology will eliminate the need for fossil fuels within 20 years ... I think global warming is real but it has been modest thus far'.[14] Kurzweil, in other words, is not a climate scientist, accepts the climate science behind global warming, but believes that continued warming is preventable due to technological progress. Another example of a 'prominent scientist' cited in the report is Steve Baskerville, a 'CBS Chicago affiliate' and 'Chief Meteorologist' who expressed 'scepticism' about a consensus on man-made global warming. Yet Baskerville's alleged qualifications in climate science amount only to a Certificate in Broadcast Meteorology from Mississippi State University.[15] Other examples include

Thomas Ring, who has a degree in chemical engineering from Case Western Reserve University, with no peer-reviewed climate science publications to his name; George Waldenberger, not a climate scientist but a meteorologist, who has repeatedly requested to be removed from the Inhofe list and reiterated his support for the anthropogenic climate change hypothesis, but who still remains on the list; Gwyn Prins and Steve Rayner, real climate scientists who, however, are misrepresented as sceptics when in fact they state: 'We face a problem of anthropogenic climate change, but the Kyoto Protocol of 1997 has failed to tackle it'; and so on. These and numerous other examples are discussed at length in an ongoing regular column, 'The "Inhofe 400" Sceptic of the Day', by Andrew Dessler, Professor of Atmospheric Sciences at Texas A & M University, which continues to demonstrate the fraudulent nature of Inhofe's list.[16]

Unfortunately, this did not stop Senator Inhofe from releasing an updated list a year later in December 2008, including the original 400, of now 'more than 650 international scientists' who 'dissent over man-made global warming claims'.[17] It was not long before the credibility of this list was also undermined. By way of example, on Inhofe's new list is IPCC scientist Erich Roeckner, a renowned climate modeller at the Max Planck Institute. Roeckner is cited in the new report as saying that there are kinks in climate models, and telling *Nature*: 'It is possible that all of them are wrong' – supposedly implying that he is questioning the validity of anthropogenic models of climate change in general. However, as the *New Republic* reported:

> But he's not! Roeckner was referring to the IPCC's emissions scenarios, which involve assumptions about the rate of growth of greenhouse-gas emissions... We already know that emissions are growing faster than the IPCC's worst-case scenario, and that's bad news, not good.
>
> Anyway, Roeckner's as far as you get from a 'dissenter'... Roeckner is quoted in multiple news stories sounding downright alarmist about the consequences of man-made warming. 'Humans have had a large one-of-a-kind influence on the climate... Weather situations in which extreme floods occur will increase,' he informed Deutsche Welle in 2004. 'Our research pointed to rapid global warming and the shifting of climate zones,' he told ABC News in 2005. Quite the heretic, that one.[18]

The pattern was the same: listing people who are not experts on climate science, and who lack peer-reviewed publications in the field; including non-scientists; and misrepresenting established climate scientists who actually do accept the reality of man-made global warming.[19] Indeed, it is instructive to compare Inhofe's fraudulent list of '650' to the official warning issued by the world's largest society of earth scientists, the American Geophysical Union, *with a membership of 50,000 scientists*, agreeing collectively that:

> The Earth's climate is now clearly out of balance and is warming. Many components of the climate system – including the temperatures of the

atmosphere, land and ocean, the extent of sea ice and mountain glaciers, the sea level, the distribution of precipitation, and the length of seasons – are now changing at rates and in patterns that are not natural and are best explained by the increased atmospheric abundances of greenhouse gases and aerosols generated by human activity during the 20th century.[20]

Unsurprisingly, Senator Inhofe was also one of the first to jump on the 2009 climate email 'scandal' bandwagon, when thousands of emails from the University of East Anglia's Climatic Research Unit from a period of more than ten years were obtained by hackers. One of the emails most cited by 'sceptics', by the head of the unit, Professor Phil Jones, reads: 'I've just completed Mike's Nature trick of adding in the real temps to each series for the last 20 years (ie from 1981 onwards) and from 1961 for Keith's to hide the decline'.[21] Inhofe's press blog commented that the email 'appears to show several scientists eager to present a particular viewpoint – that anthropogenic emissions are largely responsible for global warming – even when the data showed something different'.[22] But the Union of Concerned Scientists (UCS), analyzing this and other leaked emails, explained the language and scientific context in detail:

> Jones is talking about how scientists compare temperature data from thermometers with temperature data derived from tree rings. Comparing that data allows scientists to derive past temperature data for several centuries before accurate thermometer measurements were available. The global average surface temperature since 1880 is based on thermometer and satellite temperature measurements...
>
> In some parts of the world, tree rings are a good substitute for temperature record. Trees form a ring of new growth every growing season. Generally, warmer temperatures produce thicker tree rings, while colder temperatures produce thinner ones. Other factors, such as precipitation, soil properties, and the tree's age also can affect tree ring growth.
>
> The 'trick,' which was used in a paper published in 1998 in the science journal *Nature*, is to combine the older tree ring data with thermometer data. Combining the two data sets can be difficult, and scientists are always interested in new ways to make temperature records more accurate.
>
> Tree rings are a largely consistent source of data for the past 2,000 years. But since the 1960s, scientists have noticed there are a handful of tree species in certain areas that appear to indicate temperatures that are warmer or colder than we actually know they are from direct thermometer measurement at weather stations.
>
> 'Hiding the decline' in this email refers to omitting data from some Siberian trees after 1960. This omission was openly discussed in the latest climate science update in 2007 from the IPCC, so it is not 'hidden' at all.
>
> Why Siberian trees? In the Yamal region of Siberia, there is a small set of trees with rings that are thinner than expected after 1960 when compared with actual thermometer measurements there. Scientists are still trying to figure out why these trees are outliers. Some analyses have

left out the data from these trees after 1960 and have used thermometer temperatures instead. Techniques like this help scientists reconstruct past climate temperature records based on the best available data.

Another email from scientist Kevin Trenberth laments that 'we can't account for the lack of warming at the moment', describing this as a 'travesty' due to the fact that 'Our observing system is inadequate'. UCS points out that he is talking about short-term internal climate variability, in particular the year 2008 'which was cooler than scientists expected, but still among the 10 warmest years on record'. Yet another email by Jones, construed by 'sceptics' as evidence of scientists manipulating peer review to squeeze out legitimate climate dissenters, objects to a paper on solar variability in the climate published in *Climate Research*, and calls for scientists to boycott the journal until it effects a change in editorship. Yet as UCS clarifies:

> Half of the editorial board of *Climate Research* resigned in protest against what they felt was a failure of the peer review process. The paper, which argued that current warming was unexceptional, was disputed by scientists whose work was cited in the paper. Many subsequent publications set the record straight, which demonstrates how the peer review process over time tends to correct such lapses. Scientists later discovered that the paper was funded by the American Petroleum Institute.

Thus, UCS rightly concluded that whoever stole the emails 'could only produce a handful of messages that, when taken out of context, might seem suspicious to people who are not familiar with the intimate details of climate science'.[23]

One of the latest examples of this sort of thing, emerging shortly before this book went to press, were widely reported claims that the IPCC's Fourth Assessment Report of 2007 was deliberately promulgating fear-mongering falsehoods exaggerating the threats posed by climate change. The *Wall Street Journal*, for instance, dismissed as fraudulent the well-known IPCC 'hockey-stick' graph depicting global average temperatures over the last millennium, showing that the temperature rise of the twentieth century is 'likely' to be 'unprecedented'.[24]

In particular the *WSJ* claimed that the IPCC's 'hockey-stick' graph and others like it, depicting twentieth-century global average temperatures as anomalous, were based on questionable 'tree-ring techniques' used by scientist Keith Briffa, as well as on data gathered from these techniques. Yet as one of the scientists who contributed to the 'hockey-stick' graph study, Michael Mann, points out: 'Neither the multiple proxy-based 'Hockey Stick' reconstruction of Mann et al nor the multiple-proxy based Jones et al reconstruction used "Mr. Briffa's tree-ring techniques" let alone their data.'[25]

In fact, the IPCC 'hockey-stick' graph has been corroborated and reinforced by numerous independent scientific studies. In 2008, the *Proceedings of the National Academy of Sciences* extended the multi-proxy reconstruction of global average temperatures back nearly 2,000 years. The study, explicitly non-reliant on tree-ring data, found that: 'Our results extend previous conclusions that recent Northern Hemisphere surface temperature increases are likely anomalous in a long-term context. Recent warmth appears anomalous for at least the past 1,300 years whether or not tree-ring data are used.' With tree-ring data, this conclusion can be extended back 1,700 years.[26] In lead author Mann's words: 'You can go back nearly 2,000 years and the conclusion still holds – the current warmth is anomalous. The burst of warming over the past one to two decades takes us out of the envelope of natural variability.'[27]

'Sceptics' have also argued that the graph ignores events like the Medieval Warm Period (950–1250) – yet they fail to realize that the higher temperatures associated with MWP were only regional, and did not represent the global average temperatures illustrated in the graph. While warmer temperatures were concentrated in certain regions, other regions were even colder than during the lower regional temperatures during the ensuing Little Ice Age (1300–1850).[28]

'Sceptics' were also overjoyed when it emerged that the IPCC had promulgated the following major error within its 3,000 pages: that the Himalayan glaciers could 'completely disappear' by 2035 and 'perhaps sooner' at current rates of warming. The IPCC later conceded that this was an unjustifiable statement which relied not on the peer-reviewed scientific literature, but on a single media interview with a scientist in 1999.[29] The error was compounded by revelations that IPCC Chairman Dr Rajender Pachauri had financial links to the oil industry, and that his own company had cashed in on research grants based on the alarmist conjecture about the Himalayas.[30]

While clearly a reprehensible error that should not have made it through the IPCC's peer-review procedures (it is worth noting that the mistake was belatedly discovered not by 'sceptics' but by glacier expert Georg Kaser, himself a lead author of Volume 1, Chapter 4 of the IPCC report), no other IPCC statements have been similarly discredited. Except in the eyes of certain 'sceptic' commentators, such as the *Telegraph*'s Christopher Booker, who declared that various 'alarms were given special prominence in the IPCC's 2007 report and each of them has now been shown to be based, not on hard evidence, but on scare stories, derived not from proper scientists but from environmental activists':

> Those glaciers are not vanishing; the damage to the rainforest is not from climate change but logging and agriculture; African crop yields are more likely to increase than diminish; the modest rise in sea levels is slowing not accelerating; hurricane activity is lower than it was 60 years ago; droughts were more frequent in the past; there has been no increase in floods or heatwaves.[31]

Yet Booker – whose track record on 'science journalism' involves such 'blunders' as claiming that white asbestos is a 'non-existent' threat to human health and that passive smoking does not cause cancer – was obviously reading and replicating the earlier claims of Jonathan Leake, science and environment editor at the *Sunday Times*, in an article whose research was done by Richard North.[32] Yet North is Booker's co-author of a well-known anti-science screed, *Scared to Death: From BSE to Global Warming*, described by the *Guardian* as replete with 'egregious errors that would shame a junior reporter', including 'reporting a non-existent interview'.[33]

In fact, the IPCC's statements about African crop yields, the intensification of natural disasters and erratic weather, and the potential deforestation of the Amazon are entirely accurate, and are corroborated by the peer-reviewed literature, examples of which we will return to later in this chapter. The IPCC's statement that 'yields from rain-fed agriculture could be reduced by up to 50%' by 2020 refer to a paper by climate expert Professor Ali Agoumi. 'Sceptics' shouted that the claim is discredited because the paper is not peer-reviewed. Although this is technically correct, the paper – a report published by the International Institute for Sustainable Development and the Climate Change Knowledge Network – is 'a summary of technical studies and research', much of which *is* peer-reviewed, 'conducted to inform Initial National Communications from three countries (Morocco, Algeria and Tunisia) to the United Nations Framework Convention on Climate Change,' and is thus 'a perfectly legitimate IPCC reference'.[34] Indeed, the IPCC's projection on the potentially devastating impact of climate change on African crop yields is supported by a series of peer-reviewed scientific studies from 1994 to 2007.[35]

The IPCC's description of the link between climate change and natural disasters is also accurate, and the source of this description – a Risk Management Solutions paper by Dr Robert Muir-Wood, a former Earth Sciences Research Fellow at Cambridge University – has turned out to be entirely credible. The full paper 'was peer reviewed and accepted for publication in November 2006', a few weeks after 'the cut-off date for the IPCC 4th Assessment Report in October' – explaining why an earlier draft version of the report was referenced. But the IPCC was 'aware of the full report and that it had been accepted for publication'. Muir-Wood himself has reiterated that the IPCC did not misrepresent the conclusions.[36] Furthermore, evidence of a link between climate change and the increased risk of natural disasters, including dangerous weather, is widely explored in the peer-reviewed literature.[37]

Similarly, widely criticized is the IPCC's assertion that up to 40 per cent of Amazon rainforests 'could react drastically to even a slight reduction in precipitation', referencing another non-peer-reviewed report, this time by the World Wildlife Fund (WWF). But despite the premature proclamations of climate detractors, the IPCC's statement is entirely accurate. The WWF report itself sources a peer-reviewed 1999 paper in the journal *Nature* by Yale University tropical forest scientist Daniel Nepstad, which categorically confirms the IPCC's warning. Nepstad responded to the 'sceptic' media reports, noting 'The IPCC statement on the Amazon is correct', and cited further peer-reviewed papers written by himself and others corroborating the same conclusion.[38]

In summary, it is absurd to claim that the IPCC's Fourth Assessment Report has been discredited as a signifier of the scientific consensus that global warming is anthropogenic.

The conclusion is clear: claims that global warming is not happening, or if happening has nothing or little to do with human activities, fall outside the existing scientific consensus, and often come from people with vested political or economic interests, for whom the study of climate is outside their professional qualifications and field of expertise. The 'sceptic' strategy is simple: to misquote, quote out of context, and/or misrepresent the statements and findings of real climate scientists. Yet despite their lack of credibility, these claims and the bad science they rest on frequently receive widespread media coverage. Before examining the impacts of climate change, I critically review some of the most prominent 'sceptical' approaches to anthropogenic climate change which try to deny the role of human activities, finding them to be deeply unscientific.

Solar Activity and Climate Variation

One of the most common misconceptions cited by 'man-made climate change' sceptics (hereafter referred to simply as 'climate sceptics' or 'sceptics') is that the sun is the primary cause of contemporary global warming. The earth's climate history does evince a close correlation between solar activity and global temperature change. But while the sun plays a crucial role in climate change, scientific studies confirm that recent global warming on the earth could not be caused by solar activity. One study by scientists from Finland and Germany, commonly used by climate sceptics, concludes that the sun has been more active in the last 60 years than in the preceding 1150 years. The scientists argue that 'long-term climate variations are affected by solar magnetic activity'. Yet the same study points out that the correlation between solar activity and temperature *ceased in 1975*, after which global average temperatures escalated despite solar activity remaining at a stationary level:

> Note that the most recent warming, since around 1975, has been considered in the above correlations. During these last 30 years the solar total irradiance, solar UV irradiance and cosmic ray flux has not shown any significant secular trend, so that at least this most recent warming episode must have another source.[39]

Similar conclusions have been reiterated independently throughout the scientific literature. More recently, in 2008, a study published by *Nature* noted that 'the level of solar activity during the past 70 years is exceptional' and 'may indicate that the Sun has contributed to the unusual climate change during the twentieth century'. Yet it goes on to confirm that:

> solar variability is unlikely to be the prime cause of the strong warming during the last three decades... even under the extreme assumption that

> the Sun was responsible for all the global warming prior to 1970, at the most 30 per cent of the strong warming since then can be of solar origin.[40]

And in 2007, the *Proceedings of the Royal Society* published a paper concluding that while there is 'considerable evidence for solar influence on the Earth's pre-industrial climate and... in post-industrial climate change in the first half of the last century', over the previous 20 years, 'all the trends in the Sun that could have had an influence on the Earth's climate have been in the opposite direction to that required to explain the observed rise in global mean temperatures'.[41]

The Earth's Natural Climate Cycles

Another general misconception promulgated by climate sceptics is that contemporary global warming can be explained entirely by the fact that the earth undergoes periodic natural fluctuations in its climate. Climate change, the argument goes, is therefore simply a natural cycle, not a result of human activities. This perspective is not entirely false, but it is misleading. According to the geological record, the earth has certainly experienced long cooling and warming trends throughout the last million years. These trends adhere to an approximate 100,000-year cycle consisting of ice ages broken by shorter warm periods known as interglacials. The onset of glaciation and the subsequent interglacial periods are brought on by changes in the earth's orbit around the sun, known as Milankovitch cycles. When the Milankovitch cycle leads to an increase in the amount of sunlight reaching the earth, it can cause Antarctic sea ice and glaciers to melt. Increased ice loss would mean less sunlight reflected away, amplifying the warming process. As the ocean warms, its ability to carry CO_2 decreases, and thus it begins to increasingly emit CO_2 into the atmosphere in a process that can take up to 800–1000 years. In turn, the CO_2 amplifies and spreads warming further. Together, the Milankovitch cycle, in tandem with the positive-feedback effect of increased atmospheric CO_2, raises global average temperatures sufficiently to trigger deglaciation, bringing the earth out of its ice age.[42] However, the last 12,000-year period was the beginning of a continuing warm interglacial period, whose *current* trend in the Milankovitch cycle is toward a gradual cooling down towards an ice age. This gradual cooling phase of the current cycle does not explain the current trend of global warming. Scientists estimate that the present interglacial period is likely to continue for tens of thousands of years naturally. Indeed, so unnatural is the current phase of global warming that Andre Berger, one of the world's leading experts on Quaternary climate change and honorary president of the European Geosciences Union, calculates that current industrial fossil fuel emissions could potentially suppress the next natural glacial cycle entirely.[43]

Linked to this misconception is the claim that global warming has stopped since 1998. The argument is that due to the lapse in sunspot activity, as well as cyclical variations of global warming and cooling, the coming decades will in fact constitute a period of prolonged global cooling (also see next sub-chapter). American geologist Don Easterbrook attributes the current period of global warming to a natural global weather cycle of warming and cooling. In a 2001 paper presented at the Geological Society of America, he argues: 'Advance and retreat of glaciers in the Pacific Northwest show three distinct oscillations, each having a period of ~25 years. Glaciers retreated rapidly from ~1930 to ~1950–55 (warm cycle), readvanced from ~1955 to ~1977 (cool cycle), then retreated rapidly from ~1977 to the present (warm cycle).' This correlates with the period of global warming which therefore appears to be an outcome of such oscillations. 'If the trend continues, the current warm cycle should end soon', and global warming will be over for another 25 years.[44] By implication, global warming is nothing to worry about! In the next section, we look at this question – did global warming stop in 1998 (or 2001)? – in more detail, but before that, it is important to understand the deeper misconceptions underlying Easterbrook's approach.

Firstly, he proffers a rather eccentric argument – indeed, his is a lone voice in the scientific community, with no wider corroboration in the relevant peer-reviewed literature. Secondly, his paper itself was not peer-reviewed and remains unpublished in any recognized relevant physical science journal. These problems should obviously raise our initial suspicions. Thirdly, further examination of one of Easterbrook's own examples, the Pacific Decadal Oscillation (PDO), confirms those doubts. According to Easterbrook, the PDO, which occurs primarily in the North Pacific, along with other oscillations is responsible for most of the climate change over the last century, including the phase of accelerated warming that began in the 1970s. Although Easterbrook depicts these oscillations of warm and cool phases as occurring over clearly demarcated periods of about 25 to 30 years, in reality each phase can last for between ten and forty years. Further, these long periods can also be broken by intervals during the transition for anything between one and five years. Thus, in 1905, the PDO switched to a warm phase as global warming began. In 1946, the PDO moved into a cool phase as temperatures cooled mid-century. In 1977, the PDO again entered a warm phase around the same time as the onset of the current global warming period. These correlations, for Easterbrook, prove that global warming is nothing more than the function of natural oscillations in the earth's climate that will inevitably give way to global cooling, before giving rise to yet another warming cycle.

However, as Figure 1.2 shows, while the PDO does correlate to some degree with short-term variations in global temperature, *this correlation is starkly outweighed by the contrasts*. While the PDO oscillates between positive and negative values at roughly equivalent levels, *global average temperatures in the same period display an unambiguous long-term warming trend*.[45] To account for global temperature increases since the dawn of the twentieth century requires totalling the impact from all relevant forcings including solar, aerosols, CO_2 emissions, and so on. By itself, the PDO does not provide an adequate explanation for global warming.

Another natural phenomenon that sceptics see as contributing to imminent global cooling, rather than warming, is the Atlantic Multidecadal Oscillation (AMO), closely related to warm currents that bring heat from the tropics to

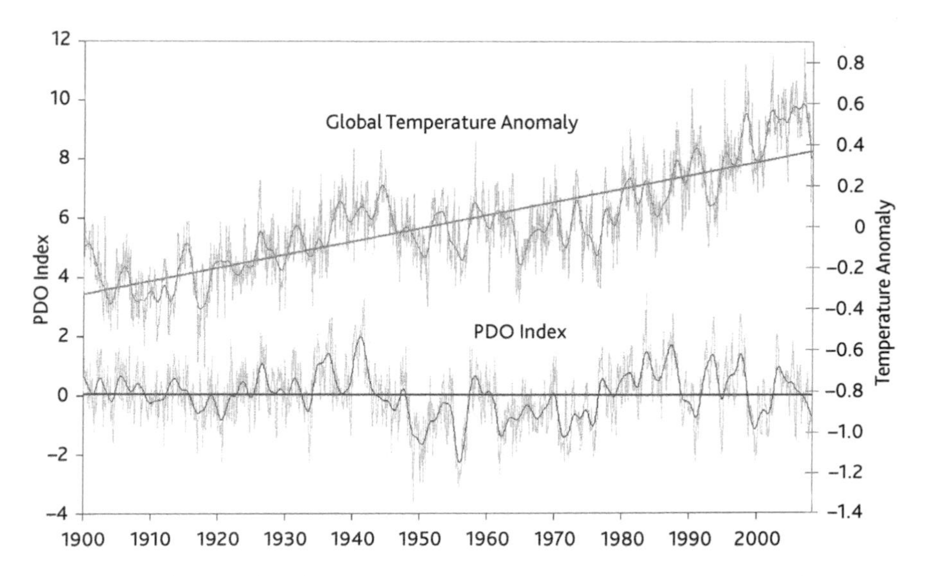

Figure 1.2 Pacific Decadal Oscillation Compared to Actual Global Average Temperature Rise.

Source: John Cook, Skeptical Science (3 May 2008), based on data from Joint Institute for the Study of the Atmosphere and the Ocean, Washington University

European shores about every 60 to 70 years. Yet when German researchers attempted to incorporate the impact of the AMO into computer models of projected climate change, they found that the AMO may temporarily ameliorate the impact of fossil fuel emissions for about a decade, but give way to rapidly rising temperatures thereafter. The scientists argued that their findings, published in *Nature*, did not contradict the consensus around anthropogenic climate change, but suggested that increased warming would occur later rather than earlier due to the AMO, and possibly with a heightened impact due to greenhouse gases accumulated in the interim. This relates to a phenomenon known as 'internal climate variability' (variations in climate trends due to internal natural fluctuations).[46]

Leading climate scientists from the online climate information network *Real Climate* – namely Stefan Rahmstorf, Michael Mann, Ray Bradley, William Connolley, David Archer, and Caspar Ammann – interrogated the study's forecast of 'no warming for a decade' and argued that it severely underestimated actual global warming trends.[47] But their criticism perhaps overlooked the most disturbing implications of the *Nature* study.

As noted by American physicist Joseph Romm, executive director of the Center for Energy and Climate Solutions, the study was widely misrepresented by the mainstream media, which interpreted it as supporting the case for global cooling. In fact, the *Nature* study confirmed exactly the opposite. Citing correspondence with the lead author of the paper, Noel Keenlyside, Romm pointed out that the researchers did not expect a rise in '*mean* temperature' between 2005 and 2015, but that this did not preclude any rise in, for example, global surface temperatures. The study, rather, explains why global average temperatures 'have not risen very much in recent years, and, perhaps, why ocean temperatures have also not risen very much in the past few years'. In the correspondence, Dr Keenlyside explicitly acknowledges that their data implies a rapid rise in global average temperature after 2010: 'However, as you correctly point out, our results show a pick up in global mean temperature for the following decade (2010–2020). Assuming a smooth transition in temperature, our results would indicate the warming picks up earlier than 2015.'

In summary, global average temperature is likely to fluctuate along a rough plateau, before rising sharply in the decade after 2010. Given that observed global mean temperatures have actually been higher than the study's simulated prediction with ocean data, it is clear that its forecast is very likely to be conservative, and overall global warming trends will be more extreme even accounting for the role of the AMO. Romm thus points out that the *Nature* study remains consistent with the following predictions:

1. The 'coming decade' (2010 to 2020) is poised to be the warmest on record, globally.
2. The coming decade is poised to see faster temperature rise than any decade since 1960, the starting point of the authors' calculations.
3. The rapid warming will likely begin early in the next decade.
4. The mean North American temperature for the decade from 2005 to 2015 is projected to be slightly warmer than the actual average temperature of the decade from 1993 to 2003.[48]

Global Cooling?

Contributing to further public confusion is the bizarre idea, mentioned above, that global warming ceased in 1998. We discussed Easterbrook's approach to this, noting that the argument is not supported by research published in the peer-reviewed scientific literature. Although Easterbrook attempted to underpin the idea of global cooling with a variety of extrapolations about the earth's various internal weather cycles in different regions, the idea that we have definitely entered a period of global cooling has been endorsed by several other non-climate scientists. For instance, in 2006 the *Telegraph* carried a piece by Bob Carter, an Australian geology professor, arguing that 'for the years 1998–2005 global average temperature did not increase (there was actually a slight decrease, though not at a rate that differs significantly from zero)'.[49] Former BBC science correspondent and fellow of the Royal Astronomical Society David Whitehouse similarly wrote in the *New Statesman*:

> The fact is that the global temperature of 2007 is statistically the same as 2006 as well as every year since 2001. Global warming has, temporarily or permanently, ceased. Temperatures across the world are not increasing as they should according to the fundamental theory behind global warming –

the greenhouse effect. Something else is happening and it is vital that we find out what or else we may spend hundreds of billions of pounds needlessly.[50]

The reason these claims find no substantiation in the actual scientific literature is simple: they are false. According to the UK Met Office's Hadley Centre for Climate Change: 'A simple mathematical calculation of the temperature change over the latest decade (1998–2007) alone shows a continued warming of 0.1°C per decade.' As noted, there has recently been a slight slowing of warming due to internal variability. This is not only due to the PDO or AMO, but more significantly due to the role of the El Niño Southern Oscillation, a periodic atmospheric and oceanic change in the tropical Pacific region, which has two phases: 'El Niño', the period when water in that region is warmer than average; and 'La Niña', the period when the water in the tropical Eastern Pacific is colder than average. The Met Office refers to 'a shift towards more-frequent La Niña conditions in the Pacific since 1998. These bring cool water up from the depths of the Pacific Ocean, cooling global temperatures.' La Niña conditions late this decade thus played a major role in cooler temperatures. In contrast, the El Niño can warm global temperatures by about 0.2°C in a single year, affecting ocean surface and air temperatures over land. The El Niño event during 1998 broke previous warming records, making subsequent temperatures appear flatter. Thus, El Niño created much warmer conditions in 1998 while La Niña generated cooler conditions toward 2008, resulting in a flatter temperature curve.[51]

Nevertheless, data sets from the National Climate Data Center (NCDC) and NASA's Goddard Institute for Space Studies corroborate the Met Office's findings, and display a continuing, if slower, global warming trend from 1998 to 2007. The Hadley (Met Office) data set is flatter because it ignores parts of the Arctic which have experienced strong warming.[52]

It is therefore clear that the apparent cooling trend witnessed from around 2007 was the outcome of several major factors including not only the PDO and AMO, but also more significantly the role of La Niña. The UN's World Meteorological Organisation (WMO) confirmed in early 2008 that the cold La Niña ocean current in the Pacific would lead to slightly colder temperatures for 2008 as compared to 2007. Yet it also noted that the 2008 global mean temperature was still 'well above the average for the last 100 years', and that 'the decade from 1998 to 2007 was the warmest on record'. According to WMO Secretary-General Michael Jarraud, 'La Niña is part of what we call "variability." There has always been and there will always be cooler and warmer years, but what is important for climate change is that the trend is up; the climate on average is warming even if there is a temporary cooling because of La Niña.'[53] It thus transpired that already by mid 2008 the La Niña cooling effect had begun to weaken.

Prior to this return to rising temperatures, the La Niña effect had contributed to a temporary drop in temperature from January 2007 to January 2008 of around 0.6°C, documented by meteorologist Anthony Watts who described the phenomenon as 'an anomaly with a large magnitude' coinciding with

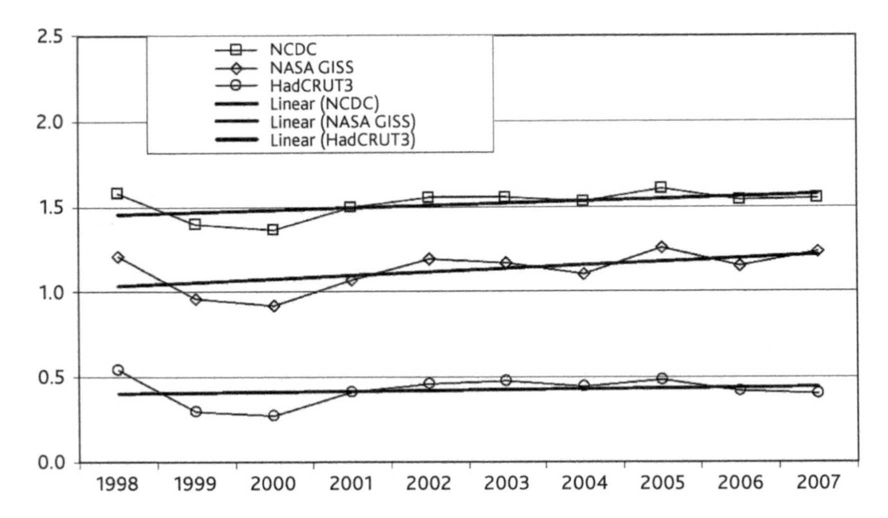

Figure 1.3 Global Warming Trend 1998–2007.

Source: Robert Fawcett and David Jones, Australian Science Media Centre (April 2008)

'other anecdotal weather evidence'.[54] Climate sceptics jubilantly reported this as yet further evidence of global cooling, rather than warming. For instance, *DailyTech* reported Watt's finding as 'a value large enough to erase nearly all the global warming recorded over the past 100 years. All in one year's time. For all sources, it's the single fastest temperature change ever recorded, either up or down'.[55] *DailyTech* then suggested that this global cooling was due not to La Niña, as documented by the scientific community, but rather to reduced solar activity. The suggestion, once more lacking substantiation from the peer-reviewed scientific literature, was that the sun is currently in its weakest cycle (23), with cycle 24 purportedly refusing to start. Sceptics point out that a similar event occurred 400 years ago in the form of a solar event known as a Maunder Minimum, precipitating a 'Little Ice Age' consisting of a sharp global temperature drop. This proves, according to *DailyTech*, that the earth is on the verge of a prolonged phase of global cooling due to yet another Maunder Minimum.[56]

Yet there is a serious problem with this idea. For diminished solar activity to be the direct cause of the 0.6°C drop in global temperature over a single year would require a dramatic reduction in Total Solar Irradiance (TSI) – that is, the total amount of radiant energy emitted by the sun hitting the top of the earth's atmosphere, measured in watts per square metre (W/m^2). According to Charles Camp and Ka Kit Tung from the Department of Applied Mathematics at the University of Washington, in a paper in *Geophysical Research Letters*, the solar cycle contributes *just under 0.2°C (to be precise, only 0.18°C)* cooling to global temperatures as the sun moves from maximum to minimum emittance – over three times smaller than the 0.6°C necessary in this case.[57] The solar cycle therefore simply cannot explain the level of cooling occurring from 2007 to 2008, which was correlated instead with aforementioned oscillations in ocean conditions.

Indeed, contradicting the speculation that solar cycle 24 would fail to kick in, leading to a period of prolonged global cooling for several decades, the solar cycle did in fact begin in early 2008 – exactly as had been predicted by leading solar scientists for some years. The scientists had already pointed out years earlier that the cycle would start later than normal, but would be between 30 and 50 per cent stronger than the previous cycle, reaching its peak in about 2012.[58] Unfortunately, this is likely to exacerbate the impact of fossil fuel emissions on global warming over the coming decades. This example illustrates the key lesson that climate change cannot be understood simply by emphasizing inevitable fluctuations and variations in weather over short periods of time, but only by analyzing the long-term trends over decades.

NATIONAL SECURITY ALERT

Abrupt, Rapid Climate Change: Plausible

So the global warming sceptics are unequivocally wrong. But how grave is the danger from climate change? One of the first explicit indications that the Western national security establishment recognized that climate change may well pose a more dangerous threat to national security than terrorism was a Pentagon study commissioned by the legendary US Department of Defense planner Andrew Marshall. Made public in January 2004, the report was authored by Peter Schwartz, a CIA consultant and former planning head at Shell Oil, and Doug Randall, a senior consultant at the Global Business Network in San Francisco. Titled 'An Abrupt Climate Change Scenario and Its Implications for United States National Security', the report for the Department of Defense's Office of Net Assessment drew on interviews and research from leading climate scientists to project a particularly dangerous global scenario that 'is plausible, and would challenge United States national security in ways that should be considered immediately'. Climate change, the report urged, 'should be elevated beyond a scientific debate to a US national security concern'. Dismissing doubts about the scientific validity of climate change, the report argued that:

> There is substantial evidence to indicate that significant global warming will occur during the 21st century... Recent research, however, suggests that there is a possibility that this gradual global warming could lead to a relatively abrupt slowing of the ocean's thermohaline conveyor, which could lead to harsher winter weather conditions, sharply reduced soil moisture, and more intense winds in certain regions that currently provide a significant fraction of the world's food production. With inadequate preparation, the result could be a significant drop in the human carrying capacity of the Earth's environment.[59]

Concerns over the national security implications of climate change were also aired in Europe. In early 2008, a high-level European Union (EU) report to 27 heads of government warned of the probability of 'significant potential conflicts' in coming decades due to 'intensified competition over access to, and control over, energy resources'. Written by Javier Solana, the EU's foreign policy chief, and Benito Ferrero-Waldner, the EU commissioner for external relations, the report argued that global warming would precipitate major security issues for Europe, such as mass migrations, failed states and political radicalization. In particular, it noted that the quickened thawing of the Arctic due to accelerating climate change would lead to intensified geopolitical contestations between Russia and NATO, and potentially even military conflict, over access to the region's large reserves of untapped oil and gas reserves. The EU report also highlighted intensified North–South tensions due to global warming, particularly the volatility of regions in the Middle East and Central Asia which hold large energy reserves and mineral deposits.[60]

Increased Probability of Resource Conflict

Such concerns were already emphasized in the Pentagon's earlier report, which had warned that rapid climate change could fundamentally 'destabilize the geo-political environment, leading to skirmishes, battles, and even war' due to three categories of resource constraint:

1. Food shortages due to decreases in net global agricultural production;
2. Decreased availability and quality of fresh water in key regions due to shifting precipitation patterns, causing more frequent floods and droughts;
3. Disrupted access to energy supplies due to extensive sea ice and stormy conditions.

Rapid climate change, leading to catastrophic droughts, famines and rioting, would thus effect mounting national and international tensions, mediated through defensive and offensive strategies that could escalate into a terrifying arc of global conflicts, oriented around deadly competition over control of increasingly scarce resources:

> Nations with the resources to do so may build virtual fortresses around their countries, preserving resources for themselves. Less fortunate nations especially those with ancient enmities with their neighbors, may initiate in struggles for access to food, clean water, or energy.

The Pentagon study thus suggests not only that the threat to national security posed by rapid climate change is potentially far worse than terrorism, but further that the future arc of conflict will be about 'resources for survival rather than religion, ideology, or national honor'.[61]

Yet the Pentagon warning is only the tip of a rapidly melting iceberg. Over the last decade alone, scientific studies have increasingly homed in on the dynamics, contours, and impacts of climate change. And the implication is not merely that climate change could dramatically undermine national security, but that it could endanger the very survival of civilization itself.

Existential Threat: Fatal Disruption to Industrial Civilization by End of Twenty-First Century

Outlining six climate change scenarios, the landmark February 2007 Fourth Assessment Report by the United Nations Inter-Governmental Panel on Climate Change (IPCC) shows that even the best-case scenario would be greatly destabilizing. But the worst-case scenario presented by the report demonstrated more bluntly than ever that climate change could mean the end of life on earth, in our own era. The report projected that by the year 2100, the average global temperature could rise by 6.4°C, leading to drastic ecological alterations that would make *life throughout most of the earth impossible*.[62] Even a rise of 3°C produced by a doubling of CO_2 production from pre-industrial levels to 550 parts per million – not a worst-case scenario, and cited by the British government's former chief scientific adviser Sir David King as a realistic upper limit at which CO_2 levels could be stabilized – would generate conditions unsupportable by society.

The IPCC report sounded alarm bells around the world about the gravity of climate change and its potentially fatal impact for life on earth. Going further, British ecologist Mark Lynas translated the IPCC's temperature rise scenarios into a detailed analysis of the impact of global warming for each degree of temperature increase. In his book *Six Degrees: Our Future on a Better Planet* – winner of the Royal Society Science Book Prize – Lynas reviewed thousands of peer-reviewed scientific studies and climate models, to try and show how each degree level increase in the global average temperature is likely to change the face of the Earth:[63]

> 1°C Increase: Ice-free sea absorbs more heat and accelerates global warming; fresh water lost from a third of the world's surface; low-lying coastlines flooded
>
> 2°C Increase: Europeans dying of heatstroke; forests ravaged by fire; stressed plants emitting carbon rather than absorbing it; a third of all species face extinction
>
> 3°C Increase: Carbon release from vegetation and soils speeds global warming; death of the Amazon rainforest; super-hurricanes hit coastal cities; starvation in Africa
>
> 4°C Increase: Runaway thaw of permafrost makes global warming unstoppable; much of Britain made uninhabitable by severe flooding; Mediterranean region abandoned
>
> 5°C Increase: Methane from ocean floor accelerates global warming; ice gone from both poles; humans migrate in search of food and try vainly to live off the land like animals

> 6°C Increase: Life on earth ends with apocalyptic storms, flash floods, hydrogen sulphide gas and methane fireballs racing across the globe with the power of atomic bombs; only fungi survive.[64]

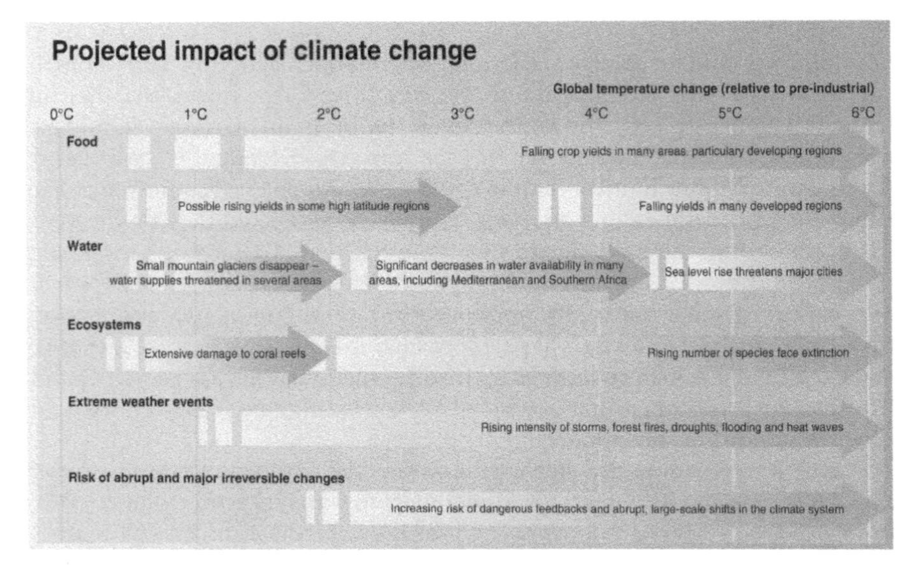

Figure 1.4 Graphical Presentation of UN IPCC Scenarios for Global Warming by Degrees Celsius.

Source: IAASTD/Ketill Berger, UNEP/GRID-Arendal based on Stern Review of the Economics of Climate Change <http://maps.grida.no/go/graphic/projected-impact-of-climate-change>

Climate change, in other words, does not simply mean more wars, social chaos, and political upheaval. In the long term, global warming threatens the survival of modern industrial civilization.

RAPID CLIMATE CHANGE

Positive Feedbacks and Runaway Climate Change by Mid Century

As dismal as these projected scenarios are, increasing scientific evidence strongly suggests that even the IPCC's worst-case scenario of a 6°C rise in world average temperature by 2100 could be overly optimistic, if current rates of increase of fossil fuel emissions continue unabated.

At the end of 2006, the Global Carbon Project (GCP) announced its findings that between 2000 and 2005, CO_2 emissions had *grown four times faster* than in the preceding ten years. GCP executive director Josep Canadell warned:

> On our current path, we will find it extremely difficult to rein in carbon emissions enough to stabilise the atmospheric CO_2 concentration at 450 parts per million [limiting global warming to 2°C] and even 550 ppm [3°C] will be a challenge.[65]

Similarly, according to a paper in the *Proceedings of the National Academy of Sciences* in April 2007, current CO_2 emissions are worse than all six scenarios contemplated by the IPCC: 'The emissions growth rate since 2000 was greater than for the most fossil-fuel intensive of the Intergovernmental Panel on Climate Change emissions scenarios.' This implies that the IPCC's worst-case six-degree scenario is a conservative underestimate of the most probable climate trajectory given current rates of emissions.[66]

Indeed, due to environmental inertia (by which the environment stores up part of the energy generated by greenhouse gas emissions, only releasing it to the atmosphere later on), even with the complete cessation of human emissions, levels of atmospheric carbon dioxide would continue to rise for up to a century, and therefore global temperatures would continue to increase for two or more centuries. So climate change could occur more rapidly than expected by many current studies.[67] This underscores the stark inadequacy of demands to stabilize CO_2 emissions at 1990 levels (the Kyoto Protocols) and to allow countries to produce their own arbitrary targets for emissions reductions (Copenhagen).

A further GCP study in 2008 found that in the preceding year, carbon released from burning fossil fuels and producing cement had increased 2.9 per cent over 2006 levels, to a total of 8.47 Gt. This output is at the extreme end of the IPCC's worst-case scenario. Similarly, in October 2008, another paper published in the *Proceedings of the National Academy of Sciences* showed that even if humans stopped generating greenhouse gases immediately, the world's average temperature would 'most likely' increase by 2.4°C by the end of this century – potentially enough to trigger irreversible and potentially even runaway climate change, as we will show below.[68]

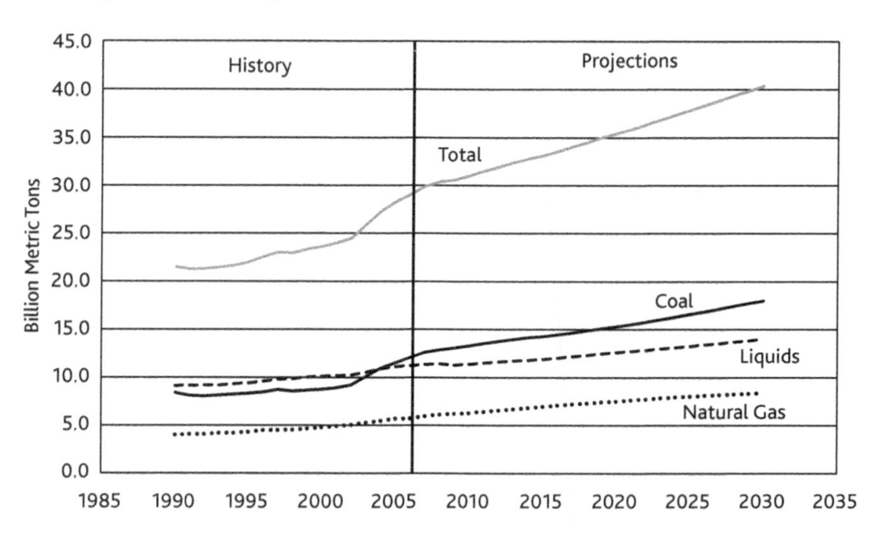

Figure 1.5 World Energy-Related CO2 Emissions by Fuel Type, 1992–2030.

Source: US Energy Information Administration (EIA), International Energy Annual 2004 (May–July 2006); EIA, System for the Analysis of Global Energy Markets (2007)

One of the most widely cited climatic models exploring the potential impact of this environmental inertia was produced by the Met Office's Hadley Centre for Climate Change. It was one of several new studies recognizing that the global climate system is in a state of unstable equilibrium, in which surface heating caused by CO_2 can act as a trigger for an accelerating process of global warming driven by amplifying 'positive feedbacks', eventually leading to a process of runaway climate change which would be completely beyond human control. The further we move away from this equilibrium due to human-generated emissions, the more powerful becomes the positive feedback system, and thus the faster the rate of climate change. David Wasdell, lead scientist on feedback dynamics of complex systems for the Global System Dynamics and Policies Project of the European Commission, has taken such findings much further. He finds that many current global warming estimates focusing on the alleged safety of the 2°C limit could be quite off the mark, and that runaway climate change could begin by mid century.[69] In this case, rather than global warming constituting a gradual, linear increase in temperature, with greenhouse gases being absorbed and retained by the atmosphere, the billions of years' worth of carbon and methane could be incontinently released in blazing surges that would drown or incinerate whole cities. Polar ice would melt rapidly, and the Amazon rainforest could collapse in a few decades. A vicious, irreversible and self-reinforcing spiral would begin which would threaten not just our way of life but the very existence of our own and every other species on Earth.[70]

The United Nations IPCC report has commendably shifted the debate on climate change by publicly affirming firstly an overwhelming scientific consensus on the reality of human-emissions generated climate change, and secondly a startling set of scenarios for how global warming will affect life on earth by the end of this century if existing rates of increase of CO_2 emissions continue unabated. But according to a growing body of scientific evidence, the IPCC's findings in 2007 were far too conservative – and dangerous climate change is more likely to occur far sooner, with greater rapidity, and with higher intensity, than officially recognized by governments.

Inaccuracies in the Intergovernmental Panel on Climate Change 2007 Report

A number of British researchers expressed grave reservations shortly after the release of the UN IPCC Fourth Assessment Report. In particular, David Wasdell, who was an accredited reviewer of the IPCC report, told the *New Scientist* that early drafts prepared by scientists in April 2006 contained 'many references to the potential for climate to change faster than expected because of "positive feedbacks" in the climate system. Most of these references were absent from the final version.' His assertion is based 'on a line-by-line analysis of the scientists' report and the final version', which was agreed in February 2007 at 'a week-long meeting of representatives of more than 100 governments'. Below we highlight three examples from Wasdell's analysis:

1. In reference to warnings that natural systems such as rainforests, soils and oceans would be less able in future to absorb greenhouse gas emissions, the scientists' draft report of April 2006 warned: '*This positive feedback could lead to as much as 1.2 degrees Celsius of added warming by 2100*'. The final version of March 2007 however only acknowledges that feedback exists and says: 'The magnitude of this feedback is uncertain.'
2. The April 2006 draft warned that global warming will increase the amount of water vapour released into the atmosphere, which in turn will act like a greenhouse gas, leading to an estimated '*40–50 percent amplification of global mean warming*'. In the final March 2007 report this statement was replaced with 'Water vapour changes represent the largest feedback'.
3. In relation to the accelerated breakup of arctic and antarctic ice sheets, the April 2006 draft paper talked about observed rapid changes in ice sheet flows and referred to an *'accelerated trend' in sea-level rise*. The government-endorsed final report of March 2007 said that 'ice flows from Greenland and Antarctica ... could increase or decrease in future.'[71]
4. The conclusion that '*North America is expected to experience locally severe economic damage, plus substantial ecosystem, social and cultural disruption from climate change related events*' was removed from the final version.[72]

In other words, the IPCC Fourth Assessment Report excluded or underplayed direct reference to the overwhelming probability of the rapid acceleration of climate change in the context of current rates of increase of CO_2 emissions and positive feedbacks. Wasdell put it down to possible political interference, and there are reasonable grounds for this conclusion. As noted by Mike Mann, director of the Earth System Science Center at Pennsylvania State University, and a past lead author for the IPCC: 'Allowing governmental delegations to ride into town at the last minute and water down conclusions after they were painstakingly arrived at in an objective scientific assessment does not serve society well.'[73]

The possible watering down of the IPCC's Report is part of a wider pattern. In the same month, a joint survey by the Union of Concerned Scientists and the Government Accountability Project concluded that 58 per cent of US government-employed climate scientists surveyed complained of being subjected to: 1) 'Pressure to eliminate the words "climate change," "global warming", or other similar terms' from their communications; 2) editing of scientific reports by their superiors which 'changed the meaning of scientific findings'; 3) statements by officials at their agencies which misrepresented their findings; 4) 'The disappearance or unusual delay of websites, reports, or other science-based materials relating to climate'; 5) 'New or unusual administrative requirements that impair climate-related work'; 6) 'Situations in which scientists have actively objected to, resigned from, or removed themselves from a project because of pressure to change scientific findings.' Scientists reported 435 incidents of political interference over the preceding five years.[74] Such widespread, systematic political interference with the work of climate scientists lends credence to the concern that they might feel unable to voice their real views about the urgency of the threat posed by global warming.

ABRUPT CHANGE THROUGH 'TIPPING POINTS'

Earth Does Not Do Gradual Change

The probability that climate change is accelerating rapidly at current rates of increase of CO_2 emissions is therefore a pivotal issue. In the last few years, the scientific evidence increasingly suggests that climate change will occur not through a long, protracted linear process of gradual intensification, but in the form of abrupt shifts through 'tipping points'. As noted by Fred Pearce, an editor of the *New Scientist* and author of *The Last Generation*, the majority of climate scientists now accept that their old ideas about gradual change simply do not represent how the world's climate system works. 'Climate change did not happen gradually in the past, and it will not happen that way in the future. Planet Earth does not do gradual change. It does big jumps; it works by tipping points.'[75]

In 2002, a comprehensive study by the US National Academy of Sciences described a 'new paradigm of an abruptly changing climatic system', now 'well established by research over the last decade'. But the new paradigm 'is little known and scarcely appreciated in the wider community of natural and social scientists and policy-makers'. The report warned that:

> Abrupt climate changes were especially common when the climate system was being forced to change most rapidly. Thus, greenhouse warming and other human alterations of the earth system may increase the possibility of large, abrupt, and unwelcome regional or global climatic events. The abrupt changes of the past are not fully explained yet, and climate models typically underestimate the size, speed, and extent of those changes. Hence, future abrupt changes cannot be predicted with confidence, and climate surprises are to be expected.[76]

There is thus 'a growing fear among scientists that, thanks to man-made climate change, we are about to return to a world of climatic turbulence, where tipping points are constantly crossed'. The last five years alone of scientific research have unearthed previously unknown tipping points that could trigger rapid climate change.[77]

The Two-Degree Limit?

At the June 2005 UK government conference on 'Avoiding Dangerous Climate Change' at the Met Office in Exeter, scientists reported an emerging consensus that global warming must remain 'below an average increase of two degrees centigrade if catastrophe is to be avoided'. It was argued that this will require ensuring that carbon dioxide in the atmosphere stays below 400 parts per million (ppm). Beyond this level, dangerous and uncontrollable climate change is likely to be irreversible.

It is commonly believed that the current concentration of CO_2 in the atmosphere is about 385 ppm. However, about two weeks after the government conference warning about the minimum threshold, the London *Independent* commissioned an investigation by Keith Shine, Head of the Meteorology Department at the University of Reading. Using the latest available figures for 2004, Professor Shine calculated that 'the CO_2 equivalent concentration, largely unnoticed by the scientific and political communities, has now risen beyond this threshold'. Unlike other calculations, Shine accounted for the effects of methane and nitrous oxide, finding that the total concentration of greenhouse gases contributing to global warming (i.e. the equivalent concentration of CO_2) is now 425 ppm and rising fast. In the absence of mitigating strategies to reduce the amount of CO_2 already in the atmosphere, these levels *guarantee that the global mean temperature will rise by two degrees before the end of this century*. Consequently, Shine argued, some of the worst predicted effects of global warming, such as the destruction of ecosystems and increased hunger and water shortages for billions of people in the South, could be unavoidable unless drastic action is taken not only to reduce emissions, but to remove CO_2 from the atmosphere.[78]

Shine's findings were corroborated by the IPCC in October 2007, which concluded that the level of greenhouse gases in the atmosphere in mid 2005 had *reached 445 ppm*, a level that had not been expected for another ten years. Macquarie University climate scientist Tim Flannery remarked:

> We thought we had that much time. But the new data indicates that in about mid 2005 we crossed that threshold. What the report establishes is that the amount of greenhouse gas in the atmosphere is already above the threshold that could potentially cause dangerous climate change.[79]

When asked about the implications, Tom Burke CBE, a former British government environment adviser for 14 years,[80] told the *Independent*:

> The passing of this threshold is of the most enormous significance. It means we have actually *entered a new era – the era of dangerous climate change*. We have passed the point where we can be confident of staying below the 2 degree rise set as the threshold for danger. What this tells us is that we have already reached the point where our children can *no longer count on a safe climate*.[81]

The two-degree limit has been adopted by the European Union as the maximum average temperature increase that humanity can risk. 'Beyond that', notes Paul Brown, 'as unwelcome changes in the earth's reaction to extra warmth continue, it is theoretically possible to trigger runaway climate change, making the earth's atmosphere so different that most of life would be threatened.'[82]

A 2005 joint task force report by the Institute for Public Policy Research (IPPR) in the UK, the Center for American Progress in the US, and the Australia Institute, argues on the two-degree basis that the point of no return may be reached as early as 2015. The report finds that a two-degree temperature rise would trigger an irreversible chain of climatic disasters:

> The possibilities include reaching climatic tipping points leading, for example, to the loss of the West Antarctic and Greenland ice sheets (which, between them, could raise sea level more than 10 meters over the space of a few centuries), the shutdown of the thermohaline ocean circulation (and, with it, the Gulf Stream), and the transformation of the planet's forests and soils from a net sink of carbon to a net source of carbon.[83]

However, it is now becoming clear that the two-degree, 450 ppm EU limit is far too high, a political figure adopted against sound scientific advice. After studying core samples from the bottom of the ocean to track CO_2 levels from millions of years ago, James Hansen, head of the NASA Goddard Institute for Space Studies, concluded in April 2008 that the absolute upper limit for acceptable CO_2 emissions is *350 ppm – a limit that has already been surpassed*: 'If you leave us at 450 ppm for long enough it will probably melt all the ice – that's a sea rise of 75 metres. What we have found is that the target we have all been aiming for is a disaster – a guaranteed disaster'. At levels as high as 550 ppm, Hansen's team found that the world would warm by 6°C, double that of previous estimates. In other words, the impact of CO_2 emissions is likely to be *double* the intensity and severity of the conservative scenarios outlined by the IPCC. This is due to the positive feedbacks that were insufficiently incorporated into IPCC assessments. It is necessary therefore not simply to stop fossil fuel emissions, but to apply technologies to safely extract carbon from the atmosphere and store it safely, allowing the atmospheric concentration to return to a safe level. Yet Hansen warns that even these technologies are seriously inadequate by themselves, and over the long term other methods to increase soil carbon storage capacity should be explored, such as extensive reforestation and more creative carbon sequestration techniques like 'biochar' – a charcoal produced from biomass which could provide long-term carbon storage while improving soil quality and agricultural productivity.[84]

It should be emphasized, then, that even the 350 ppm upper limit was not proposed by Hansen as a *target* for emissions reductions, as it would not prevent massive climate change with potentially uncontrollable and irreversible consequences. As noted by Philip Sutton, who teaches Global Warming Science at the University of Melbourne:

> The total loss of the Arctic sea-ice in summer, ... the loss over 100 years of all the permafrost stored carbon, the acidification of the ocean, the overheating of the oceans, the loss of the Amazon rainforest, the loss of most of the Greenland ice sheet, the destabilisation and major loss of the West Antarctic ice sheet are all issues that have severe ramifications and require lower CO_2 levels than 350 ppm.[85]

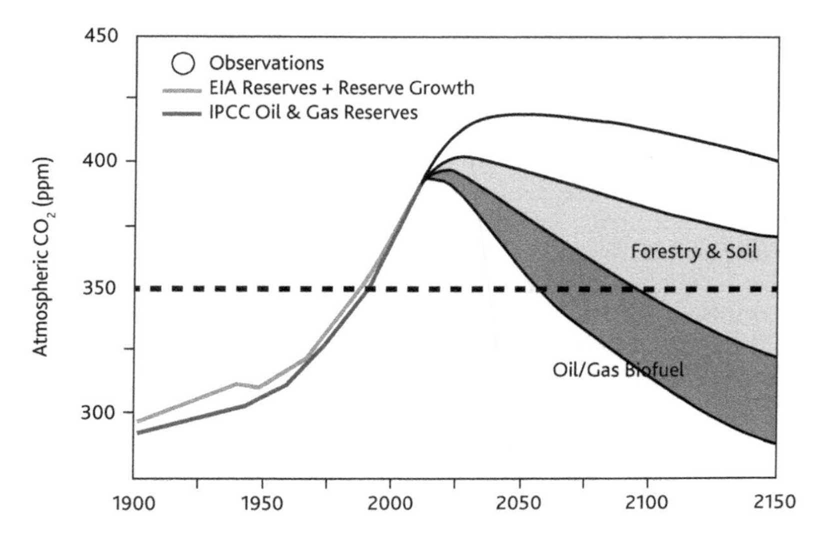

Figure 1.6 CO_2 Reductions with Coal-Carbon Phaseout by 2030.

Source: James Hansen, et al. (2008)

The problem is that each of these effects, probable even at levels below 350 ppm, have their own positive feedback impacts, each with potentially irreversible consequences. A safe level of emissions is somewhere below 330 ppm – most likely, according to Professor John Schellnhuber of the Potsdam Institute, between 280 and 300 ppm.[86] *We appear to have now passed the tipping point, and at current rates of increase of CO_2 emissions, we are well on our way to breaching temperature increases of 2°C and even 3°C at minimum.* Since 2005, increasing evidence has emerged that several major climate sub-system tipping points have thus been passed with potentially irreversible consequences. These consequences, in turn, may trigger the crossing of further tipping points, the cumulative impact of which could push the planet's climate system into a self-reinforcing, runaway warming process.

The Gulf Stream and the Arctic Ice Cap

In May 2005, climate scientists working under Peter Wadhams, Professor of Ocean Physics at Cambridge University, announced that they had found signs of a slowdown in the Gulf Stream, otherwise known as the thermohaline circulation (THC) – a huge convection system that transports warm water from the tropics to the poles and sends cool water back through the depths of the oceans. One of its driving 'engines', the sinking of supercooled water in the Greenland Sea, had 'weakened to less than a quarter of its former strength' due largely to global warming, a change likely to precipitate a drop in temperatures in the UK and northwest Europe. Wadhams and his team also predicted that the slowing of the Gulf Stream might have other effects, such as the complete summer melting of the Arctic ice cap by 'as early as 2020 and almost certainly by 2080'.[87] By December 2005, scientists on an expedition to the Atlantic Ocean, measuring the strength of the current between Africa and the American east coast, found that its circulation had slowed by 30 per cent since an expedition twelve years earlier.[88] More recent research suggests that this may not be as serious as originally suspected, and that the downward trend is less pronounced than hitherto believed.[89]

The fear is that higher temperatures caused by global warming might add fresh water to the northern North Atlantic by increasing precipitation and by melting nearby sea ice, mountain glaciers and the Greenland ice sheet. This influx of fresh water could reduce surface salinity and density, potentially slowing down the Gulf Stream. The southern hemisphere would become warmer and the northern hemisphere would experience cooling, but this cooling effect would probably be counterbalanced by overall excess warming.[90]

Although a large shift in the THC was the main variable studied in the Pentagon's apocalyptic abrupt climate change scenario, climate scientists tend to view a collapse of the Gulf Stream or the THC as a very slim probability. But as time has passed, the likelihood has increased dramatically. According to Michael Schlesinger, Professor of Atmospheric Sciences at the University of Illinois at Urbana-Champaign:

> Absent any climate policy, scientists have found a 70 percent chance of shutting down the thermohaline circulation in the North Atlantic Ocean over the next 200 years, with a 45 percent probability of this occurring in this century. The likelihood decreases with mitigation, but even the most rigorous immediate climate policy would still leave a 25 percent chance of a thermohaline collapse.

He added: 'The shutdown of the thermohaline circulation has been characterized as a high-consequence, low-probability event. Our analysis, including the uncertainties in the problem, indicates it is *a high-consequence, high-probability event*.'[91] Since then, increasing evidence indicates that other climate systems are even closer to tipping points that mark the onset of serious instability.

In August 2005, scientists reported that Arctic sea ice had reached its lowest monthly level on record, dipping an unprecedented 18.2 per cent below the long-term average. Sea ice naturally melts in summer and reforms in winter, but for the first time on record this annual rebound did not occur.[92]

By March 2007, the traversal of this tipping point was no longer in doubt. Mark Serreze, then at the US National Snow and Ice Data Centre (NSIDC), warned that the Arctic would be almost totally ice-free within the next few decades, with a dramatic impact on weather patterns across the northern hemisphere. 'I think there is some evidence that we may have reached that tipping point, and the impacts will not be confined to the Arctic region', he noted. 'With this increasing vulnerability, a kick to the system just from natural climate fluctuations could send it into a tailspin.'[93]

The potential impact that will be seen within this century is not entirely predictable, but the broad contours are clear. NASA's James Hansen has noted

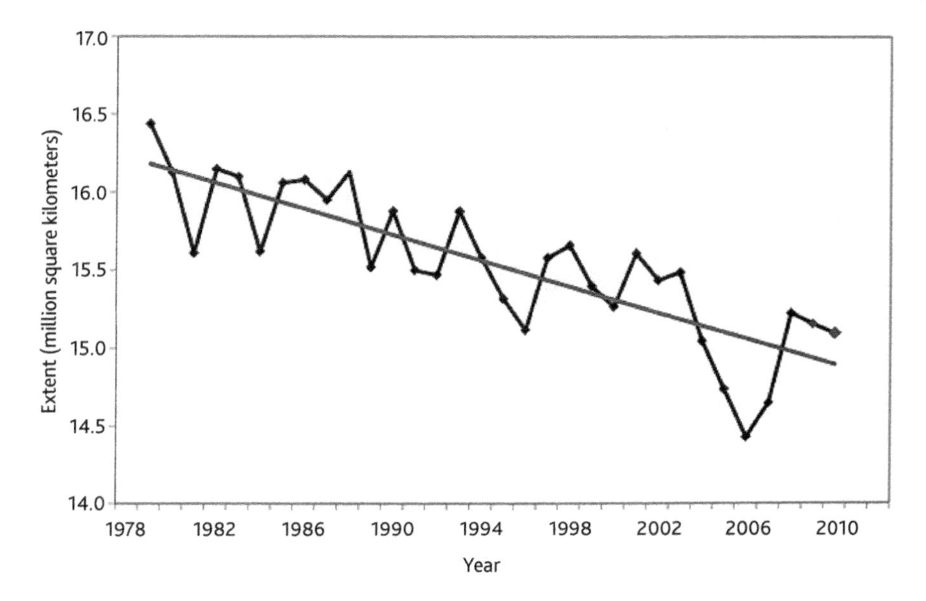

Figure 1.7 Arctic Sea Ice Loss 1978–2010.

Source: National Snow and Ice Date Center (2010)

that in the context of the rapid melting of the Greenland and Antarctic ice sheets, sea levels are already rising at an unprecedented rate, and could end up increasing by one metre every twenty years. 'That is a real disaster, and that's what we have to avoid', he warned. Eric Lindstrom, NASA's head of oceanography, reports that according to satellite data, 'If the (polar) ice sheets really get involved, then we're talking tens of metres of sea level – that could really start to swamp low-lying countries', including large areas of Britain, Western Europe and the United States.[94]

But the rate of Arctic ice's annual retreat is likely to intensify as the melting exposes the darker ocean, which absorbs more of the sun's energy and leads to the further loss of ice. Thus, with each year, the prior predictions of climate scientists are revealed not to be too alarmist, but, to the contrary, far too conservative. Thus by December 2007, NASA satellite data showed that Arctic ice was disappearing so fast that 'an irreversible tipping point has already been reached because of global warming'. Between 2002 and 2006, the volume of Arctic ice had halved, while the Greenland ice sheet had lost almost 19 billion tonnes. According to NASA climate scientist Jay Zwally:

> *At this rate, the Arctic Ocean could be nearly ice-free at the end of summer by 2012*, much faster than previous predictions. The Arctic is often cited as the canary in the coal mine for climate warming. Now as a sign of climate warming, the canary has died. It is time to start getting out of the coal mines... *It's getting even worse than the climate models predicted.*[95]

By the end of 2008, NSIDC scientists declared that the rate of Arctic melting had probably already breached the tipping point. Regional air temperatures during the autumn were higher than expected due to accumulated heat in the ocean caused by the worsening loss of sea ice. This process of 'Arctic amplification', a self-reinforcing positive feedback, was not expected for another 10–15 years, and is occurring at a pace faster than accounted for by any of the IPCC's models.[96] A summer 2009 study of the Arctic ice's thickness, rather than just surface area, found that it had thinned by 40 per cent since 2004.[97] The accelerating Arctic melt also increases the probability of a slowdown of the Gulf Stream or THC, as discussed above:

1. Surface currents carry warm, salty water from the tropics.
2. The water cools, its density increases and it sinks to the deep ocean.
3. The cold water flows back to the equator, driving the 'ocean conveyor' which in turn contributes to the North Atlantic current, the continuation of the Gulf Stream, that warms northern Europe.
4. As ice melts, freshwater dilutes the warm salty water from the tropics.
5. The water becomes less dense so does not sink as fast, weakening the 'conveyor' and therefore possibly disrupting the Gulf Stream.[98]

While potentially contributing to cooling in northern Europe, a slowdown of the Gulf Stream would simultaneously lead to more frequent droughts in other areas. This process is already well under way and getting worse. According to the US National Center for Atmospheric Research (NCAR), the percentage of earth's land area stricken by serious drought more than doubled from the 1970s to the early 2000s, from about 10–15 per cent to 30 per cent, largely due to rising temperatures. Widespread dryness occurred over much of Europe and Asia, Canada, western and southern Africa, and eastern Australia.[99] Global warming is not only melting the Arctic, it is melting the glaciers that feed Asia's largest rivers – the Ganges, Indus, Mekong, Yangtze and Yellow. Because glaciers are a natural storage system, releasing water during hot arid periods, the shrinking ice sheets might aggravate water imbalances, causing flooding as the melting accelerates, followed by a reduction in river flows. This problem is only decades, possibly only years, away, potentially resulting in water shortages for hundreds of millions of Africans and tens of millions of Latin Americans. By 2050, more than 1 billion people in Asia could face water shortages, and by 2080, water shortages could threaten 1.1 billion to 3.2 billion people. Some climate models show sub-Saharan Africa drying out by 2050.[100] By other projections, as soon as 2025 some 5 billion people globally could be suffering from serious water shortages, half a billion of them due to climate change.[101]

While Arctic sea ice is rapidly disappearing, the last 20 years have at the same time seen an apparently anomalous build-up of sea ice in West Antarctica. From 1979 to 2006, Antarctic wintertime ice extent increased by 0.6 per cent per decade. The average year-round ice level has also increased in this period.[102] Climate sceptics often claim that this is evidence against global warming. However, it actually proves the opposite. Once again, it is necessary to account for longer-term climate trends and internal variability to understand what is happening. A 2003 study published in *Science* finds that detection of long-term change in Antarctica is masked by large decade-to-decade fluctuations. These decadal variations have produced the apparent short-term increases in Antarctic sea ice from recent satellite data. The *Science* study finds that since the 1950s, however, there has been *a large overall reduction of approximately 20 per cent* in the northern extent of Antarctic sea ice in the region south of Australia.[103] Furthermore, *overall* ice loss in the Antarctic has been detected even over shorter periods. A NASA mission in 2006, based on the first ever survey of the entire Antarctic ice sheet on land, 'found the ice sheet's mass has decreased significantly from 2002 to 2005', raising global sea levels by about 1.2 millimetres – about 13 per cent of the overall observed sea level rise for this period.[104]

In addition, far from disproving global warming, the temporary increase of Antarctic sea ice has been *predicted* as its consequence. Researchers have long known that snow builds glaciers. In 2005, a team of scientists combined snow-thickness measurements with the results of modelling studies and found that snow may also build Antarctica's sea ice. Publishing their findings in the *Journal of Geophysical Research*, they argued that as global warming intensifies, more moisture has made its way to high latitudes, leading to heavier snowfalls in the Antarctic in particular. With increased snowfall, a sufficiently thick layer of snow would push the ice underwater. The seawater in the snow-ice boundary would freeze, thickening the floe. 'Some of the melt in the Arctic may be balanced by increases in sea ice volume in the Antarctic', noted lead scientist Dylan C. Powell.[105]

Another factor is ozone depletion. As global warming in the near-surface lower atmosphere (troposphere) accelerates in correlation with increased output of greenhouse gases, this warmer air becomes trapped in the lower troposphere by a CO_2 'blanket', preventing the heat from entering the upper atmosphere (stratosphere) which thus begins to cool. Thus, higher surface temperatures due to global warming are correlated with lower temperatures in the upper atmosphere. In the Antarctic, high-altitude colder temperatures intensify the impact of ozone destruction by chlorofluorocarbons (CFCs), emitted by so many industrial technologies. These lower temperatures thus counter the overall impact of global warming in the region.[106]

Adding to these findings, research published in 2008 dissolved any further ambiguity that may have remained. A new study in *Nature Geoscience* found that over the long term both the Arctic and the Antarctic are 'getting less icy because of global warming'. The study, comparing temperature records and four computer climate models, found a warming in both polar regions that could only be explained by a build-up of greenhouse gases, mainly from burning fossil fuels, rather than natural forces. In both cases, the observed warming could only be reproduced in climate models by including human influences via fossil fuels.[107]

More Positive Feedbacks

The levels of warming in the polar regions are merely isolated signifiers of the momentous scale of climate changes, among many others that scientists are racing to keep up with. Scientists have isolated a total of 12 eco-system 'hotspots' including the above, which they consider to be especially vulnerable to human intervention. According to John Schellnhuber, we have barely begun to recognize the danger of triggering large-scale, rapid and irreversible changes across the entire planet, by affecting elements of these crucial systems which act as massive environmental regulators: 'We have so far completely underestimated the importance of these locations. What we do know is that going beyond critical thresholds in these regions could have dramatic consequences for humans and other life forms.' Breaching of tipping points in these climate sub-systems could have uncontrollable consequences that could, in turn, increase the probability of breaching a tipping point in the global climate system, possibly leading to runaway warming.[108]

When accounting for the potential impact of positive feedbacks, the two-degree Celsius limit is simply too high. Even with changes greater than 1°C, the probability of runaway climate change due to positive feedbacks is greatly increased, and above 2°C, where we are currently headed at existing rates of emissions, the probability is magnified immeasurably. According to James Hansen and colleagues:

> If the [additional] warming is less than 1°C, it appears that strong positive feedbacks are not unleashed, judging from recent Earth history. On the other hand, *if global warming gets well out of this range*, there is a possibility that *positive feedbacks could set in motion climate changes far outside the range of recent experience*.[109]

Among the most dangerous positive feedbacks is the melting of Arctic permafrost, which is far more abundant in carbon than was previously believed. Arctic permafrost contains more carbon in the form of methane than is found in the entire atmosphere today, three times that from all industrial emissions.[110] Once the permafrost melts, the increased run-off of warmer water melts sub-ice methane clathrates, releasing the gas into the atmosphere. For the past five years, Alaskan and Siberian permafrost has been melting, releasing five times more methane than was previously estimated.[111] Thus, the more global warming melts the permafrost, the greater the amount of methane released. Methane is 20 times more powerful as a greenhouse gas than CO_2. In September 2008, an expedition organized by the International Siberian Shelf Study, prepared for publication by the American Geophysical Union, confirmed that millions of tonnes of sub-sea methane are being released due to rapidly melting Arctic permafrost.[112] An estimated 1400 Gt of carbon is trapped as methane under the Arctic permafrost, 5–10 per cent of which has been punctured due to melting. Scientists say that a 'release of up to 50 Gt of predicted amount of hydrate storage [is] highly possible for abrupt release at any time' – a quantity equivalent to a doubling of current levels of CO_2.[113] According to former US Energy Department geologist John Atcheson, the release of this methane signals the impending danger of a runaway greenhouse effect with extinction-level consequences:

> A temperature increase of merely a few degrees would cause these gases to volatilize and 'burp' into the atmosphere, which would further raise temperatures, which would release yet more methane, heating the Earth and seas further, and so on. There's 400 gigatons of methane locked in the frozen arctic tundra – enough to start this chain reaction – and the kind of warming the Arctic Council predicts is sufficient to melt the clathrates and release these greenhouse gases into the atmosphere.
>
> Once triggered, this cycle could result in runaway global warming the likes of which even the most pessimistic doomsayers aren't talking about... The most recent of these catastrophes occurred about 55 million years ago in what geologists call the Paleocene-Eocene Thermal Maximum (PETM), when methane burps caused rapid warming and massive die-offs, disrupting the climate for more than 100,000 years.
>
> The granddaddy of these catastrophes occurred 251 million years ago, at the end of the Permian period, when a series of methane burps came close to wiping out all life on Earth... *If we trigger this runaway release of methane, there's no turning back. No do-overs. Once it starts, it's likely to play out all the way.*[114]

Another potential source of positive feedbacks are trees and forests, which are not only capable of acting as sinks to absorb carbon from the atmosphere, but instead can also act as net sources of carbon. In the case of the pine forests of western North America, global warming created a perfect climate for bark beetles, leading to a massive explosion in the beetle population. This in turn weakened the ability of the northern forests to act as a carbon sink, as documented in *Nature*:

> During outbreaks, the resulting widespread tree mortality reduces forest carbon uptake and increases future emissions from the decay of killed trees. The impacts of insects on forest carbon dynamics, however, are generally ignored in large-scale modelling analyses... This impact converted the forest from a small net carbon sink to a large net carbon source both during and immediately after the outbreak... Insect outbreaks such as this represent an important mechanism by which climate change may undermine the ability of northern forests to take up and store atmospheric carbon.[115]

Moreover, the tropical forests of the Amazon, the Congo and Borneo are nearing critical resiliency because of decreased rainfall due to global warming.[116] A study in Amazonia by the Massachusetts-based Woods Hole Research Centre concluded that the rainforest cannot sustain three consecutive years of drought without breaking down. In year three, sample trees studied by the scientists started dying, literally came crashing down, exposing the forest floor to the drying sun. By the end of the year the trees had released more than two thirds of the carbon dioxide stored over the course of their lives, thus accelerating climate change. The Amazon rainforest contains 90 billion tonnes of carbon, enough in itself to *increase the rate of global warming by 50 per cent.*[117] Recent research shows that intensifying droughts in the Amazon are linked to warmer sea surface temperatures in the tropical Atlantic Ocean. As global temperatures rise, the Amazon is thus increasingly at risk. Hence climate models increasingly forecast a dire future for the Amazon rainforest. As the tropical Atlantic warms, the southern Amazon is likely to see higher temperatures and less rainfall.[118] Further, due to the impact of deforestation, logging, fires, and drought, the Amazon could be reduced by 55 per cent by 2030. The impact of this massive loss of rainforest will result in emissions of 15–26 Gt of carbon in less than three decades – *that is, 15–26 billion metric tonnes of carbon.*[119] The massive influx of carbon into the atmosphere could accelerate global warming by as much as 1.5°C on top of the preceding temperature increase. With each rise in temperature, positive feedbacks such as this would be intensified with irreversible impacts. These in turn would increase the probability of an unstoppable escalation of global temperatures.

Just as trees may well end up exacerbating climate change as global warming accelerates, the same also applies to soil. One quarter of our carbon emissions are now being absorbed by the soil, but its capacity to do so is decreasing. As global warming increases, soil is increasingly liable to release its stored carbon. There is some 300 times as much carbon trapped in the soils as is released each year from burning fossil fuels. From 1978 to 2003, 13 Mt of carbon held in UK soils has been released each year.[120] As global warming accelerates microbial activity in the soil, large quantities of carbon will be added to the atmostphere.[121]

Another major positive feedback is from water vapour, which in fact is the dominant greenhouse gas. Because water vapour enters the atmosphere via evaporation, the level of water vapour correlates with temperature. Temperature rise increases the rate of evaporation, and thus the quantity of water vapour in the atmosphere. Water vapour absorbs heat and this also warms the air, which in turn can cause further evaporation. The warming effect of CO_2 emissions amplifies all this by causing yet more water to evaporate. Water vapour can roughly double or, in tandem with other positive feedbacks, even triple the impact of CO_2 warming.[122]

The cumulative impact of positive feedbacks like melting permafrost, failing forests, activated soil, and water vapour on climate change is difficult to imagine, but it is clear that they pose the danger of triggering rapid, irreversible changes to key climate sub-systems once greenhouse gases are in the atmosphere at a concentration above 350 ppm for a prolonged period (we are currently approximately 75 ppm above this level).

Conventional climate models tend to omit the impact of these positive feedbacks. When incorporated, the findings are disturbing. Scientists at the Massachusetts Institute of Technology (MIT), in a study published by the

Journal of Climate, projected that between 2091 and 2100 global average temperature would rise by 5.1°C.[123] Similarly, a December 2008 Met Office study concluded that the world could warm by between 5–7°C by 2100 at the current rate of emissions increases.[124] On the way, we could reach an average global temperature of 3–4 degrees before 2060.[125] One of the most comprehensive yet least publicized studies, by climate scientists at Lawrence Berkeley National Laboratory and the University of California at Berkeley, concluded that 'global temperatures at the end of this century may be significantly higher than current climate models are predicting', *rising by as much as 8°C*. 'If the past is any guide', said Margaret Torn from the Berkeley team, 'then when our anthropogenic greenhouse gas emissions cause global warming, it will alter earth system processes, resulting in additional atmospheric greenhouse gas loading and additional warming'.[126] Given that 6°C is already recognized as a wholly unacceptable level of warming implying the potential destruction of most life on the planet, the prospect that temperatures may rise by 8°C within this century signals the necessity of urgent preventive action.

A SYSTEMIC FAILURE

A Record of Early Warnings

The landmark 2007 UN IPCC report follows a long spate of diverse scientific assessments recognizing that our overexploitation of fossil fuels could lead to the demise of civilization itself. But they were for the most part ignored by policymakers. We will review a few examples here. Consider, for instance, the 1992 'warning report' produced by the Union of Concerned Scientists, signed by over 1500 members of national, regional and international science academies, representing 69 nations from around the world, including each of the twelve most populous nations and the nineteen largest economic powers, with the full list including a majority of the Nobel laureates in the sciences. The warning report announced that: 'Human beings and the natural world are on a collision course. Human activities inflict harsh and often irreversible damage on the environment and on critical resources'. These practices constitute a 'serious risk' to human society and the plant and animal kingdoms, threatening to 'so alter the living world that it will be unable to sustain life in the manner that we know'. The report condemned 'massive tampering with the world's interdependent web of life – coupled with the environmental damage inflicted by deforestation, species loss, and climate change', and noted that such practices 'could trigger widespread adverse effects, including unpredictable collapses of critical biological systems whose interactions and dynamics we only imperfectly understand'. It continued:

> No more than one or a few decades remain before the chance to avert the threats we now confront will be lost and [with them] the prospects for humanity... The developed nations are the largest polluters in the world

> today. They must greatly reduce their overconsumption, if we are to reduce pressures on resources and the global environment... No nation can escape from injury when global biological systems are damaged. No nation can escape from conflicts over increasingly scarce resources.[127]

Just under a decade on, such conclusions were reiterated in the *Global Environment Outlook 2000* (GEO-2000), launched by the United Nations Environmental Programme based on contributions from UN agencies, 850 experts and 30 environmental institutes. The report described a variety of full-scale emergencies: The world water cycle is unlikely to cope with demand in coming decades; land degradation has negated many advances made by increased agricultural productivity; air pollution is at crisis point in many major cities; dangerous global warming is inevitable. A survey for GEO-2000 conducted by the Scientific Committee on Problems of the Environment found that according to 200 leading scientists in 50 countries, water shortage and global warming constituted the two gravest problems, followed by desertification and deforestation at national and regional levels. But GEO-2000's most important finding is encapsulated in the following conclusion: 'The present course is unsustainable and postponing action is no longer an option.'[128]

The 2005 Millennium Assessment report by the United Nations – a synthesis of research by 1300 experts from 95 countries, hailed as the most comprehensive survey of the planet's natural life support systems – found that 15 of 24 global ecosystems were already in severe decline; human civilization is absorbing the earth's natural resources at unsustainable break-neck speed; and as a consequence, we are in danger of destroying *two thirds of the earth's ecosystems*.[129] The UN report warned that the earth is faced with the emergence of new diseases, sudden changes in water quality, the creation of coastal 'dead zones', the collapse of fisheries and drastic shifts in regional climate. This combination of new diseases, absence of fresh water, continuing decline of fisheries and unpredictable weather was already having increasingly fatal results. For example, half of the urban populations of Africa, Asia, Latin America and the Caribbean suffer from diseases directly associated with global environmental decline, *already leading to a death toll of approximately 1.7 million people a year*. Similarly, whole species of mammals, birds and amphibians continue to be made extinct at nearly 1000 times the naturally occurring rate.[130]

These reports showed clearly that our 'way of life' associated with modern industrial civilization is deeply implicated in the destruction of our environment. They acknowledged that the ecological damage wrought by anthropogenic climate change is intimately tied to the structure of the global political economy; that global warming, ozone depletion, species extinction, pollution, and so on, are ultimately symptoms of modern industrial civilization's disruption of the earth's life support systems, implemented in the pursuit of an amoral, unrestrained drive for economic growth.[131] Accordingly, the GEO-2000 prescribed the necessity of 'a shift in values away from material consumption. Without such a shift, environmental policies can effect only marginal improvements'.[132] This echoed one of the earliest warnings from the Club of Rome, that avoiding 'global catastrophe' required 'fundamental changes in the values and attitudes of man... such as a new ethic and a new attitude towards nature'.[133]

Market Failure from Kyoto to Copenhagen: Too Little, Too Late

The response of governmental policymakers to such warnings has been decidedly indifferent. Not only have governments displayed reluctance to implement adjustments *within* the system; they show no intention whatsoever of effecting the necessary changes *to* the system itself. Yet ever greater numbers of environmental experts, from the UN to independent NGOs and scientists, are beginning to realize that we need some kind of systemic transformation. Indeed, this is precisely the issue that remains thoroughly underinvestigated. Civilization, in other words, is in denial.

The Kyoto Protocol – an international legally binding treaty established in 1997 and coming into force in 2005 – demanded that industrialized countries reduce their emissions by only 5.2 per cent, compared to 1990 emissions levels, by the year 2012. Apart from not being ratified by two of the larger producers of these pollutants, the US and Australia, the Protocol also treated China and India, both major fossil fuel emitters, as 'non-industrialized' countries. The Protocol also lacked enforcement mechanisms for reduction targets that were already far too modest. As Professor Gwyn Prins of the London School of Economics noted, actual emissions of the EU – the leading proponent of Kyoto – had risen by at least 10 per cent, a figure which was 'massaged' partly by 'including offsets purchased under the UN Clean Development Mechanism (CDM): offsets that were not real and, in many cases, fraudulent'.[134] The CDM facilitated an international carbon-trading scheme which, although intended to reward less developed countries investing in renewable energy technologies, in practice 'is handing out billions of dollars to chemical, coal and oil corporations and the developers of destructive dams - in many cases for projects they would have built anyway'.

Basically, carbon offsets are when a corporation financially supports other projects that supposedly reduce fossil fuel emissions, such as renewable energy and energy efficiency initiatives. Thus, a polluter who registers a certain amount of emissions, such as an airline, can purchase offsets by funding 'green' projects. It works like this. The potential reduction of emissions by a particular renewable energy (or similar) 'green' project is given a monetary value. This value is the 'credit'. The airline that makes profits from its normal business of burning fuel can 'offset' its own real emissions by purchasing these 'credits'. The problem is that this effectively allows large industrial behemoths hell-bent on continuing with the carbon-intensive emitting of greenhouse gases to continue, if not accelerate, business as usual by investing heavily in carbon credits linked to the purported 'green' projects of other companies. The more money they invest in this way, the more they 'offset' their own tremendous continuing emissions. The 'offsets' are recorded, and by law count as actual effective emissions reductions. In other words, a polluter can literally buy 'emissions reductions' – or the label of 'emissions reductions' – without actually reducing one's own emissions. Compounding the problem is that the carbon trading market is unregulated and extremely confused – there are no scientifically-determined and internationally-agreed criteria for how to calculate the 'credit value' of 'green' projects. This means that all kinds of dubious operations, which have little or no impact on actual reductions, have been fraudulently valued at high levels of credit. Companies have purchased these expensive credits and used them to record vast emissions reductions on their own books. And there is also the danger that corporations might try to attach carbon credits to purportedly 'green' projects which are not truly 'green' at all.

Thus two thirds of the recorded 'emission reduction' credits tabulated by the CDM from such offsets 'are not backed by real reductions in pollution', but are rather calculated from purchases of carbon offsets 'rather than by decarbonising their economies'. For example, if a 'Chinese mine cuts its methane emissions under the CDM, there will be no global climate benefit because the polluter that buys the offset avoids the obligation to reduce its own emissions'.[135]

In theory, the idea of carbon quotas and trading – where each country is assigned an emissions quota with permits issued to the biggest carbon emitters (such as power firms) – seems workable. Those companies that reduce emissions and use less than their quota can sell leftover permits to others which produce emissions as before. Each year the quota is reduced, so market forces push higher the penalty for emitting greenhouse gases. Outside of the CDM under Kyoto, Europe has already established an emissions trading network, and at least nine US states have started trading carbon dioxide among themselves.[136] But in practice, fossil fuel emissions have continued to escalate without relief. For instance, the EU's Emissions Trading Scheme (ETS) was from the outset compromised by corporate power, which pressured states to over-allocate emissions rights to the very industries most responsible for fossil fuel pollution. Consequently, the price of carbon dropped by over 60 per cent, reducing incentives to invest in renewable energies. Britain's heaviest polluting industries together earned £940 million in profits from the first year of the scheme.[137]

Given the impotence of the Kyoto Protocol, hopes were high that the Copenhagen summit in 2009 might produce a deal with more teeth. Unfortunately, this was not to be. The resulting Accord between the US, China, India, Brazil and South Africa, although 'recognized' by the 193 countries present, was not legally binding. The document 'recognizes the scientific view that' global average temperatures should not rise more than two degrees Celsius – although, as noted, the scientific evidence supports a far lower maximum safe temperature increase range within about one degree. But despite this recognition, no year for the peak of carbon emissions is indicated, nor are countries compelled to pledge emissions reductions necessary to respect even the two-degree limit. Rather, they are expected to simply declare their own 2020 emissions reduction targets, and no penalties are established even to help enforce these.[138]

Stated emissions pledges were minimalist. US President Obama pledged to reduce emissions by only 17 per cent from 2005 levels by 2020 – 4 per cent below the more conventional 1990 benchmark, and still far too insignificant to prevent dangerous climate change. Similarly, the EU pledged a roughly 20 per cent reduction from 1990 levels – better than the US, but still ineffective. Under the Climate Change Act passed in 2008, the UK had already pledged to a 34 per cent reduction in carbon dioxide emissions by 2020, and an 80 per cent reduction by 2050 (against the 1990 baseline).[139] Yet the UK example perfectly illustrates the ineffectiveness of pledges that are subject solely to the domestic discretion of respective national governments. The Climate Change Act, for instance, grants the Secretary of State the power to *amend* the 80 per cent reduction target, and to *amend* the base year against which it is measured. It currently excludes emissions from aviation and international shipping. In June 2008, the British government dropped the stipulation that UK companies must declare their annual carbon emissions, and the requirement that 70 per cent of the targets be met by actual emissions cuts, rather than through buying carbon offsets from less developed countries, although, commendably, the five-year carbon budget scheme will limit the latter activity.[140]

Overall, the international carbon trading and offset schemes at the focus of government efforts to deal with climate change are increasingly an arena by which powerful vested interests can exploit fears about global warming in order to consolidate unprecedented profits – and London is at the centre of it. According to the *New York Times*, 'British companies were the leading global investors in carbon projects' and 'more carbon was traded in London than in any other city'. The newspaper noted that emissions management is 'one of the fastest-growing segments in financial services, and companies are scrambling for talent. Their goal: a slice of a market now worth about $30 billion, but which could grow to $1 trillion within a decade.' Louis Redshaw, head of environmental markets at Barclays Capital, predicted that: 'Carbon will be the world's biggest commodity market, and it could become the world's biggest market overall.'[141] Similarly, Andrew Ager, head of emissions trading at Bache Commodities in London, predicts that the carbon market 'could grow to around $3tn compared to the £1.5tn market there is for oil'.[142]

The problem is that by accepting neoliberal capitalist markets as a given, carbon trading overlooks the systemic origins of climate change. Such market-oriented solutions are inspired by Aubrey Meyer's 'Contraction and Convergence' (C&C) model of action for global emissions reductions. 'Contraction' requires the adoption of a safe target for atmospheric CO_2, as the basis for calculating a progressive decline in annual global emissions toward that target by a specific target year, all determined according to climate science. 'Convergence' requires the assignment of annual emissions quotas to each country which converge toward a common level of per capita emissions by the target year. Added to this is a market-based carbon trading plan to permit wealthier, high emitters to purchase emissions rights from poorer, low emitters.[143]

The problem with the C&C model is not the model itself – which certainly identifies one viable path for global action – but the lack of attention to the socio-political and economic structures that prevent policymakers from *genuinely* implementing the C&C model in the first place. Inattention to the deeper structural and systemic issues means that the current policy processes inspired by the C&C model work in practice to sustain inequitable structures, while continuing to escalate fossil fuel emissions. Indeed, a damning 2009 report from Deutsche Bank's Asset Management Division reports that carbon markets are unlikely to contribute to significant emissions reductions 'for the foreseeable future', and will not encourage sufficient investments in renewable energy. The report called instead for governments to introduce stronger incentives such as feed-in tariffs.[144] Given that market mechanisms alone are bound to fail, the question remains as to why governments are reluctant to go beyond them. Overall, Western government attitudes toward climate change – as principally a problem of reducing CO_2 emissions by introducing new market mechanisms that penalize carbon emission – are premised on a remarkable self-deception: that we can continue with hydrocarbon exploitation, and the pursuit of economic growth, while simultaneously saving the climate.

This failed paradigm is exemplified by one of the most celebrated contributions to the debate, the *Stern Review on the Economics of Climate Change*, which is the basis for much UK and Western government climate policy. It concludes that by investing one per cent of global GDP (revised upwards to two per cent in 2008) it would be possible to avoid the worst effects of climate change, while failure to do so could damage global GDP by up to 20 per cent. To his credit, report author Sir Nicholas Stern, then head of the UK Government Economic Service, recognized that this would constitute the most monumental *market failure* the world has ever seen. Yet as illustrated by the corporate co-optation of the global carbon market shown above, this would be less a market failure than an *integral function of market behaviour under neoliberal capitalism* motivated by short-term profit maximization. In this context, Stern's recommendation to introduce market mechanisms that would generate incentives for carbon cuts are widely off the mark.[145]

The problem is that neither Kyoto nor Copenhagen provides a clear plan for *how* the world is to reconfigure its energy supply to renewable sources and reduce consumption premised on hydrocarbon dependency. While offering no meaningful curtailment of our trajectory toward climate catastrophe, current policies do provide a way of piling huge costs on the public, drastically increasing state revenues, and facilitating corporate profiteering – without actually solving the causes of the problem. This unfortunately feeds the suspicion that Western governments are exploiting climate hysteria to consolidate their own questionable political and economic programmes. *In summary, current emissions reductions policies will effectively escalate* CO_2 *emissions further, at levels liable to propel global warming well past the two-degree tipping point and into the realm of increasingly dangerous and rapid climate change.*

Unfortunately, it appears that alarming strategic decisions have already been made. In 2008, the British government's chief scientific advisers publicly asserted that *a 4°C rise in global temperatures is most probably inevitable and irreversible*, and that the task for governments now is not so much preventing dangerous climate change as it is adapting to the extreme conditions that this change will unavoidably bring. The government's chief scientific adviser to the Department of Environment, Food and Rural Affairs, Professor Bob Watson, described the 2°C tipping point as an unrealistic upper limit: 'But given this is an ambitious target, and we don't know in detail how to limit greenhouse gas emissions to realise a 2 degree target, we should be prepared to adapt to 4°C.' The government's former chief scientific adviser Sir David King backed Watson's position, arguing that 'even with a comprehensive global deal to keep carbon dioxide levels in the atmosphere at below 450 parts per million there is a 50 per cent probability that temperatures would exceed 2°C and a 20 per cent probability they would exceed 3.5°C'.[146]

This indicates that Northern governments have accepted as inevitable the catastrophic consequences of global warming well beyond 2°C, after which irreversible climate change becomes increasingly probable, and have even settled for accepting the consequences of global warming of up to 4°C, *at which level rapid, runaway warming becomes a foregone and irreversible conclusion.* Some of the consequences of these scenarios were outlined by Lynas, summarizing the findings of his book *Six Degrees*:

> The impacts of two degrees warming are bad enough, but far worse is in store if emissions continue to rise. Most importantly, 3°C may be the 'tipping point' where global warming could run out of control, leaving us powerless to intervene as planetary temperatures soar. The centre of this predicted disaster is the Amazon, where the tropical rainforest, which today extends over millions of square kilometres, would burn down in a firestorm of epic proportions.
>
> Computer model projections show worsening droughts making Amazonian trees, which have no evolved resistance to fire, much more susceptible to burning. Once this drying trend passes a critical threshold, any spark could light the firestorm which destroys almost the entire rainforest ecosystem. Once the trees have gone, desert will appear and the carbon released by the forests' burning will be joined by still more from the world's soils. This could boost global temperatures by a further 1.5°C – tipping us straight into the four-degree world.
>
> Three degrees alone would see increasing areas of the planet being rendered essentially uninhabitable by drought and heat. In southern Africa, a huge expanse centred on Botswana could see a remobilization of old sand dunes, much as is projected to happen earlier in the US west. This would wipe out agriculture and drive tens of millions of climate refugees out of

the area. The same situation could also occur in Australia, where most of the continent will now fall outside the belts of regular rainfall.

With extreme weather continuing to bite – hurricanes may increase in power by half a category above today's top-level Category Five – world food supplies will be critically endangered. This could mean hundreds of millions – or even billions – of refugees moving out from areas of famine and drought in the sub-tropics towards the mid-latitudes. In Pakistan, for example, food supplies will crash as the waters of the Indus decline to a trickle because of the melting of the Karakoram glaciers that form the river's source. Conflicts may erupt with neighbouring India over water use from dams on Indus tributaries that cross the border.

In northern Europe and the UK, summer drought will alternate with extreme winter flooding as torrential rainstorms sweep in from the Atlantic – perhaps bringing storm surge flooding to vulnerable low-lying coastlines as sea levels continue to rise. Those areas still able to grow crops and feed themselves, however, may become some of the most valuable real estate on the planet, besieged by millions of climate refugees from the south.[147]

Yet even these catastrophic scenarios, now considered inevitable, may be deeply conservative predictions. Kevin Anderson from the Tyndall Centre for Climate Change Research at Manchester University argues that it is 'improbable' that the atmospheric concentration of CO_2 could be restricted to 650 ppm even with 'draconian emission reductions within a decade'.[148] At 650 ppm, according to Hansen's analysis taking account of the positive feedbacks underplayed by the IPCC, the earth's average temperature would *rise well beyond 6°C, creating an uninhabitable world before 2100*. And with governments and corporate emitters continuing their business as usual, even this is a conservative forecast, suggesting that we may trigger a runaway warming process as early as mid century.

All this raises fundamental questions about the nature of state contingency-planning that addresses the impact of climate change from a national security angle, looking at not merely the local effects of global warming, but also the wide-ranging ramifications in terms of massive migrations, food shortages, water shortages, greater propensity for competition and conflict over resources, and the magnified danger of civil unrest. From a *national security* outlook, climate change becomes an issue of mobilizing state resources to protect existing structures of political and economic power, by controlling increasingly volatile domestic and international populations. Classified studies by the US intelligence community have already red-flagged 'regional partners' of the US in key strategic regions in Africa, Central Asia and the Middle East who are likely to face severe problems, and whose identities remain confidential to avoid diplomatic friction. Climate change is thus viewed as a 'threat-multiplier' to traditional security issues such as 'political instability around the world, the collapse of governments and the creation of terrorist safe havens'. By implication, climate change will serve to amplify the threat of international terrorism, particularly in regions with large populations and scarce resources.[149]

Their focus, then, is not on avoiding dangerous or even catastrophic climate change, but on maintaining business as usual while developing new security frameworks that sustain the fundamental structures of the global political economy, despite the massive human and social costs. The climate change discourse endorsed by Western officialdom 1) evades the mounting evidence of the exponential *acceleration* of climate change within the next few decades, long before the 2100 mark; 2) overlooks the question of our civilization's relations of energy production, in other words our fundamental *dependence on hydrocarbon energy sources*, like oil, gas and coal; and 3) ignores the global political economy's structural imperative for unlimited economic growth leading to unsustainable levels of fossil fuel exploitation.

For while the worldwide consumer demand for energy will continue to rise, the necessity of slashing our CO_2 emissions obviously requires a corresponding drastic drop in energy consumption. How governments will continue to sustain the booming energy requirements of their societies while still reducing hydrocarbon energy consumption to reduce emissions, without rearranging the structure of the economy, remains a mystery. Without addressing the specific mechanisms by which societies will eventually cease to rely on the exploitation of hydrocarbon energies through a multi-billion dollar crash programme to transform the global energy system, meaningful CO_2 reduction is wholly impossible.

Correspondingly, the question of reducing world consumption of hydrocarbon resources is also tied to issues around sustaining industrial agriculture, which is fundamentally dependent on supplies of *cheap petroleum*. The imperative to reduce oil dependency and cut CO_2 emissions raises the dilemma of how *world food production* can simultaneously be maintained in order to feed a growing planetary population. Given that climate change threatens to generate intensifying water shortages and droughts affecting the world's leading agricultural regions, a business-as-usual approach suggests that a permanently altered climate will involve a future of grossly inadequate food and water supplies. There is no doubt, then, that the fallout of this approach would be catastrophic – indicating the need to dispense with the neoliberal model of unlimited growth. *Ignoring the instrumental role of the 'growth imperative' as a systemic pressure rooted in the structure of the global political economy guarantees the continuation of global warming.*

As British environmentalist George Monbiot points out, to avoid dangerous climate change the entire world must dramatically reduce greenhouse gases by no less than 90 per cent by 2030 (in fact, it should be *before 2020*, with further efforts to safely capture and store carbon from the atmosphere through carbon sequestration, among other methods, according to Hansen). Doing so, he shows, would require large-scale changes in the infrastructure of Western societies to downsize energy consumption and revert to renewable energies. Monbiot focuses, for instance, on a combination of state-led technocratic solutions designed to transform markets and impose strict regulations on all emissions-generating activities, including: finding improved ways to build homes and other buildings based on an optimal combination of renewable and non-renewable energy sources; radically changing land transportation but without reducing mobility; firmly curtailing air travel; and massively curbing greenhouse gas emissions of the retail and cement industries; all backed up by a comprehensive emissions trading plan and international carbon rationing scheme.[150]

Whatever the shortfalls of some of these policy solutions, the overarching obstacle to their implementation is that both states and markets concurrently operate in the context of the *unequal structure of the global political economy;* the 'psychological grip of capitalist consumption patterns'; and are *both subject to the pressures of powerful vested interests* – like the 'fossil fuel lobby, heavy industry, airlines', and so on. Government policies are currently tied closely to these structures, and cannot change course sufficiently unless they are transformed.[151] It is therefore essential to account for the central necessity of fundamental transformation in the socio-political, economic, ideological and ethical structure of the global system – starting with the extent to which vested financial interests are tied to the global hydrocarbon energy system: a hierarchical structure of geopolitical domination by core Northern states over a network of peripheral states in regions like the Middle East, Central Asia and West Africa containing the bulk of the world's strategic oil and gas reserves.

2
Energy Scarcity

Over-dependence on hydrocarbon energy exploitation is a defining feature of modern industrial civilization. This chapter seeks to identify the reasons behind the contemporary global political economy's over-dependence on cheap petroleum in particular, as well as other carbon-based energy forms. The fundamental principles involved in oil exploration and exploitation are set out in order to assess the validity of the concept of 'peak oil' – the idea that at some point the world will run out of cheap oil. In exploring the major debates around this subject, I confront the variety of sceptical perspectives that deny that 'peak oil' is a problem, by reviewing the alternative theories as well as interrogating the most prominent arguments presented by the oil industry. A review of the available data reveals not only that 'peak oil' is a reality, but that by the best estimates we are already inhabiting a 'post-peak' world where conventional oil will become increasingly difficult to find, with major geopolitical ramifications for the stability of the international system in coming years.

I extend this argument to examine the prospects for finding alternative sources of hydrocarbon energy, including the short-term and long-term viability of coal, natural gas, unconventional oil and gas, as well as nuclear power. I find not only that these forms of energy are unlikely to be of much use after the mid twenty-first century, but also that continuing hydrocarbon energy exploitation in any form will have a deep impact on climate change. Without urgent mitigation and preventive actions, the future therefore holds the possibility of converging crises of climate catastrophe and energy scarcity. Further, I argue that any preventive actions must be rooted ultimately in global systemic change if we are to successfully make a transition to clean and renewable forms of energy. This means that human society after the twenty-first century will need to go beyond carbon-based forms of industrial social relations – necessarily heralding the dawn of a post-carbon civilization.

ENERGY AND SOCIETY

The Political Economy of Energy Exploitation

As CO_2 emissions increasingly alter the earth's ecological balance with devastating consequences, the process underlying emissions – the exploitation of cheap conventional oil – is simultaneously running up against its own internal limits in the form of 'peak oil'. Peak oil occurs when global oil production reaches its absolute maximum capacity, after which production inevitably declines. In the decline period, available oil becomes increasingly expensive and difficult to produce, leading to worsening energy crises.

Why is peak oil relevant? Energy is the bedrock of society. The manner in which a society derives and makes use of energy defines its relationship to nature. And this in turn affects the way in which a society conceptualizes that relationship, and thus understands both itself and nature. In turn, these conceptualizations can affect the ways in which a society attempts to obtain its energy.

How this has played out historically in the distinction between agrarian and industrial forms of energy exploitation is explained succinctly by M. Shahid Alam, Professor of Economics at Northeastern University in Boston. He notes that before the eighteenth century, all economic systems were premised on an agrarian system that drew its energy largely from plants, the source of our food, fuel, fibres and other raw materials; this constrained the supply of energy since the land necessary for growing plants was available only in finite quantities. In addition, this system only used organic instruments, men and animals, for converting the energy captured by plants into mechanical energy. Once all the land was in use, the limits on growth were more acutely felt. It was the extent of the land, or its ability to store solar energy through plants, that appeared to impose the final constraint on growth in this organic economy.

But as capitalism increasingly took hold of Europe, beginning in England in the sixteenth century, this began to change. In the context of English capitalism, the industrial revolution – from around the middle of the eighteenth century to the early nineteenth century – gave rise to a new inorganic economy. Alam points out that the new economy of industrial capitalism drew its energy and raw materials more and more from relatively large stocks of hydrocarbon minerals, seemingly lifting the cap on energy flows available to the economy:

> Energy from fossil fuels was converted to mechanical energy by machines: the steam engine and, later, internal combustion engine, which quickly outstripped the organic instruments for converting energy to work. They were applied in transportation, manufacturing, and, eventually, agriculture. No longer bound by the land *per se*, industrial capitalism premised on the exploitation of fossil fuel energy enabled unprecedented economic growth. Unlike the muscle-driven, plant-based, land-constrained agrarian economy, the industrial economy increasingly drew upon minerals for its energy and raw materials, employed engines to convert fossil fuels to mechanical energy, and used this energy to mechanize work in manufacturing, transportation, construction, and agriculture. The productive capacity of the industrial economy was not constrained by energy, as in the old agrarian economy, but by its ability to deploy machines that converted energy to work. The engine of growth in this economy was capital accumulation, since this determined how fast it could expand the stock of energy-converting and energy-using machines available to the economy.

The first countries to adopt the new energy system would have a near-lock on the global economy. It created a set of cumulative forces that concentrated manufactures, capital, technology and power in the countries that took a lead in the energy revolution – the Northern core countries. Simultaneously,

> the new energy system created a Southern periphery, economic regions that were restructured by core capital to supply food, agricultural raw materials and minerals to the core.[1]

Our dependence on fossil fuels was thus integral to the international division of labour forged by the Western European powers in the nineteenth century. Today, the overwhelming bulk of the world's energy, raw materials and mineral resources are owned and controlled by transnational corporations (TNCs) based in the developed core, and it is here that the most striking continuities between the colonial and postcolonial orders can be seen. TNCs operate across national borders, planning, producing, and marketing on a global scale, assigning various functions to different regions of the world – wherever the most considerable profits can be made. Many such corporations have more power than the states across whose borders they operate, and thanks to Western-brokered free-trade agreements, this means that they often elude national laws.[2]

The majority of TNCs are based in the North. More than half come from only five nations: France, Germany, the Netherlands, Japan and the United States. In 1970, there were about 7,000. By 1995, there were 40,000. And they truly dominate the world: today, 51 of the largest 100 economies belong to TNCs – the other 49 are countries. TNCs hold 90 per cent of all technology and patents worldwide, and monopolize 70 per cent of world trade – 30 per cent of which is 'intra-firm' (occurring between different units of the same corporation).[3]

But the most important point to remember from the perspective of control over energy and resources is that TNCs manage, mine, refine and distribute most of the world's oil, gasoline, diesel and jet fuel, and extract most of the world's minerals from the ground. This *de facto* ownership of the world's resources translates into the domination of production and consumption processes worldwide. TNCs build most of the world's oil, coal, gas, hydroelectric and nuclear power plants; harvest much of the world's wood, making most of its paper; grow many of the world's agricultural crops, while processing and distributing much of its food; and manufacture and sell most of the world's automobiles, planes, communications satellites, computers, home electronics, chemicals, medicines and biotechnology products.[4]

'The World Has Never Faced a Problem Like This'

Our civilization is thus based fundamentally on the geopolitical domination of a specific kind of resource base, hydrocarbons, on which the contemporary structure of the global political economy is dependent – so dependent that this very global structure and the political economic interests vested in it generate massive systemic pressures that constrain our civilization's ability to switch over to alternative energy sources. This degree of systemic dependence is conveyed here using the term *over-dependence*. Today, the two most prominent hydrocarbon resources are oil and gas, and it is predominantly the former that supplies the world's energy needs. Oil constitutes the driving force behind almost every dimension of contemporary human life and the world economy. Despite efforts in the UK, for instance, to invest in some renewable energy sources, fossil fuels such as oil, gas and coal supply approximately 90 per cent of all the energy that runs the UK today. Without these energy supplies, civilized life in the UK could not continue. The same applies to the US and most Western European countries. Transportation, agriculture, modern medicine, national defence, water distribution, and the production of even basic technologies would be impossible.[5] The International Energy Agency forecasts that between 2005 and 2030, the world's primary energy needs will grow by 55 per cent, and that due to the lack of investment in alternative renewable energies, exploitation of hydrocarbon energies such as oil will account for 84 per cent of this increase.[6]

The basic rules for the discovery, estimation and production of petroleum reserves were first laid down by geophysicist Dr M. King Hubbert, who pointed out that as petroleum is a finite resource, its production must inevitably pass through three key stages. Firstly, production begins at zero. Secondly, production increases until it reaches a peak which cannot be surpassed. This peak tends to occur at or around the point when 50 per cent of total petroleum reserves are depleted. Thirdly, subsequent to this peak, production declines at an increasing rate, until finally the resource is completely exhausted.[7]

Estimates of the size of extant petroleum reserves are based on the following concepts: Cumulative Production, which denotes the known; Reserves, which denotes the knowable; Undiscovered, which is predictable based on past trends; and the Ultimate, which denotes the total estimate of oil. Therefore:

Ultimate = Cumulative Production + Reserves + Undiscovered[8]

There has been considerable debate about how much oil is actually available in the world, based on different calculations derived from this equation, mainly due to disagreement over how much 'undiscovered' oil really remains – and, in this context, when exactly world oil production is likely to peak. The intensity of the debate is exacerbated in light of the fact that in the post-peak period, the increasing decline of oil production will eventually mean that world demand will outpace the ability to supply, thus resulting in an energy crisis that will escalate as resources deplete further and further.

Although Hubbert's work laid the foundations for contemporary oil exploration, discovery and production, it has been challenged by a relatively minor theory of the 'abiotic' origins of oil. In stark contrast, abiotic theory posits that oil is *continuously generated* from inorganic matter within the earth's mantle. Proponents of this theory often suggest that this means there are no intrinsic limits on the amount of oil available for extraction, and that oil wells continuously replenish themselves. Apart from contradicting the recorded history of oil reserve discovery and exploitation, this theory has no peer-reviewed scientific support except in Russia where it is accepted in some geological circles and used as the basis for significant oil exploration pursuits.

Indeed, the 'Hubbert curve' depicting the rise, peak and decline of oil production has been repeatedly, empirically proven in the concrete case histories of countries like the US and Britain (formerly major oil producers and exporters who, after the peak in production of their domestic oil reserves, eventually became net importers of oil). However, the main problem of abiotic oil theory arises when its proponents cite it to undermine the idea of an impending peak in world oil production, at which point its internal incoherence becomes obvious. If the theory were true in this strong sense, we would expect to be drowning in oil, which is empirically false; and if true in the weak sense – that although abiotically formed, oil still appears at the same rates that geologists conventionally assume– then it is simply irrelevant with respect to peak oil. This has been explained by Ugo Bardi, Professor of Chemistry at the University of Florence and author of several studies on oil depletion:

> There are, really, two versions of the abiotic oil theory, the 'weak' and the 'strong':
>
> - The 'weak' abiotic oil theory: oil is abiotically formed, but at rates not higher than those that petroleum geologists assume for oil formation according to the conventional theory. (This version has little or no political consequences).
> - The 'strong' abiotic theory: oil is formed at a speed sufficient to replace the oil reservoirs as we deplete them, that is, at a rate something like 10,000 times faster than known in petroleum geology. (This one has strong political implications).
>
> Both versions state that petroleum is formed from the reaction of carbonates with iron oxide and water in the region called 'mantle,' deep in the Earth. Furthermore, it is assumed... that the mantle is such a huge reservoir that the amount of reactants consumed in the reaction hasn't depleted it over a few billion years (this is not unreasonable, since the mantle is indeed huge).
>
> Now, the main consequence of this mechanism is that it promises a large amount of hydrocarbons that seep out to the surface from the mantle. Eventually, these hydrocarbons would be metabolized by bacteria and transformed into CO_2. This would have an effect on the temperature of the atmosphere, which is strongly affected by the amount of carbon dioxide (CO_2) in it. The concentration of carbon dioxide in the atmosphere is regulated by at least two biological cycles; the photosynthetic cycle and the silicate weathering cycle. Both these cycles have a built-in negative feedback which keeps (in the long run) the CO_2 within concentrations such that the right range of temperatures for living creatures is maintained (this is the Gaia model).
>
> The abiotic oil – if it existed in large amounts – would wreak havoc with these cycles. In the 'weak' abiotic oil version, it may just be that the amount of carbon that seeps out from the mantle is small enough for the biological cycles to cope and still maintain control over the CO_2

> concentration. However, in the 'strong' version, this is unthinkable. Over billions of years of seepage in the amounts considered, we would be swimming in oil, drowned in oil.
>
> Indeed, it seems that the serious proponents of the abiotic theory all go for the 'weak' version. [Thomas] Gold, for instance, never says in his 1993 paper that oil wells are supposed to replenish themselves. As a theory, the weak abiotic one still fails to explain a lot of phenomena, principally (and, I think, terminally): how is it that oil deposits are almost always associated to anoxic periods of high biological sedimentation rate? However, the theory is not completely unthinkable.
>
> At this point, we can arrive at a conclusion. What is the relevance of the abiotic theory in practice? The answer is 'none.' The 'strong' version is false, so it is irrelevant by definition. The 'weak' version, instead, would be irrelevant in practice, even if it were true. It would change a number of chapters of geology textbooks, but it would have no effect on the impending oil peak... To be sure, Gold and others argue that even the weak version has consequences on petroleum prospecting and extraction. Drilling deeper and drilling in areas where people don't usually drill, Gold says, you have a chance to find oil and gas. This is a very, very weak position... [D]igging is more expensive the deeper you go, and in practice it is nearly impossible to dig a commercial well deeper than the depth to which wells are drilled nowadays, that is, more than 10 km... So, the abiotic oil theory is irrelevant to the debate about peak oil and it would not be worth discussing were it not for its political aspects.[9]

Thus, the Hubbert-based projections of an impending peak and subsequent decline in world oil production are far more concordant with available data on oil production rates. Early versions of the Hubbert curve for world oil production indicated that it would begin to decline around 2000. However, these were based on insufficient data. In the last half-decade, geologists and oil industry experts extending Hubbert's model have forecasted that world oil production would peak between 2005 and 2013. More recently, there is a growing consensus that the peak was reached sometime between 2005 and 2008 – economic recession having reduced the investment in new production capacity, preventing a return to 2008 levels.[10]

A report commissioned by the US Department of Energy has thrown new light on the debate. The report, 'Peaking of World Oil Production: Impacts, Mitigation and Risk Management', authored principally by US energy expert Robert Hirsch, was released by the Science Applications International Corporation (SAIC) on behalf of the US Energy Department in February 2005. Hirsch was not asked to investigate the precise timing of peak oil, but to examine the impact of peak oil in the context of three potential scenarios: one in which there are no attempts to mitigate the impact until the very moment of peak global oil production; a second in which mitigation efforts begin ten years before the peak; and a third in which such actions commence twenty years prior. Assuming for each possibility a 'crash program rate of implementation',

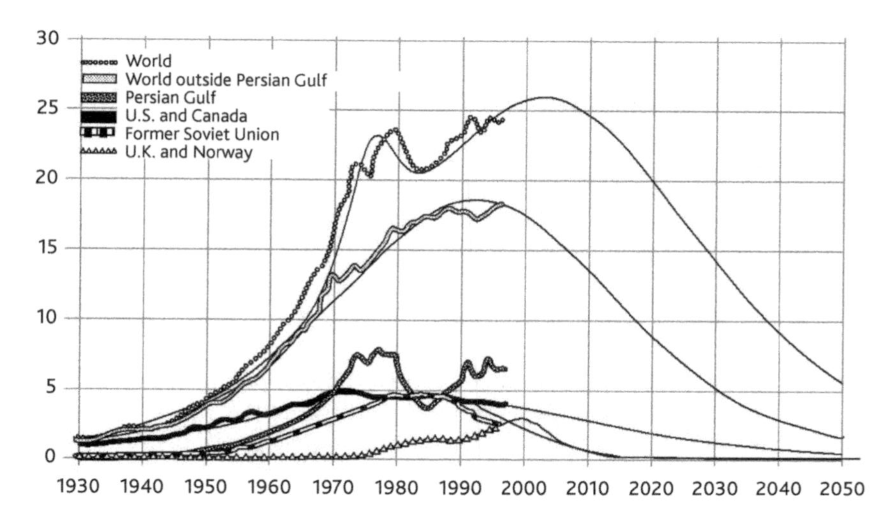

Figure 2.1 US Army Projection of Peak for World Oil Production.

Source: Army Logistician (July–August 1999, Vol. 31, No. 4)

the Hirsch Report concludes that in scenario one, the world will suffer from a 'significant liquid fuels deficit for more than two decades' that 'will almost certainly cause major economic upheaval'. In scenario two, the world will still suffer from a ten-year fuel shortfall. Even in scenario three, this will only offer a mere 'possibility' of bypassing a drastic fuel shortfall. Any mitigation programme would cost 'literally trillions of dollars' – but one is necessary because 'the world has never faced a problem like this'.[11] The report notes:

> Oil is the lifeblood of modern civilization. The peaking of world oil production presents the US and the world with an unprecedented risk management problem...
>
> As peaking is approached, liquid fuel prices and price volatility will increase dramatically, and, without timely mitigation, the economic, social, and political costs will be unprecedented. Viable mitigation options exist on both the supply and demand sides, but to have substantial impact, they must be initiated more than a decade in advance of peaking... The problems associated with world oil production peaking will not be temporary, and past 'energy crisis' experience will provide relatively little guidance. The challenge of oil peaking deserves immediate, serious attention.[12]

Increasingly, US government agencies are quietly recognizing the imminence of peak oil even as most private sector energy corporations, and the US administration, officially deny its reality. A 2006 report by the US Army Engineer Research and Development Center (ERDC) attributes great weight to warnings of peak oil's imminence, from research organizations including the Swedish Association for the Study of Peak Oil and Gas (ASPO) and the

Oil Depletion Analysis Centre (ODAC) in London. The report quotes French petroleum engineer Jean Laherrere's explanation of why official US Geological Survey statistics espousing the abundant availability of future oil deposits are implausible:

> The USGS estimate implies a five-fold increase in discovery rate and reserve addition, for which no evidence is presented. Such an improvement in performance is in fact utterly implausible, given the great technological achievements of the industry over the past twenty years, the worldwide search, and the deliberate effort to find the largest remaining prospects.[13]

Reviewing the available data, the US Army report thus argues that a shift to renewable energies is essential. The report points out that the 'doubling of oil prices from 2003–2005 is not an anomaly, but a picture of the future. Oil production is approaching its peak; low growth in availability can be expected for the next 5 to 10 years'. Disruption of world oil markets is likely to also affect world natural gas markets 'since most of the natural gas reserves are collocated with the oil reserves'. The verdict is alarming:

> The days of inexpensive, convenient, abundant energy sources are quickly drawing to a close... World oil production is at or near its peak and current world demand exceeds the supply. Saudi Arabia is considered the bellwether nation for oil production and has not increased production since April 2003. After peak production, supply no longer meets demand, prices and competition increase. World proved reserve lifetime for oil is about 41 years, most of this at a declining availability. Our current throw-away nuclear cycle will consume the world reserve of low-cost uranium in about 20 years. Unless we dramatically change our consumption practices, the Earth's finite resources of petroleum and natural gas will become depleted in this century. Coal supplies may last into the next century depending on technology and consumption trends as it starts to replace oil and natural gas.[14]

The US Army Corps of Engineers study thus tends to confirm that petroleum will be increasingly scarce over the next five years, followed by natural gas in the next twenty, and coal by the end of the century.

Whence Peak Oil?

So the main question is, when? One of the most authoritative studies on peak oil was conducted by Dr Colin Campbell, former BP petroleum geologist, and Jean Laherrere, on behalf of the Geneva-based Petroconsultants, which later became the IHS Energy Group. The Petroconsultants database, used by all international oil companies, is the most comprehensive for data on oil resources outside North America – and is considered to be so significant that it has been kept out of the public domain. Campbell and Laherrere concluded in their report, priced at $32,000 a copy and written for government and corporate insiders, that 'the mid-point of ultimate conventional oil production would be reached by year 2000 and that decline would soon begin'. They also projected that 'production post-peak would halve about every 25 years, an exponential decline of 2.5 to 2.9% per annum'.[15]

The Petroconsultants report pinpointed global oil production's peak as at earliest 2000 and at latest 2005. Murdoch University's Institute for Sustainability and Technology Policy describes the overall analysis as the most accurate available, based as it is on performance data from thousands of oil fields in 65 countries, including data on 'virtually all discoveries, on production history by country, field, and company as well as key details of geology and geophysical surveys'. Due to their unprecedented access to such data, Campbell and Laherrere, unlike other oil industry commentators, are in 'a unique position to sense the pulse of the petroleum industry, where it has come from and where it is going to. Their report pays rigorous attention to definitions and valid interpretation of statistics.' A review of the research by senior industry geologists in *Petroleum Review* indicated, apart from minor disagreement over the scope of remaining reserves, 'general acceptance of the substance of their arguments; that the bulk of remaining discovery will be in ever smaller fields within established provinces'.[16]

Other analysts predict that the peak will occur at a slightly later date than do Campbell and Laherrere, but the overall conclusion is inescapable. Reviewing the debate, Richard Heinberg, a Senior Fellow at the Post-Carbon Institute, observes that 'the clarity and logic of the analysis, and the depth of expertise, of the petroleum pessimists – Campbell, Laherre, Ivanhoe, Deffeyes, Youngquist, et al. – seem impressive'. Summarizing their work, he argues that 'global conventional oil production will peak some time during this first decade of the 21st century', and will occur no later than 2015. On the other hand, Heinberg lends credence to Campbell's view 'that the first global production peak has already happened... and that the next decade will be a "plateau" period', in which recurring economic recessions resulting in lower energy demand will 'temporarily mask the underlying depletion trend'. Of course, the peak point will only be fully discerned in its aftermath. Heinberg notes that 'one year we will notice that gasoline prices have been climbing at a rapid pace, and we will look back at previous few years' petroleum production figures and note a downward slope'.[17]

Yet one of the most prominent viewpoints challenging this picture of peak oil's imminence comes from Cambridge Energy Research Associates (CERA), based in Massachusetts. In a 2005 study, CERA argued that worldwide oil production capacity could rise by as much as 16 million barrels per day (mbd) between 2004 and 2010 – a 20 per cent increase over the period. Unconventional oils will play a much larger role in this increased capacity than hitherto recognized, providing up to 35 per cent of supply by 2020. CERA thus rejects the idea of a 'peak' in world oil production, positing instead that in the third or fourth decade of this century, the world will experience an 'undulating plateau' that itself will persist for several decades, before commencing a slow decline. In November 2006, a further CERA report argued that peak oil is a faulty theory which ignores that the global oil resource base of total conventional and unconventional reserves is three times as large as pessimistic estimates routinely made by peak oil theorists. World oil reserves, the report suggested, will last for another 122 years at current rates of consumption. In January 2008, CERA followed up with another report insisting that decline rates for global oil production are much lower than claimed by peak oil theorists, at only 4.5 per cent a year.[18] CERA continues to generate similar optimistic reports on a regular basis.

CERA's independence is questionable. According to Canada's *Globe and Mail* Inside Energy Blog, 'Underlying CERA's analysis is a likely assumption that Middle Eastern reserves are what those countries say they are', although, as noted above, these official reserve estimates are routinely over-inflated. The *Globe and Mail* adds that Dan Yergin, the founder of CERA, 'is also close with top officials in Saudi Arabia – some industry players argue he's too close – and in November received a special award for his work from the kingdom (and, it is rumoured, $100,000 in cash) when it held a summit of OPEC leaders'.[19]

There are other problems. It is universally recognized that over the long term, rates of consumption are rising. CERA's projection that world reserves will last for more than a century at 'current rates' is therefore misleading. As world population rises over the coming decades, propelled by the rapid growth of industrial economies like China, India and others, rates of consumption are set to increase dramatically. CERA's recalculation of the size of global oil reserves is also deeply questionable, primarily because it is based on projections of undiscovered conventional and unconventional reserves, whose accuracy can only be proven in the hindsight of actual discoveries, the rates of which are declining. As energy consultant and former US government energy adviser Matthew Simmons points out of the CERA figures:

> [They] create the illusion that there's a precision there that they have that is far beyond anybody's ability to grasp. They talk about "yet to discover" oil that they have a specific number for. And I just think people shouldn't basically do these sort of reports, and claim to be so precise, without a huge caveat saying, "by the way, this is just our own hunch".[20]

Simmons's scepticism is borne out by the statements of senior officials at the International Energy Agency who reported in late 2009 that:

> The IEA in 2005 was predicting oil supplies could rise as high as 120m barrels a day by 2030 although it was forced to reduce this gradually to 116m and then 105m last year. The 120m figure always was nonsense but even today's number is much higher than can be justified and the IEA knows this. Many inside the organisation believe that maintaining oil supplies at even 90m to 95m barrels a day would be impossible but there are fears that panic could spread on the financial markets if the figures were brought down further. And the Americans fear the end of oil supremacy because it would threaten their power over access to oil resources.

Another IEA source observes: 'We have [already] entered the "peak oil" zone. I think that the situation is really bad.'[21]

Moreover, CERA's purportedly optimistic 2008 estimate of the annual rate of decline of global oil production at 4.5 per cent tends to confirm, rather than disconfirm, the scenario of peak oil. According to Randy Udall and Steve Andrews, directors of the Denver-based Association for the Study of Peak Oil & Gas (ASPOG), the implications of even a 4.5 per cent decline rate are 'enormous'. They argue:

> If you start with 85 million barrels a day in 2007, but lose 4.5% each year, by 2017 you've lost 31 mbd. That's the equivalent of losing the world's four largest oil producers: Saudi Arabia, Russia, the USA and Iran. By 2030, you've lost 55 mbd, or as much as all the non-Opec nations now provide. Remarkably, CERA finds this to be 'good news.'

Udall and Andrews also critique CERA's methods of data sampling and analysis:

> One 'Iran' is what we are now losing to depletion each year... Of CERA's 811 fields, only half have entered their decline; the rest are new fields that are still ramping up or on their maximum production plateau. In other words, what CERA calls its 'aggregate global production decline' is an average of new deepwater fields, young pups, mature giants, and sclerotic geriatrics.
>
> When CERA looks just at fields that have passed peak, its results resemble those so often quoted on peak oil web sites. To wit, of 308 Non-OPEC 'post-plateau' fields, the average decline is 8%. Of 209 post plateau offshore fields, the average decline rate is 10%; 29 deepwater fields are declining at 18%.[22]

CERA's own figures for the rate of decline suggest that by 2017 the world's currently known oil fields will be producing about 33 million fewer barrels a day than they are now. Yet CERA insists that by that date, world capacity will be capable of reaching a production level of 112 million barrels a day. This would require adding 59 million barrels a day in new capacity, 'more than six times today's daily output from Saudi Arabia, the world's largest oil exporter'. This, of course, suggests that there are large numbers of giant oil fields, equivalent in size to the biggest Middle Eastern reserves, which have yet to be discovered and have thus been inexplicably overlooked by the world's best oil field technologists. Such a notion is simply not supported by oil industry experts of any persuasion. CERA attempts to overcome this impasse by suggesting that 'nearly half of that output will come from non-conventional sources', without yet specifying how such non-conventional sources will be successfully exploited to meet world demand at projected rates of consumption. As Robert Hirsch asks, 'How is this credible without a worldwide crash program, which has not yet been seriously considered, let alone initiated?'[23]

The Export Land Model (ELM) most clearly demonstrates the fallacy of CERA's approach to questions surrounding peak oil. According to Dallas petroleum geologist Jeffrey J. Brown, the decline of global oil production is not necessarily the central question; an equally significant angle of inquiry is the corresponding supply-demand nexus and its impact on when oil-exporting nations begin to *cut off their exports to international markets*. Thus, Brown argues that to fully understand the import of peak oil requires not merely inspecting production data, but also examining how production rates interact with the rate of local consumption and the effect of that consumption on the willingness of a country to export its oil. As Brown's work shows, ELM projects on very conservative assumptions that after a country's oil production peaks, production declines at a rate of 5 per cent annually, while local consumption increases by 2.5 per cent. Using these figures, the ELM foresees that exports will reach zero in a maximum of nine years. In practice, however, this occurs much faster than ELM's assumptions would indicate. While the ELM forecast hypothesizes nine years between peak and the end of exports, the historical record shows that Indonesia's exports ceased seven years after the peak of its oil production, and the UK's exports stopped only six years after its peak. The critical import is that after this period, the global market is deprived of those exports. The ELM thus forecasts an imminent oil supply-demand crunch due to the impact of declining exports in the context of increasing numbers of post-peak oil-producing countries.[24]

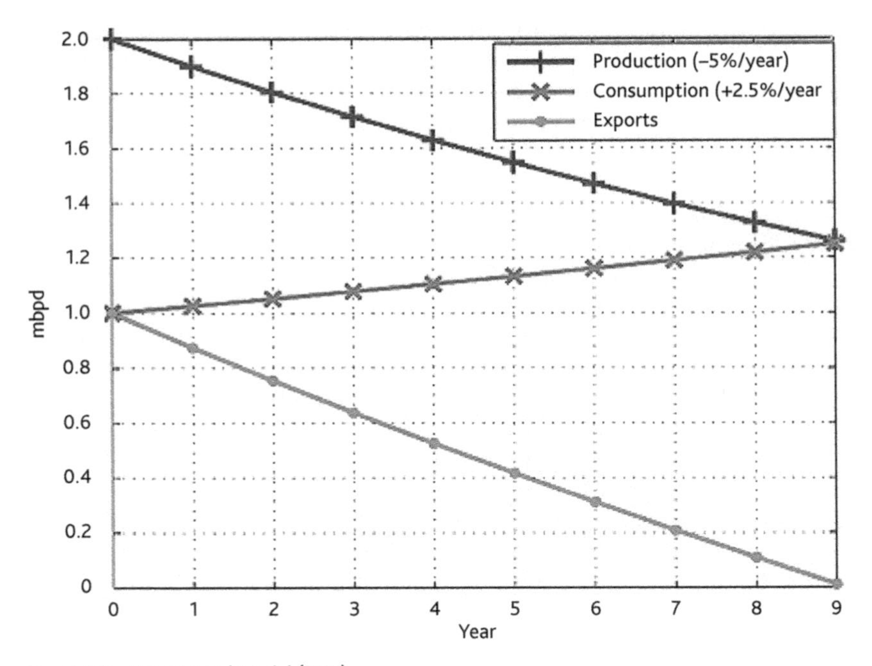

Figure 2.2 Export Land Model (ELM).

Source: Jeffrey J. Brown, US petroleum geologist

In 1993, China became a net importer of oil as its domestic production reached a plateau, then declined, against skyrocketing consumption. Subsequently, China's oil consumption has almost doubled to about 8 million barrels a day, about half of which is now imported. Over the next ten years, as domestic production continues to decline and consumption to rise, oil imports are projected to rise another 50 per cent. China's massive energy requirements should be compared to those of the US, which currently imports about 14 per cent of its oil from Mexico. In total, the US meets up to 70 per cent of its oil needs through imports. Brown's ELM model forecasts that Mexico's oil exports to the US and global markets will cease as of 2014, as over the years it has consumed more of its own oil. As demand balloons from emerging markets, major powers like China and the US still need to find viable sources of oil to sustain their swelling domestic consumption needs vis-à-vis their declining rates of production. In the months and years ahead, international competition and conflict over control of available oil supplies will become increasingly desperate. The severity of this situation will only increase with time. On Brown's estimate based on his analysis of global oil production data, *world oil production already peaked in 2005*: 'our model and case histories show that the decline rate accelerates, year by year. Using the Lower 48 in the United States as an example, you can see the annual declines going 2%, 3%, 5%, 7%, 10%, 15%, 20, on and on. So it's an accelerating decline rate.'

Underscoring these concerns is the fact that several major oil producers have effectively gone off-line. On 15 April 2008, Russia, the world's second largest oil exporter, peaked in its oil production, with a decline in the first quarter of that year – incidentally putting to rest the last vestiges of hope, voiced by abiotic theorists, for self-replenishing reservoirs. ELM forecasts that within six to nine years, the world will gradually lose Russia's current 7 million barrels a day in exports. Iran, the world's fifth-largest oil exporter, is either at or near peak output according to Iranian oil analysts, and by 2015 may consume the same amount as it exports. In April 2008, King Abdullah announced that the Saudi Kingdom would no longer allow oil production to rise above 12.5 million barrels a day, a phenomenon described as 'practical peak oil' by the Merrill Lynch vice-president who noted that the Saudis are attempting to prolong the petroleum exploitation and depletion cycle to forestall an impending peak. In Brown's words: 'The reality is that this thing is coming so much faster and so much harder than even most pessimists were expecting.'[25]

A Post-Peak World

Indeed, there is good reason to believe that the peak has already been surpassed. Until 2004, world oil production had risen continuously, but thereafter underwent a plateau all the way through to 2008. Then from July to August 2008, world oil production fell by almost one million barrels per day. It continues to decline.[26] The *New Scientist* reported at around that time:

> Based on the best information we have on global oil supply, there is growing evidence that global production of crude oil actually peaked in 2005, and

> for the past three years has struggled to remain at an undulating plateau of some 73 to 74 million barrels a day. Since the world's total petroleum consumption is about 88 million barrels per day, we currently bridge this gap through the intensive use of liquefied natural gas, refining processing gains and tapping into the world's oil inventory.[27]

Corroborating this assessment is US energy analyst Tony Eriksen, who cites US Energy Information Administration production data showing that:

> World crude oil, condensate and oil sands production peaked in 2008 at an average of 73.78 million barrels per day (mbd) which just exceeded the previous peak of 73.74 mbd in 2005... Production is expected to decline further as non OPEC oil production peaked in 2004 and is forecast to decline at a faster rate in 2009 and beyond due mainly to big declines from Russia, Norway, the UK and Mexico. Saudi Arabia's crude oil production peaked in 2005. By 2011, OPEC will not have the ability to offset cumulative non OPEC declines and world oil production is forecast to stay below its 2008 peak... Additional reserves and the related production from prospective areas such as the arctic, Iraq, and Brazil's Santos basin are highly unlikely to produce another peak but should decrease the production decline rate after 2012.[28]

At best, then, future oil production will remain flat until 2012, and thereafter fluctuate along an undulating and accelerating decline. Rapidly rising oil prices and accumulating reports of declining oil production for specific fields corroborate this conclusion (see Figure 2.3). London's *Petroleum Review* published a study toward the end of 2004 concluding that in Indonesia, Gabon, and fifteen other oil-rich nations supplying about 30 per cent of the world's daily crude, oil production is declining by 5 per cent a year – double the rate of decline one year prior to the report. Chris Skrebowski, the *Review*'s editor and a former BP oil analyst, noted that:

> Those producers still with expansion potential are having to work harder and harder just to make up for the accelerating losses of the large number that have clearly peaked and are now in continuous decline. Though largely unrecognized, [depletion] may be contributing to the rise in oil prices.[29]

A 2004 report by the US Office of Petroleum Reserves similarly concluded that 'world oil reserves are being depleted three times as fast as they are being discovered'.

> Oil is being produced from past discoveries, but the reserves are not being fully replaced. Remaining oil reserves of individual oil companies must continue to shrink. The disparity between increasing production and declining discoveries can only have one outcome: a practical supply limit will be reached and future supply to meet conventional oil demand will

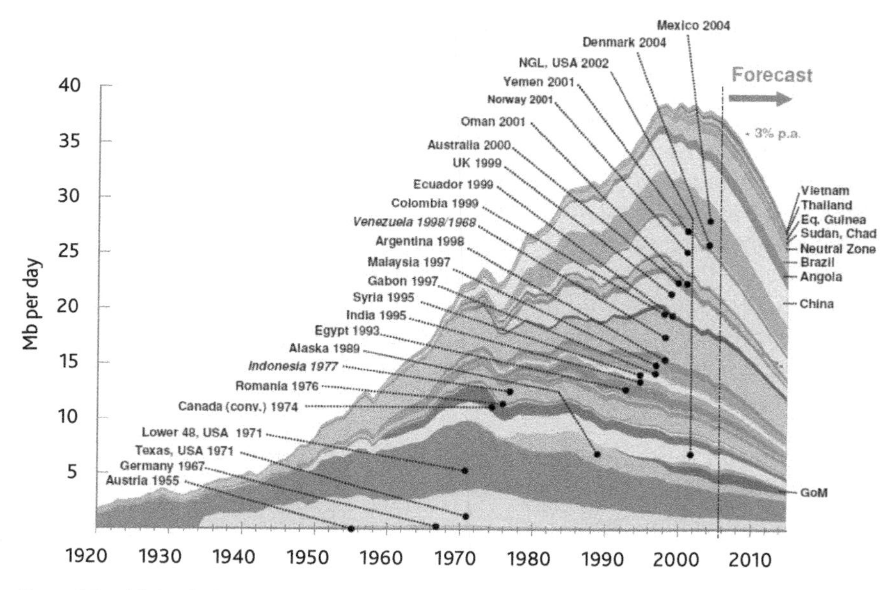

Figure 2.3 Oil-Producing Countries Past Peak.

Source: Energy Watch Group (October 2007)

> not be available... Although there is no agreement about the date that world oil production will peak, forecasts presented by USGS geologist Les Magoon, the *Oil and Gas Journal*, and others expect the peak will occur between 2003 and 2020. What is notable ... is that none extend beyond the year 2020, suggesting that the world may be facing shortfalls much sooner than expected.[30]

In fact, even this admission is optimistic. Bill Powers, editor of the *Canadian Energy Viewpoint* investment journal, confirms that geologists are increasingly coming to agree that world oil production 'is soon headed into an irreversible decline... The US government does not want to admit the reality of the situation. Dr Campbell's thesis, and those of others like him, are becoming the mainstream'.[31] Indeed, according to Chris Skrebowski in early 2005, conventional oil reserves were declining at about 4–6 per cent a year worldwide, including those of 18 large oil-producing countries, and 32 smaller ones.[32]

Denials of peak oil's imminence are therefore no longer credible. Even a little-noted 2005 ExxonMobil report, *The Outlook for Energy: A 2030 View* states that non-OPEC oil producers are due to peak no later than 2010. As Princeton-based energy consultant Alfred Cavello observes, 'once non-OPEC production reaches a peak, conventional world oil production could peak shortly thereafter', leading to price hikes in accordance with the laws of supply and demand. Thus, he argues, 'the petroleum industry is approaching a turning point. Conventional petroleum production will soon... no longer be able to satisfy demand.'[33]

Increasingly, however, it is agreed that peak oil production will not occur simply along a smooth curve after which production will neatly decline. Rather, before the onset of a steep decline in world oil output, there will indeed be a transition period consisting of a 'bumpy plateau' – yet unlike CERA's claims, the undulating plateau period is unlikely to last longer than a few years before it is followed by a sharp drop in production. According to Colin Campbell, the 'bumpy plateau' period will begin with massive oil price hikes, followed by upward and downward price swings in accordance with economic recession and volatility of demand. He identifies five basic stages of undulation which can be traversed repeatedly before the onset of irreversible decline (see Figure 2.4):

1. Price shock (as the capacity limit is breached)
2. Economic recession cutting demand
3. Price collapse (the market overreacts to small imbalances between surplus and shortage)
4. Economic recovery [followed by increased demand]
5. Price shock (as the falling capacity limits are again breached)[34]

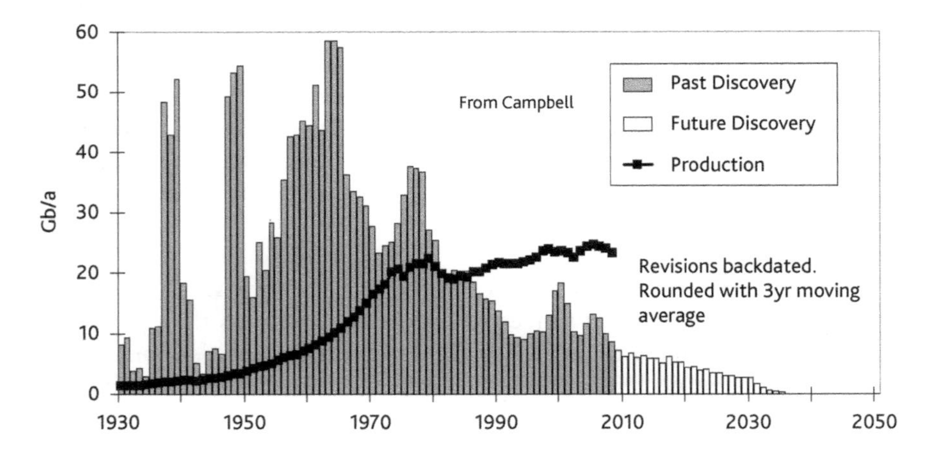

Figure 2.4

Source: Colin Campbell

In this scenario, as production capacity limits are breached the impact of oil price shocks is to exacerbate economic recession, leading to decreased consumption and, therefore, the slashing of production. As the economy contracts and demand slumps, prices correspondingly decline leading to prolonged deflation, eventually permitting rising consumption and production, and thus temporary economic recovery which, nevertheless, again hits the limits of falling oil production capacity. In other words, *oil price volatility consisting of large upward and downward price swings is precisely a symptom of the undulating plateau in the post-peak period.*

We are therefore most likely already on this undulating plateau, and it is unlikely to last more than a decade. One of the most detailed earlier warnings was an October 2007 oil market report by the German-based Energy Watch Group (EWG), run by an international network of European politicians and scientists, arguing that world oil production had peaked in 2006. This conclusion corroborates geologist Jeffrey Brown's estimate, based on his own (unpublished) analysis of the data, that the peak occurred around 2005; predictions by Campbell, Laherrere, the US Army, among others; and the more recent assessments by Simmons, Eriksen and others cited above. For past production data, and partly also for reserve data for certain regions, the EWG report relies primarily on oil production data from an industry database. The reliance on concrete production data rather than questionable estimates of total world reserves means that the report is far more reliable than others so far, such as those by CERA. The EWG report posits that the rate of world oil production will start to decline at 7 per cent annually, rather than CERA's optimistic 4.5 per cent. By 2020, and even more by 2030, global oil supply will be dramatically lower, creating 'a supply gap which can hardly be closed by growing contributions from other fossil, nuclear or alternative energy sources in this time frame'. Oil producers outside the former USSR and OPEC increased their total production until around 2000, but since then it has declined. Despite repeated claims to the contrary, only very few countries can expand production capacity, for instance Brazil and Angola. Saudi Arabia almost certainly cannot. Growth of production has come to a standstill and output now is more or less on a plateau despite historically high oil prices. The historical maximum of oil discoveries was passed in May 2005, and now 33 of the 48 largest oil-producing countries have already peaked. The EWG study also roughly corroborates the findings of a separate study by Fredrik Robelius, whose doctoral thesis at Uppsala University in Sweden projects future oil production on the basis of studies of global oil reserves, historical production, and new finds, focusing on the world's largest reserves. Robelius sees the peak as having occurred in 2008.[35] Essentially, the different studies are viewing the same production data from different ends, and the overall picture is clear: from 2005, world oil production stopped rising, remained roughly flat, and began to decline in 2008, and now continues to decline. While there is scope for future fluctuation for about five more years maximum, the trend thereafter will be downwards.

All these studies indicate that the peak of world oil production is at hand, and is fundamentally responsible for the volatility of oil markets, including rapid oil price hikes and drops, in the early years of this century. The extent of price rises was only marginally related to market speculation and purchasing of oil futures, particularly as widespread fears of a coming oil-supply crunch were voiced even by international institutions like the IMF, and while geopolitical tensions in the Middle East remain unresolved. As noted by Lord Meghnad Desai, Professor Emeritus of Economics at the London School of Economics

(LSE), independent research on financial data confirmed that the high oil prices had far more to do with 'the fundamentals of rising demand and constrained supply'.[36] Thus, with the deepening of economic recession in late 2008, world consumption slumped, relieving the pressure on oil suppliers and thus precipitating a seemingly dramatic oil price drop. By 2009, they eventually began to rise again as lower prices permitted renewed economic activity and world demand once again grew, impinging on oil supply. This phenomenon of fluctuating rise and decline, further accentuated by changing investment trends in oil hedging, may persist for some years, but this is merely a temporary phase along the undulating plateau period before we enter the permanent era of oil supply scarcity and, thus, permanently high oil prices.

Given the magnitude of the energy crisis, the official response of Western governments is worth noting. Malcolm Wicks, the UK Prime Minister's Special Representative on International Energy Issues, in 2008 denied that there was any reason for the government to prepare for an energy crisis:

> The hydrocarbon resources around the World are sufficiently abundant to sustain likely growth in the Global energy system for the foreseeable future... On the balance of the available analysis and evidence, the Government's assessment is that the World's oil resources are sufficient to sustain economic growth for the foreseeable future.[37]

How will peak oil play out for major oil-producing countries? The 2008 collapse in oil prices was linked directly to the deepening of global recession and the corresponding collapse in demand (also see chapter 4). As world demand hit the floor, leading to deflation and plummeting prices, the lower prices encouraged the renewal of world consumer spending, permitting a temporary economic recovery. Renewed consumer activity would also create escalating demand, placing renewed pressures on production – and greater demand for energy. It is likely that a temporary economic recovery would therefore breach global oil capacity limits, once again triggering spiralling oil prices. This means that even though oil will become more expensive, it will also become more difficult to export as less and less is being produced. Over time, decreasing oil production will have to be used to meet growing domestic demand. The ELM projects that it takes a *maximum* of nine years after an oil-producing country's fields peak before its exports become zero and domestic consumption of its own oil reaches a maximum, as its rates of production continue to decline.

The implications are severe. Global economic policies could generate another artificial (albeit far less effective and shorter) 'boom'. States with remaining oil production capacity would at first experience this very favourably. But any such boom, if it does occur, would precipitate an even larger global financial bust around 2017 *at the latest* (nine years after 2008), after which the economy would enter an unavoidable long-term deep depression far worse than even the Great Depression of the 1930s. At that point, the world would once again witness rocketing oil prices, generating immediate financial dividends in the short term, but in the long term (again, by around 2017 at latest) leading to exponentially declining oil exports. These would be unable to provide sufficient revenue to continue large-scale expansion, creating serious repercussions for the economy and perhaps even contribute to permanent contraction.

QUICK ALTERNATIVES TO CONVENTIONAL OIL?

Coal and Natural Gas

'Clean coal', a technology promising the burning of coal while removing its dangerous emissions, is being eagerly explored as an alternative to conventional oil. This would make sense if there was sufficient coal to meet projected demand. The most widely accepted view, endorsed for instance by the World Coal Institute, is that coal supplies will last another 155 years. As countries face difficulties securing oil and gas supplies, global coal consumption rates have already increased drastically since 2000. Yet according to a recent report by the Institute for Energy (IFE) prepared for the European Commission Joint Research Centre, future abundance of coal supplies is by no means a foregone conclusion. The IFE report warns that 'coal might not be so abundant, widely available and reliable as an energy source in the future... the world could run out of economically recoverable (at current economic and operating conditions) reserves of coal much earlier than widely anticipated'. The key findings of the report that support this observation are as follows:

1. 'world proven reserves (i.e. the reserves that are economically recoverable at current economic and operating conditions) of coal are decreasing fast';
2. 'the bulk of coal production and exports is getting concentrated within a few countries and market players, which creates the risk of market imperfections';
3. 'coal production costs are steadily rising all over the world, due to the need to develop new fields, increasingly difficult geological conditions and additional infrastructure costs associated with the exploitation of new fields'.

Thus, over the coming decades global prices of coal are likely to rise, such that 'the relative gap between coal prices and oil and gas prices will most likely narrow'.[38]

A report by the Energy Watch Group was even more specific, warning that global coal production would peak at around 2025, at 30 per cent above 2007 levels of production. The fastest depletion of reserves is occurring in China and the US. Although volumes of coal are still considerable, high-quality coal production in the US already peaked around 2002, leaving the bulk of production to concentrate on lower quality subbitumous coal – which has lower energy content. The EWG estimates that US coal production in terms of energy will only be able to maintain current levels for another 10–15 years.

While input from reserves in Australia, China, Russia, Ukraine, Kazakhstan and South Africa will allow for a temporary increase in production, this too will plateau shortly after 2020.[39]

The situation looks similar for the future of natural gas production. According to Jean Laherrere, who has conducted one of the most comprehensive surveys of the available gas reserve and production data, global natural gas production will also peak at almost the same time as coal, sometime around 2025.[40] Some of the world's largest gas reserves are already in decline. Canadian gas reserves have been declining for decades, losing 43 per cent of proven reserves between 1985 and 2005. Canadian natural gas production is forecasted to steadily decline until 2025.[41] In contrast, US natural gas production relies on a variety of conventional and unconventional sources. While conventional gas production has certainly declined, unconventional sources have increasingly been used, ensuring that overall production has remained at essentially the same level for some years. So far technological limitations remain the primary obstacle to exploiting the large resources of unconventional forms of natural gas.[42] However, recent developments have permitted deeper drilling and better exploitation of unconventional gas, potentially permitting many decades of further use at *current* rates of production and consumption – up to 118 years by one estimate.[43] Of course, at projected levels of demand, the amount of time is likely to be far less. Future developments in this regard will depend on the interplay between technology and prices. Of course, the prospect of ongoing exploitation of natural gas for several decades, while seemingly cause for optimism from an energy perspective, also raises difficult questions about the problem of escalating fossil fuel emissions and the danger of triggering rapid climate change.

Elsewhere, evidence suggests in any case that the overall prospects of global reliance on natural gas are deeply uncertain, given that the long-term trend for many of the world's largest gas fields is continuing decline. A clear example is Russia, which produces about 22 per cent of world gas supply, and is believed to hold 30 per cent of the world's remaining gas reserves. In late 2008, Dr Pierre Noel, Acting Director of the Electricity Policy Forum at the University of Cambridge, presciently warned: 'Over the next 15–20 years, Gazprom faces serious supply challenges, and the international gas market is likely to experience considerable tightening.' He noted that the coming decades could see Europe facing 'a gas supply crunch, leading to stagnant or even declining consumption'. Although the Russian company controls 'the world's largest gas reserves, Gazprom will find it difficult to maintain its current supply levels'. Noel reports that production from the 'super-giant' west Siberian gas fields, accounting for most of Gazprom's production, 'is now in steep decline'. Maintaining production depends on the development of new fields on the Yamal Peninsula in northwest Siberia, which are set to come online in 2010. Yet most of the European gas industry argue that 'this is highly unlikely', putting 2015 as a more realistic date. But the problem goes deeper than this:

In fact, Gazprom's production is already insufficient to meet all the company's commitments. It depends on two other sources of gas – 'independent' Russian producers and imports from Central Asia, especially Turkmenistan – to make up the shortfall. This 'bridge' is supposed to supply Gazprom's needs until the Yamal fields come online. But there is uncertainty over whether Gazprom will be able to source sufficient volumes from Turkmenistan, while independent Russian producers have little incentive to increase their production in the absence of access to Gazprom's transmission network, which would enable them to reach consumers directly. Moreover, domestic gas consumption in Russia is growing, driven by economic expansion and a gas-intensive electricity mix. So there is at least a risk that Gazprom's 'bridge' to Yamal could collapse. Industry assessments vary from a tight but manageable supply situation to an impending crisis.[44]

This analysis places in sharp focus the EU–Ukraine–Russia gas crisis of early 2009, where Russia halted its gas supply to Europe through Ukraine, which had signed a strategic accord with the US in December 2008. The accord called for a US diplomatic post in Crimea, where Russia's Black Sea Fleet is based, as well as for 'enhanced cooperation' in defence, security, trade, and energy security.[45] The incident must be understood as signifying the opening of a major fault line for future geopolitical competition in the region for claims over access to increasingly scarce hydrocarbon resources. An analysis of data from the *BP Statistical Review of World Energy 2008*, by the open source intelligence firm Sanders Research Associates, found that Russian natural gas could well have already peaked in 2006, suggesting that the prolonging of delays in bringing online the new Russian gas fields could imply an EU gas supply crunch as early as 2015.[46]

Unconventional Oil

Two sources of energy that governments and industry frequently point to as viable alternatives to conventional oil and gas are unconventional oil and nuclear energy. Because these sources are abundant, the argument goes, there really is nothing to fear from the peaking of conventional oil production. Recent estimates suggest that total unconventional reserves, such as oil shales and tar sands, amount to around 3600 billion barrels worth – a seemingly colossal figure. Moreover, reserves are widely dispersed, with large resources near consumers in North America (especially Canada) and China. But how much of this is geophysically recoverable is another question, and certainly is a far lower figure.[47] Oil companies such as ExxonMobil, Shell, and BP, among others, routinely advertise such huge deposits of unconventional oil to reassure consumers of the supposed lack of resource constraints for the foreseeable future. But intriguingly, ExxonMobil's own world oil production forecast shows no contribution from 'oil shale' even by 2030. And as for Canada's oil production from tar sands, it projects only 4 million barrels of oil per day by 2030, accounting for a mere 3.3 per cent of the predicted total world demand of 120 million barrels per day.[48]

This lack of confidence in unconventional oil sources is no accident. Canadian tar sands are actually deposits of bitumen (tar), which are the result of conventional oil degradation by water and air. Making tar sands usable is a capital-intensive venture that requires special procedures such as heating to separate the tar from the sand, mixing the tar with a diluting agent for pipeline transport, and constructing specially equipped refineries for processing. It is also, ultimately, extremely inefficient and even illogical. Despite intensive efforts by energy corporations to develop better technologies, to produce just *one barrel of oil* from tar sands requires between *400 and 1000 cubic feet of natural gas*, depending on the extraction method used. And natural gas production, despite a near doubling of drilling activity, is flat or decreasing both in Canada and in the United States, prompting a tripling of prices over the last few years.[49]

The extraction of oil shale is hardly easier. In fact, there is no oil in shale, but rather an organic material called kerogen, a precursor of petroleum. To extract oil, the shale (typically between 5 and 25 per cent kerogen) must first be mined, then transported to a plant where it is crushed, then heated to 500 degrees Celsius, which pyrolyzes, or decomposes, the kerogen to form oil. After processing, most of the shale remains on the surface in the form of coarse sand, so large-scale mining operations will produce immense amounts of waste material. An estimated 1–4 barrels of fresh water are required for each barrel of oil produced, both for cooling the products and stabilizing the sand waste.[50] Take the Albion Sands site run by Shell. The National Energy Board of Canada (NEB) estimates that the mining-extraction-upgrading process requires about *500,000 cubic feet of gas* for every *single barrel of bitumen*. In the *in situ* recovery process in the Peace River and Cold Lake oil sands regions, gas is used to make steam which is injected underground to force bitumen into producing wells. The NEB estimates that *in situ* generally takes *1 million cubic feet of gas to produce a single barrel of bitumen*. The total projected requirements for use of natural gas with *in situ* recovery are anywhere from 1.2 billion cubic feet (bcf) per day to 1.8 bcf per day by 2015, according to the NEB report. This is utterly unsustainable. Already, enough gas is used up daily in the Alberta oil sands to heat 3.2 million Canadian homes.[51] And the costs of this enterprise are skyrocketing, far beyond Shell's stated plans to produce unconventional oil at $30 a barrel. For instance, at Athabasca Oil Sands Project in Alberta, as of 2006 Shell faced a 'cost explosion' amounting to $11 billion, almost double its original budget, to add 100,000 barrels daily to their 155,000 barrel-per-day output.[52]

The verdict from independent experts is now that unconventional oil sources are simply irrelevant. A study by the Hydrocarbon Depletion Study Group at Uppsala University in Sweden, accepted by the journal *Energy Policy*, investigated the viability of a crash programme for the Canadian tar sands industry within the time frame 2006–2018 and 2006–2050, finding 'serious difficulties'. In the first instance, 'there is not a large enough supply of natural gas to support a future Canadian oil sands industry with today's dependence on natural gas'. Even adopting 'a very optimistic scenario Canada's oil sands will not prevent Peak Oil'.[53]

Of course, the idea that we should escalate exploitation of unconventional oil does not factor in the exacerbation of climate change due to rising CO_2 emissions – the imperative now is not to increase investment in CO_2-emitting energy sources, but to reduce it immediately. As oil companies attempt to increase the production of unconventional oil, they will only accelerate global warming. That is why, while quietly allowing oil companies to continue projects to exploit tar sands and oil shale, Western states proposing climate-sensitive solutions have publicly touted nuclear energy as a viable, 'clean' alternative to oil and gas which will avoid CO_2 emissions while successfully meeting the world's energy demands for the foreseeable future.

Nuclear Power

It is in this context that the British government's 2006 Energy Review, *Our Energy Challenge* proposed on ecological and energy grounds to build a new generation of mainland nuclear power plants. Many other Western states, along with the UK, see nuclear power as a viable long-term alternative to conventional oil, necessary to keep up with rising demand for energy. In Britain, the argument goes that an investment of at least £2 billion would be enough to build a new nuclear processing plant at Sellafield that would be capable of converting 60,000 tonnes of nuclear waste into reactor fuel, which could provide 60 per cent of the UK's electricity until 2060.[54] As a generic model, the idea is that nuclear power based on reprocessing of nuclear waste eliminates CO_2 emissions and would allow the country to retain relative energy self-sufficiency for about five more decades. This means that at best, and relying predominantly on reprocessing of nuclear waste, billions of pounds of investment in the nuclear industry would only be capable of providing just over half the country's energy needs up to around the mid twenty-first century.

Unfortunately, the British nuclear power lobby also *severely underestimates* the massive energy and economic costs of the nuclear waste reprocessing model. According to Dr Edwin Lyman, Scientific Director of the Washington-based Nuclear Control Institute and a participant in the Processing Needs Assessment conducted by the US Energy Department's Nuclear Material Stabilization Task Group, 'a reprocessing plant with an annual capacity of 2,000 metric tons of spent fuel would cost up to $20 billion to build'. The US currently has 55,000 tonnes of nuclear waste – less than Britain, which has about 60,000. Lyman notes that the US nuclear industry's own estimates show that the US would need not one but 'two of these' reprocessing plants 'to reprocess all its spent fuel'. Furthermore, the estimated ongoing cost premium for reprocessing the nuclear waste 'would range from 0.4 to 0.6 cents per kilowatt-hour – corresponding to an extra $3 to $4.5 billion per year'. This would have to be paid either 'through increased taxes or higher electricity bills'.[55]

Lyman also points out that the reprocessing of nuclear waste would drastically increase the risk of nuclear proliferation and terrorism. Separated plutonium produced by reprocessing is not highly radioactive even if blended with other elements like neptunium, and stored in a powdered form – yet it remains weapon-usable, therefore making it an attractive target for theft by terrorists. Further, because 'commercial-scale reprocessing facilities handle so much of this material', it is simply 'impossible to keep track of it accurately in a timely manner'. It would therefore be feasible for 'the theft of enough plutonium to build several bombs' to go 'undetected for years'.[56]

Reprocessing also means more, not less, nuclear waste which needs to be safely disposed of, and in fact would only make waste management more energy-intensive and costly. Indeed, achieving reasonable safety and efficiency would depend on technological processes potentially requiring centuries – time we don't have. In detail, Lyman argues:

> [A] geologic repository would still be required. Plutonium constitutes only about one percent of the spent fuel... After reprocessing, the remaining material will be in several different waste forms, and the total volume of nuclear waste will have been *increased by a factor of twenty or more*, including low-level waste and plutonium-contaminated waste. The largest component of the remaining material is uranium, which is also a waste product because it is contaminated and undesirable for reuse in reactors. Even if the uranium is classified as low-level waste, *new low-level nuclear waste facilities would have to be built to dispose of it*. And to make a significant reduction in the amount of high-level nuclear waste that would require disposal, the used fuel would need to be reprocessed and reused many times with an extremely high degree of efficiency – *an extremely difficult endeavor that would likely take centuries to accomplish*.[57]

Past experiences with nuclear waste reprocessing technologies have also proven to be dismal failures, patently unsafe, extremely damaging for the environment, and up to *six times less efficient* than claimed by the nuclear power industry. In the US, of the three commercial nuclear waste reprocessing facilities built, only one actually became operational, in West Valley, New York. In the first year of operation, the plant was supposed to complete 27 'runs' – yet it took six years to meet that target. Fires, high exposure of workers to radiation, and continued radioactive releases into air and water created contamination effects that lasted decades, and according to the US Energy Department cost $6 billion to clean up. This was the massive fallout from a *minimal* reprocessing operation.[58] Applied to more recent US, UK and EU plans to ramp up nuclear waste reprocessing, this analysis suggests that: 1) in practice, such nuclear facilities would be deeply inefficient and be capable of generating only a fraction of the electricity currently being promised; 2) security is relatively easy to compromise making these facilities obvious targets for theft by terrorist groups; 3) continued transport, management, reprocessing and disposal of nuclear waste would be necessary, and increasingly costly, requiring increased financial and construction inputs, eliciting ongoing high taxes and electricity bills; 4) overall costs would therefore be several billion dollars at inception, and several billion dollars annually for several decades – over a period of five decades taking us up to mid century, this would amount to a total cost of *at least* $100 billion, to maintain just one facility for the duration (not including potential costs incurred by environmental damage).

There are further problems when we attempt to explore the ramifications of nuclear power for the world, with or without reprocessing of nuclear waste. One is that nuclear power is simply not carbon-free. Two independent experts, chemist Jan Willem Storm van Leeuwen and nuclear scientist Philip Bartlett Smith, point out that CO_2 is released at every stage of the nuclear fuel cycle except the actual fission in the reactor. Fossil fuels are involved in the mining, milling, conversion and enrichment of the ore, in the handling of the mill tailings, in the fuel can preparation, in the construction of the station and in its decommissioning and demolition, in the handling of the spent waste, in its processing and vitrification and in digging the hole in rock for its deposition. It is not even clear whether there are real energy gains overall to be made in this process, or whether it is as self-defeating as the process of extracting and refining unconventional oil. Van Leeuwen and Smith find that generating electricity from nuclear power emits 20–40 per cent of the CO_2 per kilowatt hour (kWh) of a gas-fired system when the whole system is taken into account. According to John Busby, former Director of the Centre for Industrial Innovation at the University of Strathclyde:

> When the energy inputs, past, present and future are totalled up and set against the actual energy derived from the entire nuclear power programme and its waste handling, it may well be that the overall energy gain has been negative. This has been masked by the availability of cheap fossil fuels, but as that era passes it behooves energy professionals to make an honest assessment of the energy and monetary economics of proceeding further with a failed technology.[59]

Further, with high quality first-grade ore in decline, future nuclear power stations will have to rely on second-grade ore, which requires even larger amounts of conventional oil input to refine it. For each tonne of poor-quality uranium, some 5–10,000 tonnes of granite that contains it will have to be mined, milled and then disposed of. Even nuclear waste reprocessing technologies as well as associated extensive nuclear waste transport, management and disposal methods would continue to require ongoing inputs of conventional oil. It would not be long before the nuclear industry emits as much CO_2 as it is supposed to save. Currently, 440 nuclear reactors supply about 2 per cent of total demand worldwide. *But the Massachusetts Institute of Technology calculates that another 1000 are needed to raise this figure to even 10 per cent of total energy needs*, at which point the question of uranium ore availability becomes critical. To meet total world energy demand, therefore, we would need 10,000 new nuclear power plants, all of them supplied regularly with fresh ore – and these would need to be built within the next few decades. The costs of constructing and continually maintaining so many nuclear plants without waste reprocessing facilities would easily amount to several trillion dollars worldwide for the duration of their operation over a few decades. With waste reprocessing facilities this cost would be tripled – yet nuclear power has already received subsidies of 0.5–1 trillion dollars since the 1950s. And by the time we would be less than halfway through building these plants, peak oil will have come and gone. It is difficult to avoid the conclusion that nuclear power would never be capable of meeting the world's total energy needs, and is not worth the costs, dangers, or inefficiencies.[60]

Separate from the dismal prospects regarding reprocessing nuclear waste, therefore, is the question of new uranium supplies, which are obviously finite. Despite some debate about available quantities of uranium ore, the 'general consensus' among nuclear scientists is that there is only sufficient uranium to fuel 'the current generation fission reactors... for decades, not centuries'. Although there are plans to convert 'fertile' material into a 'fissile' material to extend the useable fuel supply (known as 'fast breeders'), this option was long ago dismissed by the US, UK and France for being 'too costly and of doubtful value'.[61] Thus fast-breeders are unlikely to solve the problem. Van Leeuwen, who produced technology-assessment studies of nuclear power for the Centre for Energy Conservation in the Netherlands, concludes in a study published by the Institute of Physics in London:

> The breeder system has proved to be unfeasible. After 50 years of intensive research it seems extremely doubtful if the theoretical and technical problems can be overcome. Even if the breeder system would operate flawlessly from this year on, the share of breeders of the world energy supply will remain marginal.[62]

Indeed, an analysis of world data on uranium resources by the Energy Watch Group concludes 'that discovered reserves are not sufficient to guarantee the uranium supply for more than thirty years' at current annual demand. With demand rising, this means that known supplies will decline even faster. Based on all estimated possible reserves, 'a shortage can at best be delayed until 2050' – but in practice, the study warns, even this is a speculative assumption. The probability that all estimated reserves will translate into producible quantities is very small, and as such, 'they are no basis for serious planning for the next 20 to 30 years'.[63] This analysis is consistent with that of the US Army Corps of Engineers, which sees world uranium supplies as becoming effectively exhausted in the next 20 years. The thrust of this assessment is also confirmed by other authoritative sources. As early as May 2001, the International Atomic Energy Agency warned:

> As we look to the future, presently known resources fall short of demand... Lead times to bring major projects into operation are typically between eight and ten years from discovery to start of production. To this total, five or more years must be added for exploration and discovery and for the potential of completing even longer and more expensive environmental

> reviews. Therefore it would most likely be no earlier than 2015 or 2020 before production could begin from resources discovered during exploration started in 2000.... Over-reliance on an ever diminishing secondary supply could lead to a major supply shortfall in the future. Complacency resulting from overconfidence in the merits of impressive (but unproved) undiscovered resource totals could have the same effect.... Future exploration will be more difficult as the remaining targets are either deeper, located in difficult terrain or inhospitable climates, or in geologic terrain where geophysical prospecting is very difficult.[64]

Thus, given how long it takes for new nuclear power stations to be constructed, it will be well after 2027 before actual production can begin via new supply sources. By that time proven reserves may well be seriously diminished, and world oil production will have long ago peaked according to the most probable estimates. Despite tremendous efforts to secure new uranium supplies, energy analysts are now reporting that the amount of uranium available to fuel the world's 440 reactors, never mind those planned or under construction in emerging economies like India and China, is 'dwindling'. As of 2005, total annual global demand for uranium was 178 million pounds, while the total supply from mines was 105.5 million pounds. Nuclear utilities were scrambling to buy uranium, 'worried they might not be able to find enough uranium to keep their plants running'. With demand growing at 1.1 per cent a year, few new sources have been found. Worse still, according to Cameco Corp., the world's largest uranium producer, world uranium production may well have already peaked. Cameco predicts that global demand will 'outpace existing supply over the next decade by more than 400 million pounds'.[65] Thus, by mid century, new supplies of uranium will be exceedingly scarce while extant supplies of nuclear waste will have been exhausted in costly reprocessing – an investment of trillions of dollars that ultimately comes to nothing.

In other words, the much-vaunted hype around unconventional oil and nuclear energy, supported by Western states, is simply unjustified. These energy sources have little chance of safely ameliorating the impact of peak oil and accelerated climate change. Yet in 2007, according to leaked official documents, the British government 'u-turned' on pledges to meet EU targets of generating 20 per cent of Europe's energy from renewable sources by 2020. The decision was reportedly made under the influence of then Business Secretary John Hutton, who argued that the pledges would entail an 'excessive' cost of £4 billion, would be opposed by the Ministry of Defence, and would generate conflict with the nuclear power lobby. This is in spite of the *real* long-term costs and dangers in creating new nuclear facilities discussed above but concealed by the nuclear power lobby, and the fact that the UK government had already been perfectly happy to spend £205 billion up to 2007 on the Iraq War alone.[66]

Renewable Energy: A Primer

In contrast, renewable energies offer a safer, environmentally cleaner, economically cheaper, and over the long term a far more efficient and productive outlet for investment.

According to Dennis Anderson, Professor of Energy and Environmental Studies at Imperial College London, over vast areas of the developing world the incident solar energy is 2000–2700 kWh per square metre of ground occupied per year. Solar-thermal power stations can convert more than 20 per cent of this to electricity, and photovoltaic solar cells about 15 per cent. Anderson thus concludes that:

> All of the world's future energy demands could, in theory, be met by solar devices occupying about: 1 per cent of the land now used for crops and pasture; or the same area of land currently inundated by hydroelectric schemes, the electricity yield per unit area of solar technologies being 50–100 times that of an average hydro scheme.[67]

New technological developments, however, can nearly double this output. Solar cells produced by Spectrolab, Inc. in the US designed to capture energy from a broader spectrum of light are doubly as efficient as common photovoltaic cells; and breakthroughs continue to occur.[68]

Similarly, scientists at Stanford University have calculated that the global long-term technical potential of wind energy is five times total current global energy production, or 40 times current electricity demand. While this could require large amounts of land in areas of higher wind resources for installation of wind turbines, offshore wind resources offer an increasingly attractive and even more productive alternative, providing average wind speeds about 90 per cent greater than on land. This means large-scale offshore wind resources could substantially increase the technical potential of wind energy, and avoid complaints that large wind farms are an eyesore on the landscape.[69]

The earth's waters provide another renewable energy source. The marine environment stores enough energy in the form of heat, currents, waves and tides to meet total worldwide demand for power many times over. However, while the *World Offshore Renewable Energy Report 2002–2007* reports the availability of 3000 Gigawatts of tidal energy, less than 3 per cent is located in areas suitable for power generation. Even so, it is estimated that at least 34 per cent of all the UK's electricity demand could be generated from tidal currents.[70]

These are just a few examples. The technical potential of renewable energy sources is over 18 times current global energy use, several times higher even than projected energy use levels for 2100.[71] Unfortunately, under the influence of powerful corporate lobbies rooted in the fossil fuel and nuclear industries, Western governments are making insufficient efforts to develop viable plans for a comprehensive move to renewable energies. Yet such plans are viable and economical.

Scientists from the Institute for Solar Energy Supply Systems (ISET) at the University of Kassel, Germany, have proven in an ongoing experiment the viability of a Combined Renewable Energy Power Plant that would allow the whole of Germany to generate 100 per cent of its electricity from diverse renewable sources. The Combined Power Plant links together 36 biogas plants, wind, solar and hydropower installations in a distributed network of jointly controlled, small and decentralized plants, and proves 'just as reliable and powerful as a conventional large-scale power station'. The current test project for the plant is scaled to meet one ten-thousandth of electricity demand in Germany, roughly equal to the requirements of a town with 12,000 households. According to projections, Germany has sufficient domestic renewable resources to eventually replace all fossil fuels and nuclear power, achieving full energy independence by the mid twenty-first century:

> The Combined Power Plant therefore shows in miniature what is also possible on a large scale: 100 per cent provision with renewable energy sources at all times. The Combined Renewable Energy Power Plant adjusts itself to the nearest minute to meet daily needs. It covers peak loads, such as at midday, and stores electricity that is not needed during quiet periods.

Eleven wind turbines, four combined heat and power (CHP) units based on biogas, twenty solar power systems and a pumped storage power plant are linked to one another via a central control unit. This model allows the advantages of different renewable energy sources to be optimally combined:

> Wind turbines and solar modules help generate electricity in accordance with how much wind and sun is available. Biogas and hydropower are used to make up the difference: they are converted into electricity as needed in order to balance out short-term fluctuations, or are temporarily stored.

The use of advanced information technology permits active control of renewable energy power plants in real-time operation, thus ensuring 'needs-oriented electricity generation': 'Variations in the various underlying conditions, such as electricity needs or the wind availability, immediately change the interaction between the networked plants. The project thus demonstrates powerfulness and ease of control of renewable energy.'[72]

But there are still major obstacles to adopting these technologies. In countries like the US, the entire electric grid infrastructure would need to be overhauled to install such a system. In general, there remain serious questions about identifying ideal locations for the tapping of renewable energy sources, and ensuring that the energy is viably transported to national electric grids. And the global recession has seriously discouraged the urgent scale of investment required to establish a new renewable energy system. The question of location, and the imminent pressures to reduce carbon-intensive travelling, mean that reducing hydrocarbon over-dependence will also require a drastic reorganization of production in general; a reconfiguring of priorities to cut down on excess activities such as paper and packing production for commercial advertising; and even the changing of patterns of settlement to minimize commuting. Yet these changes also require a concomitant shake-up of dominant consumerist culture, values and work practices.

The good news is that an assessment by the Energy Watch Group of current cost trends and investment needs in the global energy sector predicts that renewable energy's share of total energy consumption will be as high as 30 per cent by 2030. By that time more than half the electricity demand (54 per cent) in OECD countries could be covered by renewable sources. The scenarios explored explicitly do not describe possible developments from a technological perspective, but rather suggest 'that much can be achieved with even moderate investments'. The analysis shows that 'renewable energy technologies have huge potential to help in solving the climate change problem, lowering dependence on fossil fuels, and making it possible to phase out nuclear energies', creating *permanent* energy sustainability – all with *levels of investment considerably lower than current state military expenditures*.[73] But even this is too little, too late, given the scale of reduction in fossil fuel consumption required according to the latest evidence on climate catastrophe and energy depletion. Currently, global public and private sector investments in renewable energy infrastructure remain far below 'moderate' level, meaning that without far greater investment a majority of the world's population, especially those in large urban centres, will experience significant and potentially catastrophic energy shortfalls well within the first quarter of the twenty-first century, and perhaps even within the coming decade.

Post-Carbon Civilization

The long-term trend of steady depletion of hydrocarbon and other conventional energy sources, particularly oil, portends nothing less than the end of industrial civilization as currently understood. The twenty-first century represents a critical transition period in which hydrocarbon energies will become increasingly scarce. Conventional oil most likely has already peaked, with the early years of this century constituting a stage of the undulating plateau period. This period will perhaps last no more than a few years, about a decade at most. As conventional oil production begins to slide, states may resort increasingly to exploitation of natural gas and coal, neither of which are available in sufficient quantities to sustain rising world demand. They may also choose to waste monumental investments in new nuclear and coal power stations. In any case, both coal and natural gas production will peak globally around 2025 – although depending on technology and price fluctuations, this could be prolonged by a few decades at most.

In any case, long before the end of this century, hydrocarbons will be virtually useless as a viable source of energy for the sustenance of modern industrial civilization in its current form. The continuity of civilization therefore necessitates that we embark on massive socio-political and economic changes, tied to a fundamental transition to the use of alternative renewable forms of energy, premised on a vision for the development of a new post-carbon form of civilization. This should be seen not as a negative regression, but rather as an inevitable step forward that is necessary not only for our well-being and prosperity, but for the very survival of our species and of all life on earth.

The only question remaining is whether this transition's inevitability is recognized and acted upon as peacefully as possible – by transitioning as immediately as possible to more localized forms of self- and community-resilience, reskilling of individuals, and renewable energies; or vehemently denied, resulting in increasingly intense competition and conflict over the world's last remaining hydrocarbon resources. The point has been made well by Richard Duncan, Director of the Institute on Energy and Man. According to Duncan's 'Olduvai theory', the life of industrial civilization will be a 'horridly short' pulse lasting approximately 100 years (from 1930 to 2030 maximum). 'Industrial civilization doesn't evolve', he argues. 'Rather, it rapidly consumes the necessary physical prerequisites for its own existence. It's short-term, unsustainable.'[74] Duncan has worked on his theory for over a decade, and so far the data underlying it remains unchallenged. In his 2000 address to the Geological Society of America, Duncan noted that: 'World energy production per capita increased strongly from 1945 to its all-time peak in 1979. Then from 1979 to 1999 – for the first time in history – it decreased from 1979 to 1999 at a rate of 0.33 %/year (the Olduvai "slope")'. He then predicted that:

> [F]rom 2000 to 2011, according to the Olduvai schema world energy production per capita will decrease by about 0.70 %/year (the 'slide'). Then around year 2012 there will be a rash of permanent electrical blackouts – worldwide. These blackouts, along with other factors, will cause energy production per capita by 2030 to fall to 3.32 [barrels per capita per year], the same value it had in 1930... The Olduvai 'slide' from 2001 to 2011 may resemble the 'Great Depression' of 1929 to 1939: unemployment, breadlines, and homelessness. As for the Olduvai 'cliff' from 2012 to 2030 – I know of no precedent in human history. Governments have lost respect. World organizations are ineffective. Neo-tribalism is rampant. The population is over six billion and counting. Global warming and emerging viruses are headlines. The reliability of electric power networks is falling. And the instant the power goes out, you are back in the Dark Age.[75]

Duncan updated his analysis in 2006, when he published a paper predicting that the tipping point would most likely be breached in 2008 – which is precisely when the massive oil price hikes were followed by a massive economic recession triggered by an unprecedented global banking crisis. His postulates were:

1. The exponential growth of world energy production ended in 1970.
2. Average *e* [the ratio of world energy production and population] will show no growth from 1979 to circa 2008.
3. The rate of change of *e* will be steeply negative from 2008.
4. World population will decline along with *e*.[76]

It is unlikely that events will proceed exactly according to this model. Studies of the collapse processes of past civilizations show that declines tend to be drawn out over centuries, rather than transpiring quickly over decades.[77] This suggests that the decline period will not be a 'cliff' but rather a fluctuating and protracted slope. Duncan's model is also based on worst-case assumptions – that is, on what might happen without any mitigating action at all. However, as the undulating plateau period culminates in more permanent oil price shocks, it will elicit worldwide consumer demands for serious state action to diversify energy-supply sources. High hydrocarbon resource prices will also generate unprecedented new incentives to invest in alternative energy sources. Duncan's model does not account for growing inputs from such sources.

The Energy Watch Group's forecast of renewable energy scenarios based on economic trends vindicates this critique, and points to a more mitigated decline. Unfortunately, however, we should also remind ourselves of the Hirsch Report's findings, which proved that even with a crash programme for adopting renewables the world would be unable to avert seriously debilitating impacts on modern industrial civilization, forcing widespread efforts to downsize the economy. Although such efforts would probably ameliorate some of these impacts and create space for the emergence of new and more sustainable forms of living, their sheer belatedness in the post-peak period means that potentially catastrophic and destabilizing supply shortfalls will most likely be unavoidable. Duncan's model is also highly Eurocentric, by focusing entirely on how *core Western* industrial states will likely be affected by the growing scarcity of energy for industrialized forms of production and consumption. Yet the *majority* of the world's population, inhabiting the South, do not rely on the benefits of Western-dominated forms of modern industrial life, but live in conditions of near-subsistence and intermittent electricity supplies. For many such communities, hydrocarbon energy and its technological fruits are barely available to begin with, and thus transitions to post-carbon forms of life – although no less fraught with struggle and deprivation – will be somewhat more manageable than in the North.

Yet Duncan's model does illustrate that the dangers should be understood. Certainly one important issue for both North and South is that of food production, which relies predominantly on industrial technologies dependent on large supplies of oil and water. Growing hydrocarbon resource scarcity threatens to destabilize the capacity of the global food industry to sustain production at pace with demand. Furthermore, the impact of peak oil in terms of the exhaustion of our civilization's fossil fuel resource base must be seen in tandem with the impact of climate change. Both processes are accelerating in parallel, and are liable to converge shortly before the middle of the twenty-first century if preventive action is not taken. Societies will likely respond to peak oil by reverting to other hydrocarbon energy sources that will continue to accelerate fossil fuel emissions. Most governments will be initially resistant to other options. Hence, if resources like coal and gas remain available in significant quantities for further decades, it is possible that they will continue to be exploited on a large scale, leading to ongoing rises in the rate of emissions. Hence, the prevalence of resource scarcity and conflict in the era of peak oil will be indelibly connected to the problem of accelerating climate change. At current rates of increase of CO_2 emissions, climate change is increasingly likely to create conditions on earth which threaten the survival of millions of people, and if left unchecked may well lead to the devastation of all the planet's ecosystems before the end of this century. Hence, it will be the people, rather than governments, who will be on the frontline of worldwide demands for meaningful social change, and who will need to take the initiative by transforming their lives at the grassroots.

There are perhaps two primary reasons for government reluctance to make the shift to more sustainable forms of energy, which will be more fully substantiated in later chapters.

The first, captured in the idea of *over-dependence*, is that the current hierarchical structure of the international system is premised on geopolitical domination by the Northern states of Southern states in regions containing strategic energy reserves, raw materials and sources of labour. The massive levels of overconsumption in the North derive from over-exploitation of these predominantly hydrocarbon resources, the bulk of which are located in areas of the Middle East and Central Asia. The centrality of hydrocarbon energy, especially petroleum, to Western – particularly Anglo-American – global pre-eminence therefore cannot be underestimated. Effective control of oil- and gas-producing regions along with international transshipment routes is a fundamental pivot of world order under US hegemony. *Transferring to an alternative renewable energy system, fundamentally challenging the primacy of oil in the international system, would potentially undermine the parameters of US pre-eminence by rendering its present hydrocarbon-based geopolitical strategy irrelevant.*

The second reason concerns the driving force of consumerism in the North itself, which ultimately is co-extensive with neoliberal capitalism, whose dynamic tends toward unlimited economic growth, rationalizing unlimited material consumption and over-exploitation of the earth's natural resources. Immediately vulnerable as a result of these tendencies, in the context of peak oil and climate change, is industrial civilization's ability to produce sufficient food to meet world demand.

3
Food Insecurity

In 2008, the global food crisis grabbed headlines, as world prices skyrocketed for staples like rice and wheat. The rises coincided with a similar escalation of oil prices. As the global banking crisis swept across the world's financial system, triggered by the US subprime mortgage crisis, world attention was diverted. Yet even when food prices dropped to some extent, they did not return to their previous levels – and, unnoticed by much of the mainstream media, the poor populations of less developed countries in Africa, Asia and South America continued to face serious food shortages.

This chapter unearths the origins and probable trajectory of the global food crisis by examining the international politico-economic structure of the food industry. In particular, it examines the role of transnational corporate power, as well as the neoliberal industrialization of agricultural production, in configuring the highly unequal dynamics of the global food production and distribution system. These dynamics mean that despite abundant food production, it remains inaccessible to large numbers of people in the South – a situation that drastically worsened in 2007 and after.

I argue that the ongoing global food crisis is partly a direct consequence of the unsustainability of current modern industrial farming techniques, exacerbated by the simultaneous impacts of energy depletion and climate change. Scientific studies prove that those impacts are not something that will be felt sometime far in the future, but they are being felt *now* – and on a scale that was largely unanticipated by experts. In this context, I seek to explore the potential ramifications for food production if global ecological and energy crises continue to be ignored by policymakers, and find that the long-term outlook is of declining food production leading to increasing supply shortages – unless the global food production and distribution system is radically overhauled to adopt more sustainable methods based on smaller, more localized organic farms and/or innovative new sustainable farming methods.

THE GLOBAL FOOD EMPIRE

Centralization for Corporate Profit

The acceleration of neoliberal globalization over the last few decades has entailed the increasing concentration and vertical integration of the food industry, with giant agricultural conglomerates increasingly taking over the more profitable functions that were formerly performed by small farmers. The expansion of industry and trade under the logic of neoliberal capitalism is bringing more and more uniformity to the management of the world's land,

and is a growing threat to the diversity of crops, ecosystems, and cultures. As corporate agribusiness takes over, farmers who have a personal stake in their land – and who are often the most knowledgeable stewards of the land – are being 'forced into servitude or driven out'. In the United States, for instance, the share accrued by the farmer of every consumer dollar spent on food has declined from over 40 cents before 1950 to about 7 cents today.[1]

Indeed, vertically integrated corporations now monopolize almost every aspect of farm production and distribution – from seeds, fertilizers, and equipment, to processing, transporting, and marketing. Through ownership of the grain elevators, rail links, terminals, and transportation needed to move grain around the world, one company, Cargill, controls 80 per cent of global grain distribution. Four other companies control 87 per cent of American beef, and another four control 84 per cent of American cereal. Five agribusinesses (AstraZeneca, DuPont, Monsanto, Novartis and Aventis) account for nearly two thirds of the global pesticide market, almost one quarter of the global seed market, and virtually 100 per cent of the transgenic seed market. Control over food has become so concentrated that in the US, 10 cents of every food dollar now goes to one corporation, Philip Morris; another 6 cents goes to Cargill.[2]

Under neoliberal capitalism, today's economic 'winners' include investors who scour the planet for the highest return, moving capital from country to country at electronic speeds; corporate middlemen who make use of subsidies, currency swings, and 'free trade' agreements to profit by transporting everyday needs – including food – many thousands of miles; and transnational corporations that relocate to wherever they are offered tax breaks, cheap labour, and lax environmental and workplace legislation – and then move on again when a better deal is offered elsewhere. Individual farmers, however, cannot simply move their farms. Once they are hooked into the global economy, farmers are easily victimized by an economic and technological 'juggernaut' that systematically destroys the smallest and most localized enterprises. Nonetheless, the precise aim of agricultural policy almost everywhere is to pull farmers into an export-led global economy – that is, contracted food production on behalf of global agribusiness for sale on international markets, thus privileging corporate profits over those of local farmers and over local food needs. The globalization of food production impels every region to specialize in whichever commodity its farmers can produce most cheaply, and to offer those products for sale in the world market. Frequently, this means growing plants in unsuitable environments, requiring greater investment to adapt the farm for economic survival. All foods consumed locally, meanwhile, must be brought in from elsewhere. The highly specialized farms this system favours are most 'efficient' when they are large, monocultural, and employ heavy machinery. Attaining the scale and equipment required can drain the capital reserves of all but the biggest farmers, leaving others with a repressive debt burden. These smaller farms are forced to cut costs to survive, by using less fertilizer, reducing weeding, and cutting back on other practices which further impinge on overall productivity – seemingly demonstrating the superiority of industrial agriculture. Eventually, smaller farms are driven under, their properties consolidated into those of the largest and wealthiest farmers.[3]

The net result in countries where corporate farming has become dominant is that thousands of farmers have given up. In the US breadbasket states of Nebraska and Iowa, where the farmer population has plummeted, less than a quarter still remain and these are being driven out. In Poland, 1.8 million farms could disappear in the next few years. In Sweden, half of all farms are expected to go out of business in the coming decade. In the Philippines, half a million farms in the Mindinao region alone are expected to go out of business.[4]

The social costs in the US alone of this process of corporate agricultural consolidation have been staggering. By the beginning of the twenty-first century, family farming has been all but decimated by the imperative of profit-maximization culminating in the domination of corporate agribusiness, which is now the second-most profitable national US industry after pharmaceuticals, with domestic annual sales exceeding $400 billion. The significance of this is underlined by a 2003 paper by the Pentagon's National Defense University describing agribusiness as 'to the United States what oil is to the Middle East', and thus characterizing it as a 'strategic weapon in the arsenal of the world's only superpower'.[5]

The Green Revolution

Much of the groundwork for the current structure of global agribusiness was established through the 'Green Revolution' of the 1960s and 1970s, during which the US invested heavily in agricultural development in India and other countries of the South, including the introduction of new seeds, modern industrial agricultural techniques, fertilizers and pesticides, heavy water usage, and so on. Yet at the centre of this process was the integration of crop and livestock production into large-scale, transnational agri-food complexes promoting commercial agricultural exports – i.e. cash crops. Thus, subsistence farming based on the production of staple foods such as corn, beans, and so on for local consumption was largely absorbed by transnational agribusiness, often transformed and integrated into a form of contract farming. Consequently, basic food staples produced in the South are transported across vast distances to supply Western supermarket chains, while hundreds of millions of Southern residents are compelled to purchase expensive imports from US agribusiness to meet their own needs.

World Bank and International Monetary Fund structural adjustment programmes have increasingly subordinated and marginalized traditional farming by and for local communities, again privileging large-scale industrial farming in the service of transnational agribusiness. These structural reforms forced Southern governments to open markets, land and other resources to the predatory exploits of private agribusiness, and compelled governments to cut subsidies to small farmers in rural areas and cut tariffs on food imports, while instead supporting high-value export agriculture. Simultaneously, massive government subsidies to large landowners and agribusiness corporations have permitted US exporters to sell specific agricultural products such as rice and grain at 30 to 50 per cent below their real production cost, resulting in a massive influx of US exports into Southern countries, undercutting the livelihoods of subsistence farmers and forcing them to switch to monoculture production for export under contract with transnational agribusiness. Government subsidies account for some 30 per cent of farm revenue in the OECD countries.[6] Smaller peasant farmers have thus become extraordinarily susceptible to price fluctuations in the world market.

It is no coincidence then that in the very countries where export agriculture is promoted, millions of people are hungry. For example, in 2004 India exported $1.5 billion worth of milled rice and $322 million worth of wheat. Yet over one fifth of the Indian population is chronically hungry and 48 per cent of children under five years old are malnourished. Elsewhere, farmland formerly devoted to food production for local consumption is now relegated to export-led production for luxury consumption in the North. Colombia, for example, produces 62 per cent of all cut flowers imported by the US – yet 13 per cent of its population is malnourished. Similarly, while twenty-five years ago Kenya was self-sufficient in food, it now imports 80 per cent of its provisions, while 80 per cent of its exports are agricultural products.[7]

Unsustainability of Modern Industrial Farming Techniques

As small farms are taken over by larger ones, and as all farms become more subservient to the conglomerates, the rationale often given is that the larger enterprises are more efficient or productive. But that is in large part a myth, according to Worldwatch Institute Senior Researcher Brian Halweil. While a large monoculture operation may produce more output per acre of that particular crop than a small farm does, the small farm engaging in traditional polyculture (raising more than one kind of crop, using different root depths or soil nutrients on the same piece of land) makes more efficient use of resources and can produce significantly more food per acre over the long-term.[8]

As US structural geologist Dave Allen Pfeiffer argues, the principal problem is that modern intensive agriculture as practised by the globalized corporate food industry is simply unsustainable. Approximately three quarters of the land area in the United States is devoted to agriculture and commercial forestry. It takes 500 years to replace one inch of topsoil. In soil utilized by modern agriculture, erosion is reducing productivity by up to 65 per cent each year. Former prairie lands, which constitute the breadbasket of the United States, have lost one half of their topsoil after about 100 years of farming. This soil is eroding 30 times faster than the natural rate. Erosion and mineral depletion removes about $20 billion worth of plant nutrients from US soils every year. Pfeiffer also notes that more than 2 million acres of cropland are lost annually to erosion, salinization and waterlogging. On top of this, urbanization, road building, and industry claim another 1 million acres from farmland annually.

Modern agriculture also places a strain on water resources. Agriculture consumes fully 85 per cent of all US freshwater resources. A corn crop that produces 118 bushels per acre a year requires more than 500,000 gallons of water per acre during the growing season. The production of one pound (lb) of maize requires 1400 lb (or 175 gallons) of water. Unless something is done to lower these consumption rates and/or reorganize modern agriculture itself, Pfeiffer rightly warns, the global food industry will help to bring about a US water crisis.

In the last two decades, the use of hydrocarbon-based pesticides in the US has increased 33-fold, yet each year more crops are lost to pests. This is due to the abandonment of traditional crop rotation practices. Nearly 50 per cent of US corn land is farmed continuously as a monoculture. This results in an increase in corn pests, which in turn requires the use of more pesticides. Pesticide use on corn crops had increased 1000-fold even before the introduction of genetically engineered pesticide-resistant corn. However, thereafter corn losses still rose fourfold. In summary, modern intensive agriculture is unsustainable. It is damaging the land, draining water supplies and polluting the environment. And all of this requires more and more fossil fuel input to pump irrigation water, to replace nutrients, to provide pest protection, to remediate the environment, and to simply hold crop production at a constant level.[9]

PEAK FOOD?

Impact of Climate Change on World Food Production

Climate change, environmental damage, peak oil, and the dysfunctional operation of the global political and economic system are all inextricably connected to an unfolding crisis in global food production. The link between climate change and a fall in global food productivity was firmly documented in a new peer-reviewed study by the Carnegie Institution and Lawrence Livermore National Laboratory, which found that from 1981 to 2002 human emissions-driven global warming reduced the combined production of wheat, corn, and barley-cereal grains (forming the foundation of much of the world's diet) by 40 million metric tonnes per year. The reductions led to annual losses of about $5 billion. On average, global yields for several of the crops dropped in correlation with warmer temperatures, by about 3–5 per cent for every 1-degree-Fahrenheit (F) increase. Average global temperatures increased by about 0.7°F during the study period, with even larger changes observed in several regions.[10] As Christopher Field from Carnegie observed: 'Most people tend to think of climate change as something that will impact the future, but this study shows that warming over the past two decades has already had real effects on global food supply'.[11] And the situation will only worsen. At the British Festival of Science in Dublin in September 2005, US and UK scientists working at the Met Office's Hadley Centre described how shifts in rain patterns and temperatures due to global warming could by conservative estimates lead to a further 50 million people going hungry: 'If we accept that broadly 500 million people are at risk today, we expect that to increase by about 10 per cent by the middle part of this century.'[12]

Toward the end of 2006, a study by the Hadley Centre, funded by the UK Department for Environment, Food and Rural Affairs, predicted that if global warming continues, drought which already threatens the lives of millions will spread across half the earth's land surface before 2100, and extreme drought rendering agriculture impossible will affect a third of the planet. This world-scale drought would undermine the ability to grow food, the capacity for a safe sanitation system, and the availability of water, pushing millions of already impoverished people over the precipice. Even this finding, however, is conservative, because the study excludes feedback mechanisms which would make the portended drought even worse and also hasten its arrival.[13]

A 2007 IPCC report confirmed these conclusions but was more alarming. Based on the IPCC's projected conservative scenarios for climate change, the report warned that rising temperatures would create critical water shortages in China, Australia, parts of Europe and the United States. At a 2–3°C temperature rise, up to 3.2 billion people (half the world's population in 2008) would face severe water shortages that would also endanger agricultural production, putting an additional 600 million people at risk of famine. Simultaneously, coastal flooding would hit another 7 million homes worldwide.[14] According to an analysis by scientists at the University of Washington and Stanford University, published in *Science* in January 2009, the areas hit hardest by global warming will be the tropics and subtropics, encompassing about half the world's population, including Africa, the southern United States, and much of India, China and South America. In these areas, higher temperatures will at first cause crops like corn, wheat and rice to grow faster, but over the longer term will reduce plant fertility and grain production. World crop yields will eventually fall 20 to 40 per cent.[15]

Limits of World Food Production Under Transnational Agribusiness

Irrespective of climate change's impact, artificial limits on world food production are being generated by the already existing structure of an industrialized corporate agricultural system that is exhausting the productivity of the soil across the entirety of the earth's fertile land. Maps released in December 2005 by scientists at the Center for Sustainability and the Global Environment (SAGE) at the University of Wisconsin-Madison, show that 'agricultural activity now dominates more than a third [or about 40 per cent] of the Earth's landscape and has emerged as one of the central forces of global environmental change'. The maps combined satellite land cover images with agricultural census data from every country in the world, creating detailed depictions of global land use. According to SAGE scientist Dr Navin Ramankutty, 'Except for Latin America and Africa, all the places in the world where we could grow crops are already being cultivated. The remaining places are either too cold or too dry to grow crops'. The maps thus show that the Earth is 'rapidly running out of fertile land' and that 'food production will soon be unable to keep up with global population growth'. Dr Ramankutty further notes that the real question is: how can we continue to produce food from the land while preventing negative environmental consequences such as deforestation, water pollution and soil erosion?'[16]

It is precisely this question that is answered by Lester Brown, a former international agricultural policy adviser to the US government who went on to found the World Watch Institute and the Earth Policy Institute, and currently serves as director of the latter. He notes that since 1996 world grain productivity has been at its height, and that since 2000 world grain consumption has exceeded production, such that 2003 saw a grain production deficit of 105 million tonnes. On that basis, in 2003 Brown had predicted the onset of a global grain deficit in subsequent years. Thus, when China saw sudden food price hikes toward the end of 2003, Brown warned that this was a sign of an impending world food crisis brought on by global warming and increasingly scarce water supplies. In previous months, wheat prices in northeast China had risen by 32 per cent, maize prices doubled and rice prices went up by 13 per cent. Simultaneously, China faced a 40-million-tonne grain shortfall. Addressing an audience of Chinese environmental NGOs, Brown said:

> I view the price rises as an indication, as the warning tremors before the earthquake. World grain harvests have fallen for four consecutive years and world grain stocks are at the lowest level in 30 years. If farmers can't raise production by late next year we may see soaring grain and food prices worldwide.

He also noted that in 2003 the world faced a 96-million-tonne shortfall in grain following poor harvests in the United States and India in 2002, and a poor harvest in Europe due to unusually high temperatures. Such worldwide shortfalls were made up only by drawing from dwindling grain reserves.

Pertinently, Brown emphasized the linkage between environmental damage, global warming and the food crisis, noting the role of water shortages in China, India and elsewhere, as well as global warming's exacerbation of falling grain harvests. Studies by the International Rice Institute and the Carnegie Institution in Stanford have shown that grain production can fall 10 per cent with a one-degree-Celsius increase in temperature, as the increased heat puts stress on the plants.[17]

However, Brown goes beyond the role of climate, pointing out real constraints on world food production from the supply side. Concurring with the SAGE scientists, he notes that there is scarce new land that can still be brought under agricultural production, except from 'clearing tropical rainforests in the Amazon and Congo basins and in Indonesia, or from clearing land in the Brazilian cerrado, a savannah-like region south of the Amazon rainforest'. The latter would, of course, have heavy and unconscionable 'environmental costs: the release of sequestered carbon, the loss of plant and animal species, and increased rainfall runoff and soil erosion'. Elsewhere, in scores of countries, 'prime cropland is being lost to both industrial and residential construction and to the paving of land for roads, highways, and parking lots for fast-growing automobile fleets'. Apart from land, sufficient new sources of irrigation water 'are even more scarce than new land to plow'. World irrigated area has nearly tripled, expanding from 94 million hectares in 1950 to 276 million hectares in 2000. Since then there has been almost no growth, and consequently 'irrigated area per person is shrinking by 1 percent a year'.

In this context, the ability of industrial agricultural techniques to continually raise productivity to meet rising world demand is dwindling. Although from 1950 to 1990 the world's farmers raised grainland productivity by 2.1 per cent a year, between 1990 and 2007 this growth rate slowed to 1.2 per cent a year. It is no surprise then that the cumulative impact of all these trends has been that farmers are finding it increasingly difficult 'to keep pace with the growth in demand'. During seven of eight years prior to 2008, world grain consumption exceeded production. Brown reports:

> After seven years of drawing down stocks, world grain carryover stocks in 2008 have fallen to 55 days of world consumption, the lowest on record. The result is a new era of tightening food supplies, rising food prices, and political instability. With grain stocks at an all-time low, the world is only one poor harvest away from total chaos in world grain markets.[18]

Apart from climate change's effects leading to massive crop failures in regions as far apart as Australia and Bangladesh, escalating Northern investment in biofuels is encroaching on land previously devoted to food production. The impetus for this is, purportedly, an effort to avoid the damaging effects of climate change by switching to alternative sources of energy. Yet according to studies by US scientists, the conversion of land use for biofuel production will generate increased greenhouse gas emissions for up to hundreds of years, over and above current CO_2 emissions, to an even greater extent than traditional petrol.[19]

Impact of Peak Oil on World Food Production

Added to this is the impact of declining fossil fuel production, specifically peak oil. The Green Revolution of the 1950s and 1960s, resulting in the industrialization of agriculture, led to new hybrid food plants and more productive food crops. Between 1950 and 1984, world grain production increased by 250 per cent, even as the numbers of the hungry rose due to an unequal system of food distribution. But the additional energy for this rise in production did not come from an increase in sunlight, nor did it result from introducing agriculture to new vistas of land. The Green Revolution was powered by fossil fuels in the form of fertilizers (natural gas), pesticides (oil), and hydrocarbon-fuelled irrigation. Precisely these factors are now imposing limits on the global agricultural system. Without this energy, production premised on the industrial agricultural methods of the Green Revolution simply cannot continue. As Pfeiffer warned presciently in 2003, the current peaking of global oil production (and subsequent decline of production), along with the coming peak of natural gas production 'will very likely precipitate [an] agricultural crisis much sooner than expected. The end of this decade could see spiraling food prices without relief. And the coming decade could see massive starvation on a global level such as never experienced before by the human race'.[20]

The complete dependence of the global food industry on the exploitation of oil and natural gas means that peak oil will inevitably coincide not only with rising prices, but with an escalating difficulty in sustaining world food production, leading to worsening annual shortfalls, and the prospect of large-scale and permanent famines. As noted by Richard Heinberg:

> [Peak oil] will make machinery more expensive to operate, fertilizers more expensive to produce, and transportation more expensive. While the adoption of fossil fuels created a range of problems for global food production, as we have just seen, the decline in the availability of cheap oil... will exacerbate them, bringing simmering crises to a boil.

In the US, he reports, over 400 gallons of oil equivalent are expended to feed each American annually, about a third of which goes toward fertilizer production, 20 per cent to operate machinery, 16 per cent for transportation, 13 per cent for irrigation, 8 per cent for livestock raising (not including feed), and 5 per cent for pesticide production. Added to this are the costs of transportation, which is predominantly by trucking, although this is ten times more energy-intensive than moving food by train or barge. Furthermore, processed foods constitute three quarters of global food sales by price, which has enormous energy costs. While a one-pound box of breakfast cereal, for instance, requires over 7000 kilocalories of energy for processing, the cereal itself provides only 1100 kilocalories of food energy. Overall, Heinberg warns, 'the modern food system consumes roughly ten calories of fossil fuel energy for every calorie of food energy produced'.[21]

Escalating food prices in the early twenty-first century were a consequence of the cumulative impact of peaking world oil production (affecting shipping and production costs), the limits of the corporate industrial agricultural system, the exacerbating role of climate change and conversion of land for biofuels, and state responses to these new global conditions. In order to rebuild their rice reserves in the wake of falling harvests, India and Vietnam announced in early 2008 their suspension of rice exports, prioritizing domestic markets. Together, these countries account for as much as 30 per cent of global rice exports. The world market, already straining under the limits hitherto described, became even tighter as speculators began panic-buying rice futures. This led to a cycle of rising prices as existing shortages were exacerbated by unprecedented hoarding of grain stocks and increased futures purchases. As food prices skyrocketed and world agricultural production faced increasing shortfalls, with rice rising by 15 per cent and wheat by 115 per cent between 2007 and 2008, Argentina, Chile and Ukraine also resorted to restricting their exports of wheat and other grains. Throughout 2008, the World Food Programme and the Food and Agricultural Organization confirmed that increasingly violent street protests and rioting over high food prices had broken out in Guinea, Indonesia, Mauritania, Mexico, Morocco, Uzbekistan, Yemen and West Africa. These examples illustrate that civil unrest due to proliferating famines will increase unless the global food system is transformed.[22]

The implications for regional and international stability are dire. As greater numbers of people in the South become deprived of adequate food, isolated incidents of public disorder may well undermine the integrity of states, and possibly even coalesce, precipitating counterinsurgency measures by the state and the outbreak of civil conflicts. Such events would magnify the probability and scale of mass migrations to escape drought and starvation across the most vulnerable areas of the South, in turn generating further pressure in the North for tighter immigration policies designed to 'wall-off' asylum seekers and foreigners. The 'security' implications are magnified in relation to potential Northern attempts to fortify their domestic food security. The dependence of industrial agriculture on inorganic, mined fertilizers means that control over regions where these are concentrated, such as Morocco's rock phosphate supply, will become increasingly contested by rival powers, and they may well become sites of major civil unrest and terrorist activity, by both state and non-state actors. Further, as food supply shortfalls worsen, the increased dispossession of small farmers by Northern agribusiness will accelerate regional migratory trends, lead to greater urbanization, and intensify population pressures on local regimes, which in turn will rely on police and military force to control groups and protect Northern interests. Such pressures will create fault lines of political insecurity across food-producing regions, potentially catalyzing the resort to political violence by poorer, hungry and unemployed classes (who constitute a majority).[23]

As oil prices declined along the undulating plateau in early 2009, the strain on agricultural production was loosened, permitting a corresponding relaxation of world food prices. Yet just as the decline in oil prices proved temporary, so too were declines in food prices. As oil prices rose again in the latter half of 2009, global food production was already approaching its limits under the physical, geophysical and environmental constraints of neoliberal corporate agribusiness, meaning that for vulnerable Southern populations food scarcity is an increasingly permanent feature of the world economy – it is only a matter of time before this catches up with the North (and it arguably already has). Indeed, while economic recession has effected a generalized collapse in demand, it has impacted the demand for food far less than it has for other commodities. Thus, a period of food price fluctuation over some years is likely, but the long-term trend, particularly for staple foods, will consist of higher and permanently escalating prices, with significantly less fluctuation. These circumstances will only be broken by breakthrough transformations in the social and technological organization of agriculture, on the basis of more sustainable and organic methods, linked to alternative sources of energy.

The Political Economy of Food

It is often assumed that part of the solution to a lack of sufficient food is to reduce the number of people that need feeding – in other words, to introduce measures of population reduction such as birth control. The assumption is that there are simply too many people to feed, that population growth is out of control, gone beyond the earth's inherent productive capacities. Not only governments, but also independent NGOs and international aid agencies hold fast to this perspective. But this erroneous assessment really amounts to a case of blaming the victims, the vague notion that 'they' are suffering not because of 'us', but because something is inherently wrong with 'them', specifically 'their way of life' which is so far removed from the modernity of ours. The problem is not so much related to population growth *per se*, as it is related to *demographic imbalances*. In the North, declining birthrates combined with declining death rates have led to an increasingly large ageing population and an insufficiently large young population, leading to labour shortages particularly in industries requiring physically intensive manual work. Southern countries where local agriculture has been debilitated by the imposition of Northern-controlled industrialization frequently have larger, younger (and hungrier) populations of unemployed ex-farm workers. These in turn serve as lucrative sources of cheap labour to be imported into the North largely for manual work.[24]

Indeed, the preceding analysis has shown that the primary reason so many people are starving today, largely in the South, is not because there are too many people, but because of the structure of a global food industry that is exhausting the soil. According to the United Nations Food and Agriculture Organization, global food production is sufficient to provide over 2720 kilocalories per person per day, which is 17 per cent more calories per person than was produced in the 1970s despite a 70 per cent population increase.[25] This is within the *existing* hydrocarbon-dependent industrialized global system.

While it is commonly assumed that, whatever its shortfalls, this system is the only one capable of producing enough food to feed the earth's current and projected populations, the evidence seen above indicates to the contrary that this system is destroying its own productive capacity, exhausting the earth, breaching its limits to production, and is no longer sustainable. Yet there *are* viable alternatives which do not require industrial technologies powered by extensive hydrocarbon energy inputs. An extensive study by the University of Michigan Department of Ecology and Evolutionary Biology finds that organic, low-input, low- or no-pesticide, integrated, small-scale, and sustainable agricultural production not only addresses environmental concerns, but can produce sufficient food to sustain a large and growing human population. The study finds that even under conservative estimates, organic agriculture could provide almost as much food on average at a global level as is produced today (2641 as opposed to 2786 kilocalories/person/day after losses). Furthermore, based on what University of Michigan researchers considered to be a more 'realistic' estimation, organic agriculture could actually *increase* global food production by as much as 50 per cent (to 4381 kilocalories/person/day). This would be more than enough to support an estimated population peak of around 10–11 billion people by the year 2100.[26] The solution for post-carbon civilization, therefore, is to make the transition to more localized and sustainable forms of organic farming, which are capable of sustaining even higher calorie intakes per person, for an even larger world population. Thus the fundamental problem is not immediately a question of technology to increase food production. Nor is it an issue of overpopulation. Rather, the fundamental difficulty lies in the unsustainable manner in which food is currently produced, and the socio-political structures which generate entrenched inequalities in global food distribution.

Firstly, Western agribusiness dominates global food production in its own interests, imposing industrial technologies of monocultural production which rapidly debilitate and exhaust the land. In the process, small farms raising several crops and using traditional methods of production are wiped out, although they are not only more efficient in the use of resources, but can produce more food overall per acre without devastating the land. Therefore the farming methods of the global food industry are self-defeating. It is not that we are exhausting the natural agricultural limits of the earth – it is that we are imposing upon it a specific corporate organization of agricultural exploitation that is inefficient, destructive, and ultimately unsustainable. We are reaching the limits of our own agricultural system. We therefore need to adopt new, more decentralized methods of agricultural production which respect the soil.

Secondly, the problem is not therefore that food is unavailable – it potentially is, and in abundance. But the global system of food production is geared toward a specific system of international distribution which is devastating for the environment. This system of distribution is inherently unequal, conditioned within the wider structures of the world economy, and hence results in the obscene situation where populations in the North indulge in overconsumption and generate huge amounts of waste, while populations in the South suffer from hunger and even starvation.

Thirdly, predictions of unrestrained global population expansion are greatly exaggerated. Careful scientific analysis suggests not that the world population will rise inexorably, but rather will taper off, stabilize and eventually decrease naturally, without active intervention. World population is expected to reach 9 billion by 2050, and level off at 10 to 11 billion by 2100, after which it will gradually decline.[27]

The idea that there are too many people in the world is not only conventionally applied to the problem of world hunger, but to global economic inequalities in general, of which hunger is ultimately only one dimension: people are poor, it is believed, because there are simply too many of them for Southern governments to cope with, a problem exacerbated by regional corruption. Once again, this is a very partial picture. Global Western agribusiness, of course, does not operate in isolation, but rather in the context of a global economic system, the same system which functions on the basis not only of hydrocarbon over-dependence, but in relation to an artificial international division of labour where huge sections of the world are functionally subordinated to the requirements of industrial and finance capitalism in the advanced Northern states. While corruption's role in the apparent inability of Southern states to develop cannot be underestimated, the truth of the matter is that in a fundamental but subtle way, they are no more corrupt than our own states and institutions. And there is no better illustration of this than the crisis of global finance.

4
Economic Instability

In August 2006, I interviewed a former US government adviser with high-level access to the US policymaking establishment who told me that, according to leading US financial analysts, 'a collapse of the global banking system is imminent by 2008'. Subsequent to receiving this warning, I gathered a wealth of evidence on the coming crisis, and was perhaps one of the only non-economists publicly warning of an imminent major global banking crisis to unfold in 2008.[1] Drawing on the groundbreaking work of a tiny minority of economists who had their finger on the pulse of the financial system, I further specified at a November 2007 seminar at Imperial College London that the banking collapse would likely begin in January 2008, precipitated by a crisis in the housing markets.[2]

This chapter builds on much of the material I had already gathered before the global banking crisis of 2008, and the wider financial crisis it sparked. Although the crisis brought home to Western publics how deeply fragile neoliberal capitalism in its current form could be, for the most part it led to little soul-searching about the desirability – let alone viability – of the current structure of the global political economy. However, it did bring home the dangers of completely deregulating the activities of the banking and corporate actors; while the massive government interventions in collapsing markets through stupendous 'bailouts' of insolvent banks and financial entities showed the world that many of the assumptions of neoliberal ideology – such as the unqualified superiority of the 'free market' system – are not always borne out.

Yet missing from the ruminations about looming economic recession, or even a new Great Depression, was recognition that what seemingly erupted almost out of nowhere in 2008 was not simply an unpredictable consequence of shortsighted policy failures, having to do with irresponsible lending in the housing markets, but was rather an *anticipated direct consequence* of deeper structural tendencies in the financial system. These crisis-prone tendencies of the global financial and economic system, moreover, are intimately tied to the global political economy's long-term propensity to generate apparently limitless profits for a minority based largely in the North, while increasingly marginalizing the vast majority of the world's population based largely in the South.

This chapter begins by contextualizing the global financial and economic crisis in relation to a much more long-term, escalating, yet virtually invisible crisis of impoverishment and deprivation. The intensifying bifurcation of the world economy into 'haves' and 'have-nots', although denied by the leading international financial institutions, is proven by their own data. Moreover,

I argue that this bifurcation process is also related to the world economy's subjugation to what I call the 'Dollar-Debt Standard,' a US dollar-denominated world-reserve currency system that is premised on the proliferation of debts as a mechanism to generate profits, created virtually *ex nihilo*. Based on unequal ownership structures of the world's physical energy, mineral, and human resources, the global political economy systematizes a form of exponential growth that is accelerating the fatal exhaustion of these resources and the destabilization of the earth's ecosystems, nudging global ecological and energy crises beyond the tipping point. This demonstrates beyond a doubt that global crises cannot be resolved by simply implementing minor reforms within the existing system, but only by effecting a fundamental transformation of the system itself.

THE PROBLEM OF 'GROWTH'

Northern Overconsumption

Climate catastrophe, peak oil, and peak food are converging in the context of the specific form of social organization by which we inhabit and live off of the earth: the global political economy. It is by no means a coincidence that the world's largest polluters today are not in the less developed South, but are the countries of the North which preside over an unequal global system that is destructive in the exploitation of energy, raw materials, and land. The Union of Concerned Scientists places the blame squarely on us when it warns that the developed nations 'must greatly reduce their over consumption, if we are to reduce pressures on resources and the global environment'.[3] There is thus an inherent connection between the destruction of the environment, the acceleration of climate change, and the inequalities of the global economy, a fact also noted in the UN Development Program's *Global Environmental Outlook-2000* report: 'The continued poverty of the majority of the planet's inhabitants and excessive consumption by the minority are the two major causes of environmental degradation.'[4]

According to the Wuppertal Institute for Climate, Environment and Energy, the global economy 'favours the rich, promotes concentrations of power, and with the World Trade Organization (WTO), the World Bank (WB), and the International Monetary Fund (IMF) has created institutions and rules that favour the global players'.[5] These networks of production, consumption, and finance in the global political economy serve the interests of only a third of the human population: 'Two-thirds of the world (the bottom 20 per cent of the rich countries and the bottom 80 per cent of the poor countries) are either left out, marginalized, or hurt by these webs of activities.'[6] Thus, in 2005, the wealthiest 20 per cent of the world accounted for 76.6 per cent of total private consumption, while the poorest fifth accounted for only 1.5 per cent.

There is thus a direct relationship between disparities in the distribution of wealth and environmental decline – between the escalating instabilities of an inherently unequal global economic system, and global ecological imbalance. This is partly because of the way the global economic system functions, driven by pressures to minimize costs and maximize productivity, and thereby maximize profits accruing to an elite minority without recognizing limits to economic growth, and with indifference to negative fallout on wider communities. 'The global corporation, which is programmed by its internal structures to respond to the incessant demand of financial markets to seek its own unlimited growth, behaves much like a cancerous tumor', writes David Korten, author of *When Corporations Rule the World*:

> Furthermore, the economy internal to a corporation is centrally planned and directed by top management, not to serve the whole of the society on which its existence depends, but rather to maximize the capture and flow of money to its top managers and shareholders. These characteristics – growth at the expense of the whole and centralized planning – represent serious violations of the principle of cooperative self-organization in the service of life.[7]

Therefore, perhaps it is not that there should be limits to growth, but that the very concept of growth inherent to the global economy is a dysfunctional one, if not a perverse one – given that current trajectories of growth are no longer sustainable. As the Worldwatch Institute in Washington DC puts it:

> The global economy as now structured cannot continue to expand much longer if the ecosystem on which it depends continues to deteriorate at the current rate. Just as a continuously growing cancer eventually destroys its host, a continuously expanding global economy is slowly destroying its host – the Earth's ecosystem.[8]

These are harsh words, and their radical import should not be underestimated. If the global economy is operating like a cancer, then the earth and its inhabitants are in serious need of medical attention. But not everyone seems to agree, least of all the World Bank, which in 2002 optimistically argued that:

> Globalization generally reduces poverty because more integrated economies tend to grow faster and this growth is usually widely diffused... Between countries, globalization is now mostly reducing inequality... The number of extreme poor (living on less than $1 per day) in the new globalizers declined by 120 million between 1993 and 1998... Within countries, globalization has not, on average, affected inequality, although behind the average there is much variation.[9]

To its credit, the Bank conceded that 'many poor countries – with about 2 billion people' are now 'becoming marginal to the world economy, often with declining incomes and rising poverty'. But the Bank's explanation for this is that they 'have been left out of the process of globalization'.[10]

The Bank's argument is that there may be problems, but these are nothing to do with the system itself – in fact, to solve those problems we need *more* globalization, the *greater* and more *consistent* application of World Bank policy prescriptions, and thus the consolidation of the global political economy in general. But the Bank's claim that globalization has brought 120 million people out of extreme poverty rests on the erroneous definition that population groups with *per capita* income of one dollar per day or more are 'nonpoor'. As the Canadian economist Michel Chossudovsky points out:

> The one dollar a day standard has no rational basis: population groups in developing countries with per capita incomes of two, three or even five dollars remain poverty stricken (i.e. unable to meet basic expenditures on food, clothing, shelter, health and education).[11]

Although the World Bank's revision of its previous estimates of global poverty in August 2008 was thus welcome, it was not enough. In its report, the Bank argued that its previous estimate of 985 million people living below the 1-dollar-a-day poverty line was incorrect. The new estimate suggests that as many as 1.4 billion people live in poverty, newly defined as below $1.25 a day (the average national poverty line for the poorest 10–20 countries). Under the revised estimates, the Bank was still able to suggest that the world is on track to halve 1990 poverty rates by 2015, arguing that although a larger number are poor, less live in extreme poverty as defined by the dollar-a-day standard.[12]

Yet as noted by Columbia University economist Professor Sanjay G. Reddy, formerly of Harvard and Princeton, and a longtime adviser to various UN agencies including the World Bank:

> [T]he Bank's chosen international poverty line is far too low to cover the cost of purchasing basic necessities. A human being could not live in the US on $1.25 a day in 2005 (or $1.40 in 2008), nor therefore on an equivalent amount elsewhere, contrary to the Bank's claims. Indeed, it appears to be far too low in many countries to account for the cost of purchasing basic necessities... The Bank's claim that its poverty line is sufficient in other countries, despite being insufficient in the United States, implicitly acknowledges the second problem: that the Bank uses inappropriate purchasing power parities (PPPs) to convert its poverty line across currencies... The Bank's new $1.25 poverty line is itself based on an average of poverty lines allegedly used in poor countries. However, as before many of these poverty lines have been defined by the Bank itself and they are translated into common units using the very PPPs the application of which is in question. The claim that the international poverty line (whether the new or the old one) is representative of standards prevailing in poor countries is untenable.[13]

It is therefore simply not clear that the alleged massive movement of millions of people out of poverty in India and China has actually occurred; it may rather be that millions of people are moving from extreme poverty to slightly less extreme poverty. Indeed, the Bank's application of the contentious and narrow definition of a $1.25 per day poverty line also ignores the extent to which millions of members of the middle classes in India and China have actually become poorer, but are not counted as their income is slightly above this figure. It is also worth noting that India and China are only two countries to have allegedly experienced a decrease in poverty (a highly questionable claim on its own terms) while another 80 or so less developed countries have consistently experienced worsening levels of poverty.[14]

In both countries, the Bank's designation of millions of deprived people as 'nonpoor' is disingenuous, as they continue to live in conditions of deprivation according to more realistic standards of poverty defined by credible, independent social scientists. The other problem is that the evidence does not support the World Bank's claim that the alleged rising income of millions has been a consequence of *neoliberal free-market* economic policies. Economists largely credit the phenomenal growth experienced by India and China not with neoliberal reforms but with the specific role of government intervention involving high levels of protection and strict currency controls, among other policies.[15]

It is not entirely surprising then that the vast majority of people in the South strongly disagree with the World Bank; and that the World Bank's own data contradicts its claims about the resounding success of globalization. Indeed, around the dawn of the new millennium, the World Bank decided to ask Southern civil society groups not only what they thought, but to participate in a five-year programme of collaborative research to discern the impact of World Bank policies. The collaboration solidified in the form of the World Bank Structural Adjustment Participatory Review International Network (SAPRI) and culminated in the release of *The SAPRI Report*, which concluded that the global economic system had produced

> increased current-account and trade deficits and debts; disappointing levels of economic growth, efficiency and competitiveness; the misallocation of financial and other productive resources; the 'disarticulation' of national economies; the destruction of national productive capacity; and extensive environmental damage.

Rather than poverty and inequality diminishing under the tutelage of Western-inspired economic medicine, they are now 'far more intense and pervasive than they were 20 years ago, wealth is more highly concentrated, and opportunities are far fewer for the many who have been left behind by adjustment'.[16] The response of the World Bank to the findings of its own research exercise is instructive – once it became clear what the findings would be, they were dismissed, and the Bank immediately distanced itself from SAPRI.

The Bretton Woods System

World capitalism's institutional framework came into being just before the end of the Second World War. The IMF and the World Bank were founded at the 1944 Bretton Woods conference, having been jointly conceived as early as 1941 by the United States and Britain to preserve and promulgate US global financial hegemony. The general idea was to establish international institutions capable of organizing and administering the growth in postwar debt, in order to finance the continued purchase of US goods and services by allies.[17]

Although foreign governments held some shares in these institutions, the US government held the 'dominating veto', and hence these governments could 'do little to control the use of capital of these bodies', except facilitating their own debt servicing. Thus, these institutions enabled the US economy to 'draw the finances of other governments into an international cartel directed by its own policy-makers, dominated by US officials and their appointees'. The IMF and the World Bank promoted a narrowly US-centred conception of world financial stability 'as an expansion in foreign economic life and its adjustment to the needs and capacities of America. US Government claims thus were formalized into an institutional edifice of world economic domination.'[18] US officials ensured that European loan repayments were not simply poured directly into the US economy, but were recycled to allow the IMF and World Bank to maintain a large reserve of further loanable funds – a pool of intergovernmental capital dominated by the United States and used 'to finance a worldwide Open Door policy and to facilitate the breaking up of colonial spheres of influence'. Subsequently, the IMF, World Bank, GATT (the precursor to today's World Trade Organization [WTO]) and US foreign aid programmes were used as 'a formal system for the political implementation of American economic strength'.[19]

The Open Door policy, moreover, was a product of the logic of US economic self-interest. Europe and Asia were required to dispense with protectionist measures to permit US producers to expand abroad. This was achieved by providing other countries with foreign assistance inducing them to 'adhere to free trade, stable currency parities, general dependence on American food and industrial exports, and to open their investment markets to private capital'.[20] As explained in detail by Wall Street financial analyst Michael Hudson, Distinguished Professor of Economics at the University of Missouri and a former balance-of-payments economist at Chase Manhattan Bank and Arthur Andersen, the establishment of a global financial regulatory system through the Bretton Woods institutions was integral to US government efforts to restore financial liquidity and monetary stability to the world economy after the war, precisely to support American industry and exports. The US government now recognized that due to massive debt repayments from its allies and war reparations from Germany, its balance of payments had reached a surplus unmatched by any other nation in history. The US Treasury had accumulated three fourths of the world's gold, 'denuding foreign markets of the ability to continue buying US exports at their early postwar rates'. Having amassed an 'embarrassment of riches' that had almost bankrupted Britain and Western Europe, for international trade to continue the US government required 'a payments deficit to promote foreign export markets and world currency stability. Foreigners could not buy American exports without a means of payment, and private creditors were not eager to extend further loans to countries that were not creditworthy'. The US government had to find a way to sustain US exports by replenishing the world's dollar reserves – the answer was to allow its allies to emerge from bankruptcy and thereby continue purchasing American goods and services. The line of least domestic resistance was to provide Congress with 'an anti-Communist national security hook on which to drape postwar foreign spending programs. Dollars were provided not simply to bribe foreign governments into enacting Open Door policies, but to help them fight Communism that might threaten the United States if not crushed in the bud.' The Cold War thus permitted the US to generate political support for the British Loan, Marshall Aid, 'along with most subsequent aid lending down through the present day'.[21] As historian Joseph Stromberg thus notes:

> '[D]efense of the Free World against Communism' became the most potent slogan veiling US imperial activities and justifying Open Door intervention everywhere. It did overlap reality, because the triumph of revolutionary nationalism, usually under communist leadership, could, indeed, exclude American business from certain markets.[22]

Bretton Woods heralded a new postwar US-dominated financial system under the dollar standard. But rather than US financial power being premised on its position as the world's leading creditor nation, it became tied to a new form of debt. If the United States had continued to absorb more foreign gold and dollar balances, the world's money supplies would have been depleted, constraining world trade and especially countries' ability to purchase American exports. This meant the US had to ensure that sufficient cash and assets were available in the system, by buying more foreign goods, services and capital assets than it was supplying to foreigners. In other words, the only way the world financial system could become more liquid and continue growing was for the United States to pump more dollars into it by running a *payments deficit*. It had, in effect, to spend its key currency abroad by borrowing more from other 'payment surplus' nations. This debt character of the world's growing dollar reserves was not initially recognized as they were underwritten by the Treasury's gold stock. But by the early 1960s, the US was clearly approaching the point at which its debts to foreign central banks would exceed the value of its gold reserves. This point was reached and passed in 1964, by which time the US payments deficit stemmed entirely from foreign military spending, mainly for the Vietnam War.[23]

Yet the US continued to spend abroad and at home without regard for the balance-of-payments consequences. To sustain this spending and avoid its debt obligations, in August 1971 President Richard Nixon made official the gold embargo, eliminating the key-currency standard based on the dollar's convertibility into gold, and inaugurating the new treasury-bill standard – that is, the Dollar-Debt Standard, no longer pegged to gold. Rather than being able to use their dollars to buy American gold, foreign governments found themselves able only to purchase US Treasury obligations (and, to a much lesser extent, US corporate stocks and bonds). Running a dollar surplus in their balance of payments became synonymous with lending this surplus to the US Treasury. In effect, records Hudson, the US government was enabled to borrow automatically from foreign central banks simply by overspending and running a payments deficit, leading to the extraordinary circumstance whereby the larger the US deficit, the more dollars ended up in foreign central banks, which then lent them back to the US government by investing them in Treasury securities. As the US federal budget moved deeper into deficit, it triggered a domestic spending spree on more imports, foreign investment and foreign military spending to maintain US hegemony. Foreign economies bore the brunt of financing the rising federal deficit by purchasing the new Treasury bonds being issued. 'America's Cold War spending thus became a tax on foreigners', writes Hudson. 'It was their central banks who financed the costs of the war in Southeast Asia'.[24]

In effect, America was financing its domestic budget deficit by running an international payments deficit. But why did foreign governments continue to allow this? If they chose not to buy US dollars, the dollar's value would drop vis-à-vis foreign currencies, as the available supply of dollars would then greatly exceed international demand. A depreciating dollar would make US goods cheaper, providing US exporters with a competitive advantage over other export-oriented economies, and would reduce the value of foreign dollar holdings. In other words, not buying dollars would be severely detrimental for foreign economies by making their exports too expensive for world markets, and making their substantial dollar reserves too cheap. Hence, they continued buying dollars to support their exchange rates, and hence their competitive export prices. In effect, the US Dollar-Debt Standard locked these economies into an unequal financial relationship making their ongoing support for seemingly unlimited US borrowing appear necessary for their economic survival.[25] Therefore, Hudson argues:

> While applying creditor-oriented rules against third world countries and other debtors, the IMF pursues a double standard with regard to the United States. It has established rules to monetize the deficits America runs up as the world's leading debtor, above all by the US government to foreign governments and their central banks. The World Bank pursues its own double standard by demanding privatization of foreign public sectors, while financing dependency rather than self-sufficiency, above all in the sphere of food production. While the US Government runs up debts to the central banks of Europe and East Asia, US investors buy up the privatized public enterprises of debtor economies. Yet while imposing financial austerity on these hapless countries, the Washington Consensus promotes domestic

US credit expansion – indeed, a real estate and stock market bubble – untrammeled by America's own deepening trade deficit.[26]

It is therefore no surprise that since its inception the IMF has always been dominated by the policy decisions of the US government. Membership of the IMF is a prerequisite to membership of the World Bank. Until 1956, US control of the IMF was so complete that its decisions were effectively made by the US Secretary of the Treasury, while the IMF's own staff lacked the authority to negotiate loan conditions. The IMF Managing Director by both tradition and agreement must always be European, and since 1949 the IMF's Deputy Managing Director has consistently been American. Similarly, the President of the World Bank is always American. Although US unilateral domination of the IMF has declined since the 1960s with the rise of European and Japanese economic power, it remains impossible for the IMF to promote policies that contradict major US policy interests. The IMF itself is controlled by its members in proportion to their monetary contributions. The US continues to hold the largest quota at over 20 per cent. Since key issues require 80 per cent majority support, the US effectively has veto power. Five members with the largest quotas (the US, UK, France, Germany and India) and the two countries whose currency is most drawn upon each appoint an executive director. As for the IMF's US Executive Director, s/he cannot vote at her/his own discretion, but is legally bound to follow instructions from the Secretary of the Treasury. Thus, the IMF is principally dominated by the US, and secondarily by the world's other wealthiest powers.[27]

Structural Adjustment and Market Imperialism

Although the fundamental components of the Bretton Woods system – the gold exchange standard and the par value system – were dispensed with in the early 1970s, the fundamental structure of the international economic order remained in place, but on a new footing based on the proliferation of debt.[28] IMF/World Bank structural adjustment programmes have consistently advocated essentially the same package of policies throughout the world. According to the late development economist Hans Singer, on the pretext of allowing debtor nations to perpetually service their ever-expanding debt, these programmes foster an increased 'outward orientation', involving 'devaluation, trade liberalization and incentives for exporters' – policies that 'are widely believed to lead to social hardships, increased inequality of income distribution and increased poverty'.[29] The focus of these policies, then, is not poverty alleviation, but economic restructuring conducive to the penetration of foreign markets by Northern capital.

A step-by-step account of these policies is provided by Nobel Prize-winner and former World Bank chief Joseph Stiglitz. According to Stiglitz, every country receives 'the same four-step programme' consisting of the following measures:

1. *Privatization*: The World Bank demands that state industries, such as water and electricity companies, are sold off at massively reduced prices to private foreign investors. In Russia, for instance, oligarchs backed by the US government 'stripped Russia's industrial assets, with the effect that national output was cut nearly in half'.
2. *Capital market liberalization*: Although this theoretically allows investment capital to flow in and out, often it only flows out following the first sign of economic instability. The IMF responds to the escalating capital flight by raising interest rates to 30, 50 or 80 per cent, with 'predictable' results, according to Stiglitz, including plummeting property values, drastic cuts in industrial production and massive losses to national treasuries.
3. *Market-based pricing*: This entails massive price hikes for essential commodities such as food, water and cooking gas, leading to what Stiglitz terms 'the IMF riot', yet another 'painfully predictable' result of these policies. In the World Bank's 2000 Interim Country Assistance Strategy for Ecuador (a secret report obtained by BBC's 'Newsnight'), for example, 'the Bank several times suggests – with cold accuracy – that the plans could be expected to spark "social unrest"... The secret report notes that the plan to make the US dollar Ecuador's currency has pushed 51% of the population below the poverty line'. The ensuing spiral of increasing capital flight and government bankruptcy is accompanied by the buying up of remaining assets at the lowest possible prices by foreign investors.
4. *Free trade*: This is concerned with simply 'opening markets' to Western investors, while simultaneously protecting Western markets from Southern agriculture. Less developed countries are compelled to remove tariffs and cut taxes, allowing foreign capital free reign and subjecting local populations to the influx of Western goods and services. Simultaneously, apart from particular monocultural exports, Southern goods and services are effectively debarred from Western markets through extortionate tariffs, undercutting Southern agricultural and industrial development.[30]

Thus, the logic of neoliberal economics tends towards a form of *market imperialism*. As former US Federal Reserve Board economist Martin Wolfson observes, the IMF loans issued in response to economic crisis are intended to allow Southern borrowers to repay Northern capital investors. Meanwhile, further IMF austerity measures such as continued deregulation and slow-growth policies are designed to 'restore the "confidence" of investors and to make it again profitable for them to return'. In other words, market economics is consistently focused 'on restoring the investments of the wealthy, *not on helping the people*'.[31]

For example, less developed countries worldwide participating in the IMF's Enhanced Structural Adjustment Facility (ESAF) programmes have experienced slower economic growth, and even drastic declines in per capita income, compared to those outside these programmes. Although increasing millions of Africans are suffering due to lack of appropriate health, education and sanitary facilities, IMF rules enforced a continuing reduction in public spending on such services from 1986 to 1996. In the same period, external debt escalated exponentially, rising for instance in sub-Saharan Africa from 58 per cent as a share of GDP in 1988 to 70 per cent in 1996.[32]

This has had devastating results. According to the Food and Agricultural Organization (FAO), between 1985 and 1990 the number of people living in poverty in sub-Saharan Africa increased from 184 to 216 million. By 1990, 204 million people suffered from malnutrition – a situation that is drastically worse than 30 years ago.[33] Indeed, the proportion of Africa's population living on less than a dollar a day increased from approximately 18 per cent in 1980 to 24 per cent in 1995. The numbers living on less than $1000 a year increased from 55 per cent to 70 per cent.[34] World Health Organization data shows that adult mortality rates in parts of sub-Saharan Africa are now higher than 30 years ago. In Botswana, Lesotho, Swaziland and Zimbabwe, for instance, life expectancy for men and women has shortened by 20 years. In Africa, 40 per cent of deaths occur within the first five years of life. Of the 10 million children who die needlessly worldwide every year, half are in Africa. Of the 20 countries with the highest child mortality rates, 19 are in Africa. Of the 16 countries with higher mortality rates than in 1990, 14 are in Africa. Of the nine countries whose child mortality rate is higher than those recorded over 20 years ago, eight are in Africa. The vast majority of these deaths were foreseeable and preventable, resulting largely from malnutrition, diarrhoea, malaria and infections of the lower respiratory tract. Simple investments in clean water, improved sanitation and basic precautionary health care such as insecticide-treated nets and more effective malarial drugs could easily prevent many of these deaths.[35] Indeed, over one million children die of malaria annually – 3000 every day – simply due to a lack of access to appropriate medication.[36]

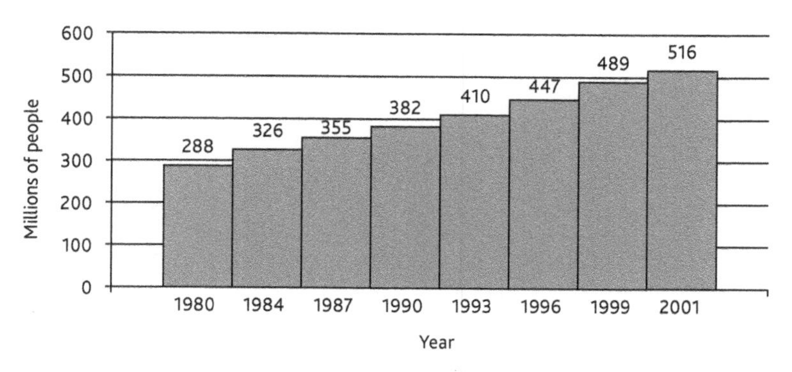

Figure 4.1 Rising Poverty in Sub-Saharan Africa, by Millions Living on Less than $1 a Day, 1980–2001.

Source: UN Africa Renewal, based on data in UN Department of Economic and Social Affairs, The Inequality Predicament (2005)

Despite the steady rise of mortality rates over decades, international financial institutions continue to impose the same policies, *slashing government expenditures on health while facilitating the penetration of transnational finance capital*. The example of Africa, moreover, is representative of a pattern of *death-by-deprivation* throughout the South, generated systematically by the North's 'development assistance' programmes.

Southern Underdevelopment

The era of neoliberal globalization from the 1980s until today has witnessed the systematic and intensifying destruction of life – not with guns and bombs, but through social policies: in the influential words of the Norwegian founder of Peace Studies, Professor Johan Galtung, 'structural violence'.[37]

Indeed, the very process of Northern economic growth has brought the retardation of many non-Western economies. A comprehensive study by the Center for Economic and Policy Research in Washington DC on the impact of neoliberal economic reforms during the 'golden age' of globalization (from 1980 to 2000) found 'a clear decline in progress as compared with the previous two decades [1960–1980]' for all indicators of economic progress:

> *Growth*: The fall in economic growth rates was pronounced and across-the-board for all groups of countries.
> *Life Expectancy*: Progress in life expectancy was also reduced for four out of the five groups of countries studied, with the exception of the highest group (life expectancy 69–76 years).
> *Infant and Child Mortality*: Progress in reducing infant mortality was also considerably slower during the period of globalization (1980–1998) than over the previous two decades.
> *Education and literacy*: Progress in education also slowed during the period of globalization.[38]

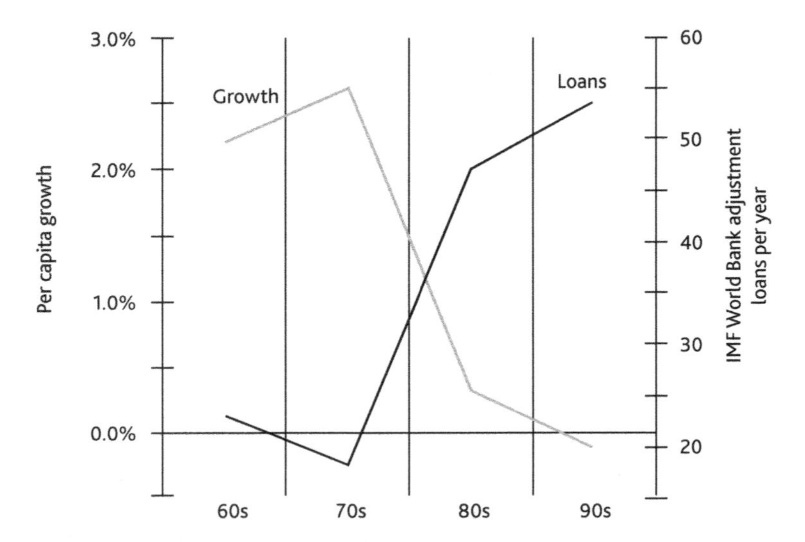

Figure 4.2 IMF/World Bank Loans Versus Growth, 1960s–1990s.

Source: UNICEF (2004)

These findings were updated in a more detailed study published by the United Nations Department of Economic and Social Affairs (UNDESA), drawing on economic data from 1980 to 2005. The UNDESA study essentially reconfirmed that:

> The past quarter-century has seen a sharp decline in the rate of growth for the vast majority of low and middle-income countries. Accompanying this decline has been reduced progress for almost all the social indicators that are available to measure health and educational outcomes... [A]t least some of the policy changes that have been widely implemented over the last 25 years have contributed to this long-term growth and development failure. In some of the financial and economic crises that took place in the late 1990s, for example, in Argentina, East Asia and Russia, it seems clear that policy mistakes contributed to severe economic losses... a long-term failure of the type documented here should at the very least... encourage scepticism with regard to economists or institutions who believe they have found a formula for economic growth and development. Indeed, some economists have recently concluded that more 'policy autonomy' – the ability of countries to make their own decisions about economic policy – is needed for developing countries.[39]

The human impact of this stunting of development can be discerned from an array of relevant data published by various international agencies since the 1970s. According to the FAO, the numbers of hungry people in the 'Third World' countries outside the Eastern Bloc and China rose by approximately 15 million during the 1970s, and by 37 million during the first few years of the 1980s.[40] The United Nations Development Programme (UNDP) reports that the gap between rich and poor nations *doubled* between 1960 and 1989.[41] Since then, it has continued to widen. In 1960, the income of the 20 per cent of the world's population living in the richest countries was 30 times greater than that of the 20 per cent in the poorest countries. By 1997 it was 74 times greater.[42]

By the late 1990s, the fifth of the world's population living in the highest-income countries had:

1. 86 per cent of world GDP – while the bottom fifth had 1 per cent.
2. 82 per cent of world export markets – while the bottom fifth had 1 per cent.
3. 68 per cent of foreign direct investment – while the bottom fifth had 1 per cent.
4. 74 per cent of world telephone lines – while the bottom fifth had 1.5 per cent.[43]

About a decade later, the global picture was still worse. By 2003, some 54 countries were poorer than in 1990. In 34, life expectancy fell; in 21, more people were hungry; 37 poor countries experienced an increase in poverty; and overall, in 54 less developed and transition countries across the South in sub-Saharan Africa, Eastern Europe and the Commonwealth states, Latin America and the Caribbean, East Asia and the Pacific, and the Middle East, per capita income growth was less than 3 per cent, and average per capita income had dropped.[44]

Figure 4.3 Population Living on Less than $2 a Day.

Country	Percent	Country	Percent	Country	Percent
Zambia	92.4	Mongolia	44.6	Honduras	35.7
Nigeria	87.2	Sierra Leone	74.5	Egypt	43.9
Mali	72.1	Laos	74.1	Tajikistan	42.8
Tanzania	89.9	Benin	73.7	Bolivia	42.2
Burundi	87.6	Pakistan	73.6	Sri Lanka	41.6
Niger	85.8	Burkina Faso	71.8	El Salvador	40.6
Madagascar	85.1	Nepal	68.5	Trinidad and Tobago	39.0
Central African Republic	84.0	Moldova	20.8	Ecuador	40.8
Rwanda	87.8	Mauritania	63.1	South Africa	34.1
Zimbabwe	83.0	Senegal	56.2	Paraguay	29.8
Gambia	82.9	Kenya	58.3	Guatemala	31.9
Bangladesh	84.0	Lesotho	56.1	Peru	30.6
Nicaragua	79.9	Namibia	55.8	Armenia	31.1
India	80.4	Indonesia	52.4	Venezuela	40.1
Ghana	78.5	Cameroon	50.6	Georgia	25.3
Mozambique	74.1	Botswana	55.5	Thailand	25.2
Haiti	78.0	Côte d'Ivoire	48.8	Swaziland	77.8
Ethiopia	77.8	Philippines	43.0	Kyrgyzstan	21.4
Cambodia	77.7	China	34.9	Brazil	21.2
Malawi	62.9	Yemen	45.2	Mexico	20.4

Source: UNDP (United Nations Development Programme). 2007. Human Development Report 2007/2008: Fighting Climate Change: Human Solidarity in a Divided World. New York: Palgrave Macmillan

Other UNDP figures provide a disturbing overview of this growing global inequality. Of the 6 billion people in the world, only 500 million live in the developed North – that is, approximately one twelfth of the world population. The rest of the world's population lives in less developed countries. Of these, about 1.3 billion have no access to clean drinking water; over 840 million suffer from chronic malnutrition (300 million of which are children); around 2 billion (a third of the world population) suffer from anaemia; and 2.6 billion (nearly half) lack basic sanitation.[45]

According to the Population Reference Bureau in Washington DC, drawing on World Bank data, 'more than one-half of the world's people live below the internationally defined poverty line of less than US $2 a day – including 97 percent in Uganda, 80 percent in Nicaragua, 66 percent in Pakistan, and 47 percent in China'. That is well over 3 billion people. The pattern of global poverty correlates with levels of energy consumption. The North 'uses over 5 times the energy per capita used by the less developed world'. In particular, North America uses over 8 times as much energy per person as does Latin America.[46]

Thus, the UNDP concludes that the opportunities presented by globalization are increasingly 'spread unequally and inequitably – concentrating power and wealth in a select group of people, nations and corporations, marginalizing the others'.[47] In China, for instance, disparities between export-oriented regions of the coast and the interior are widening. Similar patterns are visible in Eastern Europe and the Commonwealth, currently experiencing the largest ever increases in income inequality.[48] Indeed, subsequent to its opening up to global market forces, Russia now has the greatest inequality in the world. Income inequalities have also skyrocketed within other purported beneficiaries of the 'free market', including Indonesia, Thailand, and other East and Southeast Asian countries. Surprisingly, even the industrialized countries, especially Sweden, Britain and the United States, are no exception.[49]

This situation is not alleviated by the international aid system. For every $1 in aid received by a less developed country, over $25 is spent on debt repayment.[50] According to *The Economist*, the 'main motive' of aid 'has not been to end poverty but to serve the self-interest of the giver, by winning useful friends, supporting strategic aims, or promoting the donor's exports'. The result is that 'the richest 40 per cent of the developing world's population still gets more than twice as much aid per head as the poorest 40 per cent', with the vast bulk of aid accruing to 'countries that spend most on guns and soldiers, rather than health and education'. Accordingly, almost 'half of all aid is still tied to the purchase of goods and services from the donor country', which 'costs developing countries some 15–20 per cent of the value of the aid because they pay higher import prices'.[51]

All this suggests there is something fundamentally wrong with the neoliberal concept of 'growth' as such, which cannot be addressed without transforming the structure of the global political economy. Indeed, it is not only the bifurcation into haves and have-nots – into appalling extremes of overconsumption and underdevelopment – that is integral to the very character of 'growth' as currently conceived, but also the very conditions that lie behind the global financial and economic crisis.

GLOBAL FINANCIAL CRISIS

The Seeds of Instability: the Dollar-Debt Standard

The global financial crisis of 2008 is often viewed as purely a problem of the housing market, whereby American and British banks became increasingly willing over the years to grant subprime mortgages to clients who were less able to repay what they had borrowed. In turn, those debts became repackaged in the form of new financial instruments, and certified as secure, to be sold on to other investors, insurers and banking firms. As larger numbers were granted subprime mortgages, increasing volumes of bad debt were repackaged and resold, on the basis of which even larger amounts of credit and thus new loans flooded into markets. Consumers increased their spending on the basis of the security provided by their houses, while banks and investors made credit more accessible and accelerated their lending on the basis of the high long-term financial returns due from the rapidly proliferating mortgages, all of this contributing to rising prices and a mounting inflationary property and consumer bubble. This was a form of virtual growth, not indicating a real surplus derived from increases in productivity, but rather derived entirely from a monetary system based on credit and debt – that is, on the ability to continually borrow (and thus *de facto* create) cash that in real terms did not yet exist except as the expectation of a return. Thus, underlying this growing house of cards was the reality that its growth was contingent on the assumption that the mortgages and loans would be repaid. As the first spate of mortgage defaults began, and as the Anglo-American financial establishment began to realize that a substantial portion of these mortgages could never be repaid, it became clear that the entire edifice of growth, built on nothing less than bad debt and non-existent money, might suddenly collapse. Yet this system of escalating debt linked to the housing market cannot be understood in isolation from the overall structure of an economy whose very foundation is the intensification and exploitation of debt. The mounting mortgage-based debt in the US economy was and is merely one element of a parasitical international financial system.

Indeed, without the US balance-of-payments deficit, the housing market would not have been able to boom in the uncontrolled manner that it did. As Hudson shows, there is a direct link between the property and consumer booms and the international Dollar-Debt Standard:

> America's balance-of-payments deficit supplies the 'foreign' capital, as foreign central banks recycled the dollar outflows – that is, their own dollar inflows – into Treasury securities. The larger the balance-of-payments deficit, the more dollars foreign governments were obliged to invest in US Treasury securities, financing simultaneously the balance-of-payments deficit and the domestic federal budget deficit. Consequently, the stock and bond markets boomed as American banks and other investors moved out of government bonds into higher-yielding corporate bonds and mortgage loans, leaving the lower-yielding Treasury bonds for foreign governments to buy. US companies also began to buy up lucrative foreign businesses. The dollars they spent were turned over to foreign governments, which had little option but to reinvest them in US Treasury obligations at abnormally low interest rates. Foreign demand for these Treasury securities drove up their price, reducing their yields accordingly. This held down US interest rates, spurring yet further capital outflows to Europe.
>
> The US Government had little motivation to stop this dollar-debt spiral. It recognized that foreign central banks hardly could refuse to accept further dollars, lest the world monetary system break down. It was generally felt that such a breakdown would hurt foreign countries more than the United States, thanks to the larger role played by foreign trade in their own economic life. US strategists recognized this, and insisted that the US payments deficit was a foreign problem, not one for American citizens to

worry about. In the absence of the payments deficit, Americans themselves would have had to finance the growth in their federal debt. This would have had a deflationary effect, which in turn would have obliged the economy to live within its means. But under circumstances where growth in the national debt was financed by foreign central banks, a balance-of-payments deficit was in the US national interest, for it became a means for the economy to tap the resources of other countries.

All the government had to do was to spend the money to push its domestic budget into deficit. This spending flowed abroad, both directly as military spending and indirectly via the overheated domestic economy's demand for foreign products, as well as for foreign assets. The excess dollars were recycled to their point of origin, the United States, spurring a worldwide inflation along the way. A large number of Americans felt they were getting rich off this inflation as incomes and property values rose.

In other words, the inflationary rise in US property prices and incomes, which spurred the explosion of unrestrained mortgage-lending and credit-backed consumer spending, was caused directly by the escalating US federal deficit. Writing in 2002, Hudson's analysis of the unsustainable character of the debt-driven US-dominated global political economy is disturbingly prescient:

> The US deficit insists that foreign economies supply the consumer goods and investment goods that the domestic US economy no longer is supplying as it postindustrializes and becomes a bubble economy, while buying American farm surpluses and other surplus output. In the financial sphere, the role of foreign economies is to sustain America's stock market and real estate bubble, producing capital gains and asset-price inflation even as the US industrial economy is being hollowed out. Today, America's trade deficit is pumping dollars into the central banks of East Asia and Europe, to be recycled into the US capital markets creating a new form of financial bubble. The United States has drained the financial resources of its industrial Dollar Bloc allies while retarding the development of indebted third world raw-materials exporters and, most recently, the East Asian 'tiger economies' and the formerly Soviet sphere. *The fruits of this exploitation are not being invested in new capital formation, but dissipated in military and civilian consumption, and in a financial and real estate bubble that cannot last.*[52]

Early Warning Signs

The bursting of the bubble was only a matter of time. Warning signs that the global economy has teetered on the edge of an abyss for over a decade have surfaced regularly since the dawn of the new millennium. By the end of 2002, for instance, US stock markets had fallen for six consecutive weeks, to their lowest levels in five years. European markets had collapsed even further, wiping out nearly half the value of European corporations in that year alone. And after a decade of recession, bad debt and falling prices, Japan's banking system was struggling to salvage a recovery plan. Then a leaked email revealed that Germany's third largest bank, Commerzbank, had lost a quarter of its value by the end of September of that year. The revelation sparked 'increased fears of the international stock market malaise exploding into a fully-fledged banking crisis'. The news precipitated an emergency meeting of Europe's finance ministers in Luxembourg to discuss possible measures for rehabilitating the financial system.[53]

Almost exactly four years later, the world's financial leaders were meeting yet again, this time at a joint World Bank/IMF conference in Singapore to discuss measures 'to diffuse the risks posed by the world's massive trade imbalances'. The US had borrowed another $800 billion to finance its trade deficit, soaking up roughly two thirds of all global net savings, and provoking widespread concern of an impending 'massive dollar depreciation' resulting in 'a sharp global slowdown or even a devastating financial crisis'.[54] The conference followed an unprecedented IMF report in March 2006 documenting the pitfalls of 'deregulation' and 'liberalization', the central doctrines of the 'Washington Consensus' underpinning the global financial architecture. Although creating 'tremendous private and social benefits', these doctrines also hold 'the potential for fragility, instability, systemic risk, and adverse economic consequences', having 'created scope for financial innovation and enhanced the mobility of risks'.[55]

As US deficits mounted, with imports far outpacing exports, the value of the dollar declined, falling by 28 per cent against the euro between 2001 and 2005.[56] In August 2005, the problem was becoming so serious that US Federal Reserve Chairman Alan Greenspan warned that the federal budget deficit 'hampered the nation's ability to absorb possible shocks from the soaring trade deficit and the housing boom'. He criticized America's 'hesitancy to face up to the difficult choices that will be required to resolve our looming fiscal problems'. His comments were echoed a week later by David Walker, US Comptroller General and head of the Government Accountability Office:

> I believe the country faces a critical crossroad and that the decisions that are made – or not made – within the next 10 years or so will have a profound effect on the future of our country, our children and our grandchildren. The problem gets bigger every day, and the tidal wave gets closer every day.

The budget surplus of $236 billion in 2000 had become a deficit of $412 billion by 2004, more than doubling by the end of 2009 to a record $1.4 trillion.[57]

Then there is the trade deficit – the difference between what the United States imports and exports, which is now more than twice as big as it was two decades ago. US citizens who go into debt attempting to sustain lifestyles beyond their immediate financial means were in fact spending increasing amounts of the government's borrowed money to buy goods from overseas. Simultaneously, the government provided more services to the public than it could afford, subsequently incurring further debt to cover the costs. That debt was purchased by other nations, private investors and foreign banks in the form of US Treasury bonds and notes. Japan holds the greatest amount of

US debt, seconded by China, with which, incidentally, the US has its biggest trade deficit. In other words, the US economy, despite its status of *de facto* world domination, is still dependent on investment in the US Treasury by the central banks of Japan, China, and others. This maintains lower interest rates and allows US consumers to continue purchasing imported goods and services. The critical downside of this is that the value of the dollar is being undermined, which while it can encourage US exports also lowers the value of US Treasuries held by foreign banks. The decline in dollar value reduces incentives for international investment in the US. As this process continues, dollar-holding foreign banks and investors are increasingly likely to either reduce their dollar holdings or to invest less in the dollar, clearly exacerbating the problem and potentially igniting an escalating drop in the dollar's value.[58]

According to one leading expert, Clyde Prestowitz – former trade adviser to President Ronald Reagan, President of Washington DC's Economic Strategy Institute, and associate of US financiers such as Warren Buffet and George Soros – the world's central banks, now 'chock-full of US dollars', are holding dollars simply because 'at the moment there's no great alternative and also because the global economy depends on US consumption. If they dump the dollar and the dollar collapses, then the whole global economy is in trouble'. What is worse is that the process of dollar-dumping, culminating in a global economic meltdown, could be triggered by something as simple as a 'hedge-fund miscalculation':

> So picture this: you have a quiet day in the market and maybe some smart MBA at the Central Bank of Chile or someplace looks at his portfolio and says, 'I got too many dollars here. I'm gonna dump $10 billion'. So he dumps his dollars and suddenly the market thinks, 'My god, this is it!' Of course, the first guy out is OK, but you sure as hell can't afford to be the last guy out. You would then see an immediate cascade effect – a world financial panic on a scale that would dwarf the Great Depression of the 1930s.[59]

The main reason the dollar has not yet plummeted is the lack of a viable alternative world reserve currency. Former Federal Reserve Chairman Paul Volcker, Greenspan's predecessor, had already warned in 2005 that there was a 75 per cent chance of a dollar crash occurring in the following five years. While Prestowitz optimistically believed that the US economy would continue to be the world's most powerful economy for the 'foreseeable future', he agreed that it would undergo an 'inexorable decline'.[60]

American economist Douglas Casey, author of the best-selling finance book in history, *Crisis Investing* (1979), commented on the fate of the dollar in uniquely apocalyptic terms in late 2006:

> Foreign owners of the big green mountain of US dollars have become uneasy and are generally looking to sell. There's no dumping, at least not yet. When it comes, the flight from the dollar will come slowly, and then gain momentum before moving into a blow off. Like a glacier sliding toward

> a cliff, movement that seems inevitable may take a puzzlingly long time to get underway. But once it does, things speed up at a surprising rate... *Given the choice between (A) a dead housing market and a scorched earth depression in the US or (B) a collapsing currency, which at least has the virtue of reducing the real cost of paying off all those Treasury bonds, I'm forced to believe the US government will choose to sacrifice the dollar.*[61]

Yet although the dollar depreciated drastically through 2008 in the context of the global banking crisis, the dollar's position repeatedly surprised observers from later that year until the end of 2009, at times suddenly reversing and again appreciating in value. This should be understood properly as a temporary phenomenon due primarily to the nature of the global debt-system and the continuing strictures of the Dollar-Debt Standard. One underlying (but by no means the only) structural cause of dollar value fluctuation was the drop in consumption due to economic recession, weakening demand for energy, allowing fuel prices to drop, which also allowed the value of the dollar to appreciate. This may result in dollar value fluctuation for perhaps several years, despite a long-term trend of decline.[62]

Casey presciently emphasized that the trigger for a run on the dollar would be found in mounting unsustainable debts in the *consumer and housing markets*. He was not alone. In mid 2006, for instance, Stephen Roach, chief economist for Morgan Stanley, warned that the world 'has done little to prepare itself for what could well be the next crisis'. About a month earlier, Roach had already warned that a major financial crisis seemed imminent and that the global institutions tasked to forestall it, including the IMF, the World Bank, and other mechanisms of the international financial architecture, were utterly inadequate.[63]The IMF further predicted that by September of that year the risk of a severe slowdown in the global economy was greater than at any other time since 2001, mainly because of the sharp downturn in housing markets in the US and much of western Europe; also important was the decline in US labour's real income and insufficient consumer purchasing power.[64]

The Property and Consumer Credit-Debt Bubble

Just how soon a full-fledged global financial meltdown could hit was underlined by UC Berkeley economics professor Brad DeLong, who in 2007 argued that a global economic recession was already in motion due to three factors:

> 1) A Federal Reserve that finds itself with less inflation-fighting credibility than it thought it had; 2) upward pressure on inflation from rising energy and, perhaps, import prices; and 3) millions of middle-class homeowners who for too long have treated their houses as gigantic ATMs, using home equity loans and refinancing to generate extra spending money.

A key trigger for an economic crisis, he warned, could be the housing market – the unprecedented use of home loans to squeeze cash out of equity, permitting middle-class consumers to spend well beyond their means.

> Someday this spending spree has to come to an end. If it comes to an end suddenly, at a time when the Federal Reserve has raised interest rates a little too much, then we have our recession... Make no mistake about it: The US economy is close to the edge... What can be done to head off the danger? Unfortunately, very little. The bag of macroeconomic tricks is empty.[65]

Some analysts even specified a date as to when the crisis was likely to erupt. As early as July 2006, Dr David Martin – founding CEO of M-CAM, a financial institution that is the international leader in intellectual property-based financial risk management – predicted that the unsustainability of the 'mortgage house of cards' could culminate in a collapse of the US banking system, beginning around January 2008.[66] As much as 90 per cent of the growth of US GDP in 2001 and 2002 was in the housing market, and it was on this basis that the population purchased greater quantities of consumer products. In 2002, recognizing the availability of consumer credit, investors turned more and more to the housing market. This consumer-spending spree continued on the basis of capital gains on houses, sending consumers and mortgage-holders deeper into debt. By 2005, Martin reported, American savings went negative by four per cent – 'That means for every dollar you earned, you spent a $1.04. That's for those of you who were saving.' The circumstances were worse for those not saving, for whom 'we actually had 120 percent spend-over-earned in 2005. For every dollar you earned, you spent $1.20'. The response of the Federal Reserve to this overspending was simply to accelerate it by expanding consumer credit, adding $1.8 trillion in the last quarter of 2005 consisting of entirely leveraged commercial consumption. Martin pointed out that according to historical default rates, the maximum period at which debt appears to be sustainable is approximately 14 months. After 17 months, the limit of debt-sustainability is breached:

> That's when you start not making your payments. That's where you get into things like technical defaults and these kinds of little bumps in the road. You try to make a payment. You can't make a payment... [So] somewhere between the 17th month and the 24th month, all of a sudden, the fecal matter hits the rotary oscillator and it's bad. Everything that looked like a good credit starts to look like a bad credit.

By this assessment, with late 2005 as a starting point for the accumulation of these debts, and January 2006 constituting the point of overextension of credit, the 17 to 24 months debt-sustainability limit suggested the probability of massive defaults sometime after January 2008.[67]

Further, the escalating spate of mortgage defaults through the years 2006 and 2007 occurred in the context of the rising inflationary pressures on food prices, energy prices, cost of production, and cost of living. This was a consequence of rising oil prices, the massive hikes in which originated primarily due to the peak of world oil production and ensuing supply-demand constraints. There is, in other words, a clear link between 'peak oil' and the global recession – both of which converged in 2008.[68] As US actuary Gail Tverberg argued:

> Nearly all of the economic analyses we see today have as their basic premise a view that the current financial crisis is a temporary aberration. We will have a V or U shaped recovery, especially if enough stimulus is applied, and the economy will soon be back to Business as Usual... this assumption is basically incorrect. The current financial crisis is a direct result of peak oil. There may be oscillations in the economic situation, but generally, we can't expect things to get much better. In fact, there is a very distinct possibility that things may get very much worse in the next few years.[69]

In hindsight, Martin's assessment was startlingly accurate. Indeed, it is precisely the scale of the impending crisis that led financial leaders to publicly deny that such a crisis was inevitable – for they know that the moment confidence in the system is eliminated, the bubble will burst. The strategy instead has been to focus efforts on sustaining positive perceptions and confidence in the system, and while this allowed the US-dominated global political economy to continue growing beyond January 2008, the bursting of the credit-debt bubble was only a question of time, as was proven by the series of banking collapses occurring through Autumn of that year. As David Korten warns:

> Much as the cancer kills its host – and itself – by expropriating and consuming the host's energy, the institutions of capitalism are expropriating and consuming the living energies of people, communities, and the planet. And like a cancer, the institutions of capitalism lack the foresight to anticipate and avoid the inevitable deathly outcome.[70]

The Debt Spiral

The 2008 banking crisis is often explained purely as a result of financial greed and consumer recklessness, largely underplaying the pivotal role of the very structure of the global political economy in systematizing debt as a mechanism to generate inflated 'growth' purely on the basis of faith in the integrity of an increasingly fragile system. The problem is that both greed and recklessness are integral ingredients of this system. This is why opportunities to regulate the banking system were ignored by governments despite the myriad warning signs of a deep-seated global financial crisis. According to former US Treasury analyst Richard Cook, although the unhealthy granting of subprime mortgages began during the Clinton administration, the Bush administration actively 'turned these acts of reckless lending into a national program of mortgage fraud'. Citing mortgage industry insiders, Cook reports that:

> Soon after George W. Bush became president in 2001, meetings at the White House between Federal Reserve Chairman Alan Greenspan and administration officials became more frequent... [D]irection soon began to

come down from the banks to mortgage brokers to falsify borrower income information to allow them to qualify for loans that were otherwise out of reach. The FBI has investigations underway to prosecute some of these cases of mortgage fraud. But they are not reaching above the brokers' level. The FBI is not gaining access... to information about collusion at the political level or at the level of the banks which provided the leveraged funding for mortgage money.[71]

Although no US federal agency was tasked with the regulation of mortgage lending, this responsibility belonged by law to the states. Yet former Governor of New York Eliot Spitzer confirmed that when state attorneys general attempted to intervene to regulate mortgage lending several years prior to the outbreak of the 2008 crisis, the US Treasury Department's Office of the Comptroller of the Currency unilaterally blocked their efforts. Here are some excerpts from his testimonial in the *New York Times*:

> Several years ago, state attorneys general and others involved in consumer protection began to notice a marked increase in a range of predatory lending practices by mortgage lenders. Some were misrepresenting the terms of loans, making loans without regard to consumers' ability to repay, making loans with deceptive 'teaser' rates that later ballooned astronomically, packing loans with undisclosed charges and fees, or even paying illegal kickbacks. These and other practices, we noticed, were having a devastating effect on home buyers. In addition, the widespread nature of these practices, if left unchecked, threatened our financial markets.
>
> Even though predatory lending was becoming a national problem, the Bush administration looked the other way and did nothing to protect American homeowners. In fact, the government chose instead to align itself with the banks that were victimizing consumers...
>
> This threat was so clear that as New York attorney general, I joined with colleagues in the other 49 states in attempting to fill the void left by the federal government. Individually, and together, state attorneys general of both parties brought litigation or entered into settlements with many subprime lenders that were engaged in predatory lending practices. Several state legislatures, including New York's, enacted laws aimed at curbing such practices...
>
> In 2003, during the height of the predatory lending crisis, the OCC invoked a clause from the 1863 National Bank Act to issue formal opinions preempting all state predatory lending laws, thereby rendering them inoperative. The OCC also promulgated new rules that prevented states from enforcing any of their own consumer protection laws against national banks. The federal government's actions were so egregious and so unprecedented that all 50 state attorneys general, and all 50 state banking superintendents, actively fought the new rules.
>
> But the unanimous opposition of the 50 states did not deter, or even slow, the Bush administration in its goal of protecting the banks. In fact, when

my office opened an investigation of possible discrimination in mortgage lending by a number of banks, the OCC filed a federal lawsuit to stop the investigation.

The Bush administration, allied with the banks, therefore had every intention of producing the housing bubble, precisely in order to inflate the cost of homes and real estate, and thus 'pumping billions of dollars of borrowed cash into the economy through mortgage and home equity loans'. The bubble 'enriched huge numbers of executives, managers, and shareholders throughout the financial and real estate industries, and provided jobs to millions of people'. It also invited new foreign capital back into US markets.[72]

The accumulation of mortgage debt was integral to the emergence and proliferation of investments in new interrelated financial instruments providing the basis for the massive growth in financial markets on the back of derivatives trading. As Hudson notes, the economic 'growth' that emerged from this profiteering was literally fictional:

> [B]illions of dollars were devoted to keeping a dream alive – the accounting fictions written down by companies that had entered an unreal world based on false accounting that nearly everyone in the financial sector knew to be fake. But they played along with buying and selling packaged mortgage junk because that was where the money was. Even after markets collapse, fund managers who steered clear were blamed for not playing the game while it was going. I have friends on Wall Street who were fired for not matching the returns that their compatriots were making. And the biggest returns were to be made in trading in the economy's largest financial asset – mortgage debt. The mortgages packaged, owned or guaranteed by Fannie [the Federal National Mortgage Association] and Freddie [the Federal Home Loan Mortgage Corporation] alone exceeded the entire US national debt.[73]

Of particular significance were Credit Default Swaps (CDS) and Mortgage Backed Securities (MBS). For example, the Federal National Mortgage Association (Fannie Mae) – the giant US government-sponsored but privately owned mortgage finance company – began moving concertedly into the CDS market due to a 'combination of hubris, greed, and pressure to surpass Wall Street's rising expectations'. As US investment advisor Dan Amoss notes, 'The CDS business involves two parties – the sellers and the buyers of default risk. The buyers shoulder mortgage default risk in return for a future stream of mortgage insurance premiums'.[74]

Credit Default Swaps are the most widely traded type of derivative, which is an ingenious financial instrument that systematizes gambling on a global economic scale. As US attorney and financial analyst Ellen H. Brown points out, derivatives 'have no intrinsic value but derive their value from something else. Basically, they are just bets. You can "hedge your bet" that something you own will go up by placing a side bet that it will go down.' Moreover, these bets can literally be placed on anything, 'from the price of tea in China to the movements of specific markets'. Congressional hearings in the early 1990s recognized derivatives trading as an illegal form of gambling, but Alan Greenspan legitimized the practice as a form of 'risk management', although derivatives in reality socialized and globalized risk to the entire financial system. The derivatives market is now larger than the value of all the goods and services produced in the world. According to the Bank for International Settlements, in 2008 total derivatives trades exceeded 1000 trillion dollars – compare this to the total GDP of all the countries in the world, approximating a mere 60 trillion dollars. In other words, the size of the derivatives market has no foundation in the real economy, because it is based on credit creation through debt: '[G]amblers can bet as much as they want. They can bet money they don't have, and that is where the huge increase in risk comes in.'[75] Brown elaborates with some startling examples:

> CDS are bets between two parties on whether or not a company will default on its bonds. In a typical default swap, the 'protection buyer' gets a large payoff from the 'protection seller' if the company defaults within a certain period of time, while the 'protection seller' collects periodic payments from the 'protection buyer' for assuming the risk of default. CDS thus resemble insurance policies, but there is no requirement to actually hold any asset or suffer any loss, so CDS are widely used just to increase profits by gambling on market changes... [A] hedge fund could sit back and collect $320,000 a year in premiums just for selling 'protection' on a risky BBB junk bond. The premiums are 'free' money – free until the bond actually goes into default, when the hedge fund could be on the hook for $100 million in claims.
>
> And there's the catch: what if the hedge fund doesn't have the $100 million? The fund's corporate shell or limited partnership is put into bankruptcy; but both parties are claiming the derivative as an asset on their books, which they now have to write down. Players who have 'hedged their bets' by betting both ways cannot collect on their winning bets; and that means they cannot afford to pay their losing bets, causing other players to also default on their bets.
>
> The dominos go down in a cascade of cross-defaults that infects the whole banking industry and jeopardizes the global pyramid scheme. The potential for this sort of nuclear reaction was what prompted billionaire investor Warren Buffett to call derivatives 'weapons of financial mass destruction.'[76]

As noted by veteran derivatives trader Nassim Nicholas Taleb, Distinguished Professor of Risk-Engineering at New York University's Polytechnic Institute, banks routinely certified these transactions as solid and risk-free using quantitative models which, in reality, simply concealed the actual scope for risk and its potential consequences.[77] As of 2004, over 90 per cent of all financial transactions in the US were not properly recorded, such that from a financial perspective we are now 'flying blindfolded'.[78] They were thus able to proliferate risk-laden financial instruments which could be sold on for stupendous profits, resulting in the socialization of the impact of losses in the event of defaults.

Bush administration Treasury Secretary Henry Paulson's $700 billion bailout plan in 2008, and further bailout schemes around the world, had no real prospect of sufficiently restoring liquidity that would permit banks to reinject credit into the failing markets. On the contrary, as US economist Dean Baker observed, 'the bailout rewards some of the richest people in the country for their incompetence'.[79] Thus, interviewing a variety of US financial experts, Bloomberg concluded that the US economy had been headed for a major recession with or without the bailout plan. Although bailouts might restore a semblance of confidence, they would not restore confidence in the dollar, which would continue to fall in value regardless.[80]

The reason for this was simple. The bailouts took taxpayers' money – trillions of dollars of cash – out of the real economy (the exchange of tangible products and services) and pumped it into the private banking system. The banking system, at this time, had no intention of using its newfound cash to kick-start lending. On the contrary, as the banks were in fact insolvent (and in many cases declared bankruptcy), the bailouts were used to replenish their lost reserves and could not be used for new spending. The effect of this was to drastically reduce the quantity of money in circulation in the real economy, thus creating unprecedented constraints on spending – because there was literally less money available to spend. Thus, the global banking crisis was not a consequence of the lack of liquidity – insufficient credit as banks stopped lending – in money markets. Rather, the lack of liquidity was a consequence of the fact that *banks were simply insolvent*. The spate of mortgage defaults revealed that the 'structured financial' activities previously believed to be highly profitable, and the basis for further financial investments, were actually fictional. If they weren't real, then the preceding decades of financialization-driven growth granting American, British and European banks their image of solvency and safety were an illusion – one that was now inexorably fading.

But the strategies adopted by Western governments, largely consisting of 'bailouts' siphoning large quantities of public funds into the hands of private financiers, served primarily to centralize financial power and consolidate public bankruptcy. The net effect of the slowdown of the global financial system has been the slowdown of the real economy – the destruction of businesses, the layoff of millions of workers, the drop in consumption levels, and consequently further bankruptcies of businesses due to insufficient purchases of overabundant goods and services. This has meant that even corporate giants across major sectors of the real economy have collapsed. As the share values of these giants have plummeted, a tiny minority of the world's most powerful financial speculators and banks have, through short-selling and other speculative activities, been able to generate windfall profits allowing them to establish ownership of these real-economy corporate assets. As Canadian economist Michel Chossudovksy points out:

> What we are dealing with is an unsavoury relationship between the real economy and the financial sector. The financial conglomerates do not produce commodities. They essentially make money through the conduct of financial transactions. They use the proceeds of these transactions to take over bona fide real economy corporations which produce goods and services for household consumption.
>
> In a bitter twist, the new owners of industry are the institutional speculators and financial manipulators. They are becoming the new captains of industry, displacing not only the preexisting structures of ownership but also instating their cronies in the seats of corporate management.[81]

Although higher oil prices, sparked by peak oil, impacted on Western economies by promoting the onset of economic downturn, the slide into the Great Depression of the twenty-first century will not be smooth. From 2006 to 2009, the escalation of mortgage defaults accompanied the growing trends of businesses losing confidence and laying off staff, contracting investment in production, and overall declines in consumer spending. These factors induced a recession that threatened to unseat the vast trillions of dollars of debt-driven growth pushed by financial services, and they will continue to intensify over the coming years. Yet recession in itself reduces demand, which in accordance with the 'undulating plateau' model of peak oil elicits a corresponding decrease in oil consumption. The onset of deflation due to the drying up of credit contributed to the reduction of demand which in turn weakened oil prices, temporarily alleviating the strain on national economies, and granting them a brief window for a partial recovery which delays the onset of deep depression. However, as the rates of production of high-volume post-peak oilfields decline year by year, with rising consumption global supply will eventually level off until it matches global demand. Following this, fluctuating oil prices will give way to permanently elevating oil prices due to an irreversibly depleting hydrocarbon resource-base, which is most likely to occur at around 2015 (give or take, depending on actual supply-demand patterns and the impact of financial speculation).

In the absence of concerted mitigating strategies, the period after this will therefore consist of inexorable economic contraction. The 2008 banking crisis is only the beginning of this long-term process. That is why the repertoire of economic and financial policy responses actually worsened the crisis. As Bloomberg reported presciently in November of that year:

> The Federal Reserve, which has already pumped out hundreds of billions of dollars, might formally adopt a policy of flooding the world financial system with even more money. The Treasury, on course to borrow some $1.5 trillion this fiscal year, may tap global capital markets for even more to finance a fiscal stimulus package of as much as $700 billion and provide additional bailout money for banks.[82]

This was the beginning of a policy of 'massive quantitative easing', including 'letting the dollar weaken sharply, flooding markets with unlimited unsterilized liquidity; talking down the value of the dollar; direct and massive intervention in the forex to weaken the dollar' – all designed to facilitate access to credit and thus encourage increased production, consumer spending and exports.[83] Similar policies were pushed through in the UK, which borrowed a further £25.6 billion, running up a debt that would become £1 trillion by 2014.[84] Indeed, throughout 2009, US, British and Western European states flooded billions of dollars of taxpayers' cash into markets in order to bail out banks. The fundamental problem with this policy was that the cause of the problem was the insolvency of the banks – the ongoing bailouts removed even more cash from the real economy, decreasing prospects of consumer spending and aggravating deflation in the *short-term* while setting up Western economies for *long-term currency devaluations and inflation*. As almost universal insolvency was the real barrier to banks recommencing lending, the new influxes of capital assets provided no real incentive to lend to other insolvent institutions or cash-strapped citizens.

By pumping excessive credit into the financial system, governments temporarily offset the most potentially severe immediate impact of the recession, continued deflation, by perhaps a few years at most. Yet the period of minor economic and financial recovery it would permit, based purely on yet more debt-based consumer spending and associated long-term asset and price inflation, would ultimately only deepen the problem by escalating levels of unrepayable debt. Not only would any new levels of growth over the ensuing period be as unsustainable and 'virtual' as before, it would also trigger the breaching of oil capacity limits, once again precipitating rising fuel prices and exacerbating further inflation, eventually contributing to yet another round of catastrophic debt defaults. On its own terms, the economy would be unable to survive more than a couple of such cyclical fluctuations before falling into deep depression. In the context of peak oil and the insights of the Export Land Model, this 'undulating plateau' period could last only 10–15 years at the absolute maximum before giving way to a protracted economic contraction unlike anything the world has ever seen.

As the United States remains the centre of the world economy, it also remains the most deeply affected by global recession. A probable picture of how the US will fare was set out by Gerald Celente, CEO of Trends Research Institute in Rhinebeck, New York, who is renowned for his accuracy in predicting major world events – including the 1987 stock market crash, the fall of the Soviet Union, and the 1997 Asian financial crisis, as well as the subprime mortgage crisis and 2008 global banking collapse. According to Celente, by 2012 the United States will experience a major transition leading to the lowering of living standards and plummeting retail sales. This is the beginning of what the Trends Research Institute describes as the 'Greatest Depression'.[85] Over the long term, the dollar will depreciate in value by as much as 90 per cent. Eventually, the prolongation of recession will create circumstances 'worse than the great depression', which will begin with the organization of tax rebellions by American citizens as 'people can't afford to pay more school tax, property tax, any kind of tax. You're going to start seeing those kinds of protests start to develop.' These could escalate into food riots and other forms of civil unrest as the rate of homelessness rises exponentially. 'Tent cities are already sprouting up around the country and we're going to see many more', as well as 'huge areas of vacant real estate' occupied by squatters with nowhere else to go. This will also lead to skyrocketing crime rates, worse than during the Great Depression, due to the large numbers of people addicted to modern drugs. The depression will, in other words, create 'a huge underclass of very desperate people'.[86] In varying ways, these trends will come to affect Britain and Western Europe by around 2014 (when Britain's debt, for instance, will near the £1-trillion mark).

This picture is corroborated by US economist John Williams, Edward Tuck Scholar at the Dartmouth College Amos Tuck School of Business Administration, and a specialist in government economic reporting:

> The US economic and systemic solvency crises of the last two years are just precursors to a Great Collapse: a hyperinflationary great depression. Such will reflect a complete collapse in the purchasing power of the US dollar, a collapse in the normal stream of US commercial and economic activity, a collapse in the US financial system as we know it, and a likely realignment of the US political environment. The current US financial markets, financial system and economy remain highly unstable and vulnerable to unexpected shocks. The Federal Reserve is dedicated to preventing deflation, to debasing the US dollar. The results of those efforts are being seen in tentative selling pressures against the US currency and in the rallying price of gold...
>
> Before the systemic solvency crisis began to unfold in 2007, the US government already had condemned the US dollar to a hyperinflationary grave by taking on debt and obligations that never could be covered through raising taxes and/or by severely slashing government spending that had become politically untouchable. The US economy also already had entered a severe structural downturn, which helped to trigger the systemic solvency crisis.
>
> The intensifying economic and solvency crises, and the responses to both by the US government and the Federal Reserve in the last two years, have exacerbated the government's solvency issues and *moved forward my timing estimation for the hyperinflation to the next five years*, from the 2010 to 2018 timing range [previously] estimated... Accordingly, risks are particularly high of the hyperinflation crisis breaking *within the next year*.[87]

The onset of major economic depression may in turn limit the availability of surplus capital for investment in new technologies based on renewable energies. In any case, the temptation to allow the fossil fuel industry to continue operations, providing a major source of tax revenue for states, acts as a disincentive to cutting greenhouse gas emissions. Ongoing hydrocarbon exploitation, of course, would only accelerate global warming and contribute to rapid climate change with unpredictable and dangerous consequences, leading to a spiral of industrial-infrastructural collapse, accelerating famines and droughts, and intensifying conflicts between and within states. In this context, the 2008 banking crisis signified perhaps the first concrete sign of the global political economy's inability to continue growing without regard for its concrete relationship to the natural world.

It is worth considering the prospects of a worst-case scenario, based on the notion that business as usual will continue to be the official approach to economic, ecological and energy crises, in which case the decline of industrial civilization is likely to be a violent and protracted process. This exercise helps to clarify the dangers of which policymakers will need to be keenly aware in responding to global crises. Here is how such a scenario might unfold from 2009 onwards: The 2008 global banking crisis does not provoke any serious consideration among financial elites about restructuring the global political economy along more equitable and stable lines. Instead, massive government and taxpayer bank bailouts and new state-bank and bank-corporate mergers serve to consolidate a minority of banking institutions and financial elites, further entrenching global debts and centralizing financial power in a concentrated network of state-financial-banking structures. The crisis elicits a new level of disenfranchisement whose worst effects will be felt by the poorest sectors in both the North and the South, but which will also increasingly affect the middle classes. Neoliberal doctrines of 'no state intervention in the economy' give way to increased governmental intrusion into markets and financial institutions, and a reversion to policies of protectionism to prop up failing domestic industries, sparking increasingly bitter international trade wars between core powers. The dramatic slump in global demand prolongs the duration of hydrocarbon exploitation by temporarily restraining global rates of consumption. However, this leads only to a period of fluctuating demand for hydrocarbons, and population pressures in emerging industrial regions continue to provide an upward pressure on consumption. The flooding of credit into world markets as deflation bottoms out over some years will encourage increased spending, bringing consumption levels up and contributing to a temporary economic recovery consisting of yet another debt-based inflationary 'boom', which will eventually generate an unavoidable oil supply crunch. Beyond this point, price fluctuation gives way to permanently high oil prices. This is likely to occur within 5–10 years.

Simultaneously, as countries turn increasingly to gas and coal options escalating fossil fuel emissions could mean a rise in world temperatures of up to four degrees Celsius, in which case by mid century large areas of the world would become uninhabitable, culminating in hundreds of millions of deaths due to droughts, famines and intensifying conflict for control over scarce resources. Such drastic global population reduction in the latter part of the century would simultaneously register as a massive alleviation of consumption pressures, which in turn would permit business-as-usual hydrocarbon exploitation to continue, albeit for the benefit of a much more concentrated and centralized minority of the world population. Northern states and economies would be forced to contract into smaller, highly fortified and heavily militarized territories relying on the last vestiges of control over diminishing fossil fuel supplies, managed by means of violence. How long such a scenario could continue without major disruption by inter- or intrastate conflict is impossible to foresee – the time between the onset of the 1929 Great Depression and the outbreak of the Second World War was ten years in far more favourable global circumstances, suggesting that this convergence of global crises may culminate in major international conflict before 2018. In any case, in this worst-case scenario of continuing fossil fuel exploitation, runaway climate change could be triggered even before mid century, eventually culminating in an unrecognizable and almost uninhabitable planet.

This worst-case scenario is by no means inevitable. As the impact of global crises escalates and converges, states and investors will increasingly be forced to downsize their ever more unsustainable operations, as peoples and communities worldwide demand more and more a return to genuine sustainability. Indeed, massive permanent price hikes in hydrocarbon energies and the paucity of traditional avenues of financial investment – highly probable within about 10–15 years at the most – will make alternative technologies and the development of more sustainable infrastructures based on renewable energies appear far cheaper, more productive, and thus even more economically viable. Grassroots pressures could converge with this economic opening, creating the possibility of more concerted efforts to reorganize social relations in a more sustainable way. Unfortunately, the later the occurrence of such developments, the more difficult it will be to avoid certain catastrophic situations. Ultimately, the final long-term outcome is not written in stone, but depends entirely on human choices. While this period of civilizational turmoil could well bring many catastrophic consequences, this should not deter ongoing efforts to push for the success of a Post-Carbon Revolution. However, if such efforts will be successful, there must be a pragmatic awareness of the limitations inherent in the way policymakers are currently dealing with these issues.

5
International Terrorism

At first glance, the idea that international terrorism is somehow directly linked to the escalation of the global ecological, energy and economic crises discussed in the preceding chapters seems far-fetched. This chapter shows that the phenomenon cannot be understood without examining at least three intertwining issues: 1) the over-dependence of the global political economy on hydrocarbon energies in service of industrial (and 'post-industrial') capitalist social relations of production, consumption and finance; 2) the function of covert Western military intelligence policies in calibrating low-intensity warfare to control potentially recalcitrant populations in strategic regions, in order to diversify and secure hydrocarbon energy supplies; 3) the function of covert Western military intelligence policies in facilitating criminal activities, particularly drug trafficking, which play an instrumental role in shoring up the global economy. All these issues, further, make no sense without recognizing their deeper context in the structure of the global political economy as hitherto explored, and the immense financial pressures they place on states and their corresponding national security apparatuses to equate 'security' with 'prosperity' – with the latter defined, of course, in rather narrow terms relevant only to an elite minority.

I begin by first establishing the link between geopolitical control of certain strategic regions, and the 'security' of energy supplies for the centres of modern industrial civilization. I then extend this analysis to explore how Western covert operations have often been mobilized to exert this geopolitical control. Unfortunately, these covert operations – which continue today in the Middle East and Central Asia – have facilitated the activities of interconnected extremist, terrorist and criminal networks worldwide. In the context of escalating global crises, which may have amplifying effects on both the use of covert action and the scope of terrorist activity, this all has profoundly destabilizing implications for international security.

HYDROCARBON OVER-DEPENDENCE

National Security and Resource Control: The Making of Foreign Policy

The convergence of the preceding four global crises represents an imminent and catastrophic threat to the national security of all countries, and the human security of communities everywhere, far outweighing the threat currently posed by international terrorism. Yet despite this, government defence expenditures are overwhelmingly focused on fighting a 'War on Terror', in which the threat of international terrorism inspired by Islamist extremism is believed to be a danger to Western civilization, thus requiring the establishment of new and comprehensive security architectures in far-flung strategic regions, as well as at home in the form of intensified community policing and surveillance. Comparatively, government expenditures on preventing or mitigating dangerous global warming, preparing for peak oil, transforming global food production, and restructuring the neoliberal economic system represent a tiny fraction of the sums invested in sustaining a permanent state of global warfare.

Yet the threat of international terrorism is, like these other global crises, also tied to the very structure of the global political economy. Indeed, it is integral to over-dependence on hydrocarbon energy, particularly petroleum. This systemic over-dependence has led to the evolution of a very specific type of Western foreign policy strategy designed to maintain a monopoly over the world's resources.

In general, US and UK foreign policies have been guided by principles described in the 1940s in a series of planning documents of the War and Peace Studies Project, a joint initiative of the US Department of State and the Council on Foreign Relations. US policy planners, preparing for the reconstruction of world order after the Second World War, identified a minimum 'world area', control of which was deemed to be 'essential for the security and economic prosperity of the United States and the Western Hemisphere'. The US aimed 'to secure the limitation of any exercise of sovereignty by foreign nations that constitutes a threat' to this world area, which included the entire Western Hemisphere, the former British Empire and the Far East. This objective was premised on 'an integrated policy to achieve military and economic supremacy for the United States'. So the concept of 'security interests' was extended beyond traditional notions of territorial integrity to include domination of regions considered 'strategically necessary for world control'. State Department planners, recognizing that 'the British Empire as it existed in the past will never reappear', candidly argued that 'the United States may have to take its place'. 'Grand Area' planning, as it was then known, aimed to fulfil the 'requirement[s] of the United States in a world in which it proposes to hold unquestioned power'. Meanwhile, Britain would be brought in as a 'junior partner', within the 'orbit' of American control.[1]

This underlying framework of concepts, despite important variations and evolution over time, has remained fundamentally constant throughout the postwar period, as evidenced by a series of unclassified documents from the Office of the Secretary of Defense dating from the early to late 1990s. A consistent theme of these documents is that the US should maintain global 'pre-eminence'.[2] This entails ensuring that other powers recognize the established unipolar order, and do not seek to increase their power in the international system. Thus, a 'first objective is to prevent the re-emergence of a new rival' to US global pre-eminence, by working 'to establish and protect a new order that holds the promise of convincing potential competitors that they need not aspire to a greater role or pursue a more aggressive posture to protect their legitimate interests'. This world order must 'account sufficiently for the interests of the advanced industrial nations to discourage them from seeking to overturn the established political and economic order' under US hegemony. In particular, this means the US must also 'endeavour to prevent any hostile power from dominating a region whose resources would, under consolidated control, be sufficient to generate global power', these regions including Western Europe, East Asia, the former Soviet Union and the Middle East. It is paramount to maintain 'the sense that the world order is ultimately backed by the US'.[3]

This strategic framework has, then, obvious implications for specific regions, the Middle East being one where the US seeks 'to prevent any hostile power from dominating', and whose resources might 'generate global power'. The principal interest in the region, of course, is oil, the first reserves of which were discovered by the British in Persia in 1908. The UK controlled most Middle East oil until the end of the Second World War, after which the US secured its sphere of influence in Saudi Arabia.[4] Although this led to significant US-UK tension, it was eventually resolved through British compliance with American primacy, expressed in 1945 by US planners in the form of a joint approach with the UK:

> [O]ur petroleum policy towards the United Kingdom is predicated on a mutual recognition of a very extensive joint interest and upon control, at least for the moment, of the great bulk of the free petroleum resources of the world... US-UK agreement upon the broad, forward-looking pattern for the development and utilisation of petroleum resources under the control of nationals of the two countries is of the highest strategic and commercial importance.[5]

In 1947, policy planners stated that the US should 'seek the removal or modification of existent barriers to the expansion of American foreign oil operations' and 'promote... the entry of additional American firms into all phases of foreign oil operations'.[6] By 1953, the US National Security Council stated the US position as follows: 'United States policy is to keep the sources of oil in the Middle East in American hands.'[7]

It is precisely in this context that Anglo-American policy has tended to ally itself with the most authoritarian regimes in the region to maintain the supply of petroleum to the West as cheaply as possible, aided by the willingness of these regimes to control their societies through force and coercion. Thus, secret British Foreign Office documents confirm that the Gulf sheikhdoms were largely created by the British to 'retain our influence' in the Middle East. This required not only protection from external threats, but also from internal overthrow. Hence, Britain had to 'counter hostile influence and propaganda within the countries themselves'. Police and military training would help in 'maintaining internal security'. Similarly, US foreign policy planners concurred that Anglo-American regional interests would be preserved by countering challenges 'to traditional control in the area', to sustain the 'fundamental authority of the ruling groups'.[8]

Oil Addiction and National Insecurity

US strategy thus required a policy of sustaining local authoritarian regimes and countering indigenous democratic aspirations, thus maintaining the regional framework of order through which US energy interests would be protected. Yet this has had two principal unintended implications.

Firstly, it has generated entrenched grievances among Middle Eastern populations based on the recognition that official US regional policy has been opposed to genuine democratization in order to guarantee access to cheap petroleum supplies. A 2004 Zogby International poll found that less than 10 per cent of people surveyed in Egypt, Jordan, Lebanon, Morocco, Saudi Arabia, and the United Arab Emirates approved of US foreign policy. When asked to indicate their 'first thought' about the United States, the most common response was 'unfair foreign policy'. Yet simultaneously they continued to view US popular culture, science and technology, and the American people in favourable terms.[9] Grievances about US foreign policy are, however, exploited by the ideology of extremist networks like al-Qaeda to justify their terrorist activity. Osama bin Laden has repeatedly described the authoritarian regimes across the Middle East as illegitimate client states of the West that should be overthrown through violent action. Al-Qaeda ideology attempts to justify terrorism as a form of resistance against the political repression of Muslim populations by these regimes.[10]

Secondly, extensive historical and empirical evidence confirms that al-Qaeda terrorist networks have been, and continue to be, covertly sponsored by several key Muslim states in the Middle East and Central Asia, such as Saudi Arabia and the Gulf states, Pakistan, Algeria, Azerbaijan, and Turkey, among others.[11] By directly and indirectly sponsoring these regimes, the US is essentially supporting al-Qaeda's state sponsors. Yet this relationship is a direct consequence of the global political economy's over-dependence on oil. In short, industrial civilization's oil addiction has generated a structural entanglement with dictatorial oil-exporting regimes that are cultivating terrorist networks which target the core centres of power in the West. This contradictory policy was candidly described in 2006 by US Navy Commander Thomas D. Kraemer, who has served in the Pentagon for the Director of Naval Intelligence, in a paper published by the US Army War College Strategic Studies Institute:

> America is buying billions of dollars of oil from nations that are sponsors of, or allied with, radical Islamists who foment hatred against the United States. The dollars we provide such nations contribute materially to the terrorist threats facing America... In the War on Terror, the United States is financing both sides. While spending billions of dollars on US military efforts in the war, we are sending billions more to nations such as Saudi Arabia... where the cash is used to finance training centers for terrorists, pay bounties to the families of suicide bombers, and fund the purchase of weapons and explosives. Oil revenues in these countries underwrite new media outlets that propagandize hatefully against the United States. They

> pay for more than 10,000 radical madrassahs set up around the world to indoctrinate young boys with the idea that the way to paradise is through murderous terror.... Men energized by oil-revenue resources killed 3,000 American civilians on September 11, 2001 (9/11), and continue to kill large numbers of Westerners in Iraq and elsewhere. We are thus subsidizing acts of war against ourselves... America is hamstrung because any forceful action on our part... could result in the disruption of oil supplies that the world economy completely depends on. We cannot stand up to those who support our enemies because we rely upon those supporters for the fuel that is our own lifeblood.[12]

But the precise character of this oil-dependent structural relationship with terrorism does not end there, and in fact has an even more disturbing component which will constitute the main focus of the rest of this chapter. In Central Asia, the Balkans, the Middle East, North Africa, the Asia-Pacific, and the Caucasus, a wealth of government and intelligence sources confirm that militant Islamist networks affiliated to al-Qaeda have been used as mercenaries by Western security agencies in promotion of geostrategic ambitions. The nominal usefulness of al-Qaeda elements for Anglo-American geostrategy has in many cases granted them a temporary operational immunity, permitting them to expand throughout the post-Cold War period. This process began during the Cold War, but contrary to the conventional wisdom *accelerated during the post-Cold War period*, and even continued after 9/11 *until today*. In other words, *Western hydrocarbon energy security has systematically undermined Western national security.*

This has radical implications for our understanding of national security. It suggests that, like other global crises, the threat of international terrorism is directly linked to the unequal structure of the global political economy, its over-dependence on hydrocarbon energies, and the overt and covert military intelligence strategies developed by Western states to control global energy supplies. As long as this structure and associated militarization strategies remain in place, so too will the threat of terrorism, as well as the even more pressing dangers of other global crises. In this sense, the militarization strategies of the 'War on Terror' are inherently self-defeating, and doomed to fail. Ultimately, security from terrorism will require the same sort of systemic transformation of the global political economy as that demanded by other global crises.

AL-QAEDA AND POST-COLD WAR WESTERN COVERT OPERATIONS

Al-Qaeda and Unconventional Warfare Doctrine

According to the conventional wisdom, Osama bin Laden and al-Qaeda were supported by the West during the Cold War to facilitate the expulsion of the Soviet Union's occupying forces from Afghanistan. This strategy ended in 1989 after the collapse of the USSR, after which bin Laden turned against his former supporters. His extremist ideological goals led al-Qaeda to mount increasingly devastating attacks on Western targets around the world throughout the 1990s, culminating in the 9/11 terrorist attacks, and followed by other insurgent operations in Madrid, London, Bali, Istanbul, Bombay and elsewhere.

The problem with this narrative is that the West never genuinely severed its military intelligence connections to al-Qaeda. During the Cold War, Anglo-American ties to al-Qaeda were localized in a single area, Afghanistan. After the Cold War, such ties proliferated in strategic regions around the world. Indeed, the globalization of al-Qaeda terrorist networks was a function of covert Western military intelligence interventions to secure Western regional interests.

In September 1999, Graham Fuller, former Deputy Director of the CIA's National Council on Intelligence, alluded to the continuing covert use of Islamism to promote US interests while countering Russian and Chinese influence:

> The policy of guiding the evolution of Islam and of helping them against our adversaries worked marvelously well in Afghanistan against the Red Army. The same doctrines can still be used to destabilize what remains of Russian power, and especially to counter the Chinese influence in Central Asia.[13]

The policy that 'worked well' in Afghanistan, and which Fuller argued should be transplanted to counter Russian and Chinese influence, was precisely the sponsorship of al-Qaeda as a mercenary force to conduct US covert operations. The implication is that, after the Cold War, al-Qaeda operations were seen as integral to a new doctrine of covert destabilization, to be implemented in new theatres of operation strategically close to Russian and Chinese influence – namely Eastern Europe, the Balkans, the Caucasus and Central Asia. Notably, Fuller is also one of the individuals identified in the State Secrets Privilege Gallery of former FBI translator and whistleblower Sibel Edmonds – banned by the US government from publicizing information on US state collaboration with Islamist extremists released during her testimonials before various US House and Senate Committees.[14] In a recent interview, Edmonds herself confirmed:

> I have information about things that our government has lied to us about. I know. For example, to say that since the fall of the Soviet Union we ceased all of our intimate relationship with Bin Laden and the Taliban – those things can be proven as lies, very easily, based on the information they classified in my case, because we did carry very intimate relationship with these people, and it involves Central Asia, all the way up to September 11.[15]

Edmonds reveals that Turkey acted as a primary intermediary in the Central Asian operations, with assistance from Pakistan and Saudi Arabia. The idea was to sideline China and Russia, and undermine popular resistance to US influence by appealing to Central Asian aspirations for an Islamic and Turkic resurgence. She says that al-Qaeda and the Taliban were used by the US as proxies in 'a decade-long illegal, covert operation in Central Asia by a small group in the US intent on furthering the oil industry and the Military Industrial Complex'.[16] In an interview with former CIA official Philip Giraldi, Edmonds further elaborates on classified conversations she translated 'that suggested the CIA was supporting al-Qaeda in central Asia and the Balkans, training people to get money, get weapons, and this contact continued until 9/11'.

> [B]etween 1997 and 2001, [the conversations] had to do with a Central Asia operation that involved bin Laden... It was always 'mujahideen,' always 'bin Laden' and, in fact, not 'bin Laden' but 'bin Ladens' plural. There were several bin Ladens who were going on private jets to Azerbaijan and Tajikistan. The Turkish ambassador in Azerbaijan worked with them.
>
> There were bin Ladens, with the help of Pakistanis or Saudis, under our management. [A senior US government official] was leading it, 100 percent, bringing people from East Turkestan into Kyrgyzstan, from Kyrgyzstan to Azerbaijan, from Azerbaijan some of them were being channeled to Chechnya, some of them were being channeled to Bosnia. From Turkey, they were putting all these bin Ladens on NATO planes. People and weapons went one way, drugs came back.[17]

After 9/11, indications of the continuation of this strategy have surfaced from different quarters. A classified document prepared in August 2002 by the Pentagon's Defense Science Board for then US Defense Secretary Donald Rumsfeld, leaked to the *Los Angeles Times*, recommended the creation of a 'super-Intelligence Support Activity' called the 'Proactive, Preemptive Operations Group (P2OG)', bringing together 'CIA and military covert action, information, warfare, and cover and deception'. The organization would expand on an existing highly classified Pentagon covert action agency, known as the Intelligence Support Activity, which had operated in Afghanistan and elsewhere in the 1980s and 1990s. According to the *Times*, the revamped body would 'launch secret operations aimed at "stimulating reactions" among terrorists and states possessing weapons of mass destruction' – that is, '*prodding terrorist cells into action* and exposing them to "quick-response" attacks by US forces'. The Board even proposed 'creating a "red team" of diabolical thinkers to *plot imaginary terror attacks on the United States so the government can plan to thwart them*'. One key role for US counterterrorism agents would be 'duping al Qaida into *undertaking operations*' and attempting to 'stimulate terrorists into *responding* or moving operations'.[18]

According to Seymour Hersh, by early 2005 the P2OG strategy described here had been fully activated, with several pilot covert operations already under way. Under the Defense Secretary's direction, the Pentagon had reconsolidated control over, and greatly expanded the scope of, Special Forces operations. Citing a Pentagon consultant as well as former and active US military intelligence officers, Hersh reported that:

> US military operatives would be permitted to pose abroad as corrupt foreign businessmen seeking to buy contraband items that could be used in nuclear-weapons systems. In some cases, according to the Pentagon advisers, *local citizens could be recruited and asked to join up with guerrillas or terrorists. This could potentially involve organizing and carrying out combat operations, or even terrorist activities.* [emphasis added]

He refers to a series of articles by John Arquilla, a professor of defense analysis at the Naval Postgraduate School in Monterey, California, and a RAND terrorism consultant, elaborating on this strategy. 'When conventional military operations and bombing failed to defeat the Mau Mau insurgency in Kenya in the 1950s', muses Arquilla, 'the British formed teams of friendly Kikuyu tribesmen who went about pretending to be terrorists. These "pseudo gangs", as they were called, swiftly threw the Mau Mau on the defensive, either by befriending and then ambushing bands of fighters or by guiding bombers to the terrorists' camps'. He goes on to advocate that Western intelligence services should use the British case as a model for creating new 'pseudo gang' terrorist groups, purportedly to undermine 'real' terror networks. 'What worked in Kenya a half-century ago has a wonderful chance of undermining trust and recruitment among today's terror networks. Forming new pseudo gangs should not be difficult.'[19]

More recently, a restricted US Army Special Operations Field Manual, leaked in December 2008, confirms that the penetration and mobilization of criminal and terrorist networks is a fundamental pillar of US unconventional warfare (UW), defined as: 'Operations conducted by, with, or through irregular forces in support of a resistance movement, an insurgency, or conventional military operations'. The US has a 'tested capability to use UW' going back to World War II. The document clarifies that:

> UW must be conducted by, with, or through surrogates; and such surrogates must be irregular forces.... These forces may include, but are not limited to, specific paramilitary forces, contractors, individuals, businesses, foreign political organizations, *resistance or insurgent organizations*, expatriates, *transnational terrorism adversaries, disillusioned transnational terrorism members, black marketers, and other social or political 'undesirables.'*

Furthermore, the 'strategic purpose' of UW is 'to gain or maintain control or influence over the population'. The document elaborates on this as follows:

> UW generally assumes that some portion of the indigenous population – sometimes a majority of that population – are either belligerents or in support of the UW operation. *UW is specifically focused on leveraging the unwillingness of some portion of the indigenous population to accept the status quo* or 'whatever political outcome the belligerent governments impose, arbitrate, or negotiate'. *A fundamental military objective in UW*

> *is the deliberate involvement and leveraging of civilian interference* in the unconventional warfare operational area (UWOA).

Further, the document points out that although UW is distinct from irregular warfare (IW), 'UW is a component and method of prosecuting IW'. Among the variety of 'constituent activities' belonging to IW, the document lists the following: '*Insurgency*. COIN [counterinsurgency]. UW. *Terrorism*. CT [counterterrorism]... *PSYOP* [psychological operation]... *Transnational criminal activities, including narco-trafficking, illicit arms-dealing, and illegal financial transactions that support or sustain IW*.'[20] In summary, UW is a form of covert action conducted largely through surrogate criminal and terrorist networks, which seeks to mobilize political violence as a tool to influence the political choices of civilian populations.

A wealth of evidence in the public record confirms that UW doctrine has manifested in a policy of collaboration with Islamist extremist networks throughout the post-Cold War period, as was advocated by Graham Fuller. This policy was designed to destabilize strategic regions in such a way as to compel local populations to accept integration into the global political economy along lines favourable to US investors, particularly with regard to control of hydrocarbon energy resources.

The strategic complexity of UW doctrine, however, goes much further, and other outcomes may involve dominating the profit circuits of organized crime, as well as subordinating potential rivals to the US and thereby consolidating the US-dominated unipolar order. According to a confidential 2009 report to the Norwegian Foreign Ministry by Professor Ola Tunander – a defence consultant and Research Professor at the International Peace Research Institute in Oslo (PRIO) – covert Western sponsorship of terrorist networks under UW doctrine is also concerned with controlling profits from organized criminal activities, especially drug trafficking: 'In several states and not least in the USA, there are significant war elites that actually seek to introduce military conflicts in order to run them and profit from them.' As 'the economy of war (weapons and drugs) is as important as the world's oil economy', US covert forces may play a central role in 'calibrating the violence in various areas at a certain level to gain hundreds of billions of dollars in profits from weapons and drug trafficking'. In some cases, therefore, to prolong the war, US strategy is to 'support both sides in the conflict'. But apart from the objective of war profiteering, this strategy also has the deeper geopolitical objective of protecting a US-dominated unipolar order against economic multipolarity and the rising power of major rivals. 'The USA's superior military strength and intelligence hegemony could only be translated into power and real global strength if there were ongoing conflicts – wars and terrorist attacks – that threatened the multipolar power structure of the economic-political world order', continues the Norwegian report:

> Accordingly, from a European or Chinese or Japanese point of view, every US war, wherever it is fought, is not just directed against a local insurgent

> or an anti-American ruler, it is directed against the economic-political multipolar power structure that would give Europe, China and Japan a significant position in the world.

By fanning the flames on both sides in strategic regions, US forces are able to 'increase and decrease the military temperature and calibrate the level of violence' with a view to permanently 'mobilize other governments in support of US global policy'.[21]

According to the US Army special operations field manual cited above, 'UW has been conducted in... Operation ENDURING FREEDOM(OEF)/Afghanistan in 2001 and Operation IRAQI FREEDOM (OIF)Iraq in 2003'.[22] These themes will be explored below by focusing on how post-9/11 UW doctrine appears to involve the co-optation of Islamist terrorist groups in Afghanistan and Iraq.

Afghanistan and Pakistan: Endless Insurgency

Afghanistan is a crucial strategic region in terms of Central Asian pipeline politics. Osama bin Laden moved back to the country from Sudan in June 1996, after being offered protection by Pakistan on the condition that he firmly align his forces with the Taliban. The move was 'blessed by the Saudis'.[23] According to the leading expert on the subject, Ahmed Rashid – a correspondent for the London *Telegraph, Far Eastern Economic Review* and *Wall Street Journal* – the United States supported the Taliban from 1994 to 1998 as a vehicle of regional American influence. Between 1999 and 2000, US continued to support them, albeit more cautiously.[24] Radha Kumar, director of the Project on Ethnic Conflict, Partition and Post-Conflict Reconstruction at the Council on Foreign Relations, points out that this was because the Taliban:

> was brought to power with Washington's silent blessing as it dallied in an abortive new 'Great Game' in central Asia... Keen to see Afghanistan under strong central rule to allow a US-led group to build a multi-billion-dollar oil and gas pipeline, Washington urged key allies Pakistan and Saudi Arabia to back the militia's bid for power in 1996.[25]

US sponsorship of the Taliban was confirmed as late as 1999 and 2000 in Congressional hearings, confirming that the policy was 'based on the assumption that the Taliban would bring stability to Afghanistan and permit the building of oil pipelines from Central Afghanistan to Pakistan.'[26] US officials held several 'track-two' meetings with the Taliban from 2000 to summer 2001, where they tried to pressure the Taliban to join a federal government with the Northern Alliance as a condition for financial aid and international legitimacy. By then, they knew the Taliban would never bring the stability needed for the pipeline project. According to one meeting participant, then Pakistani Foreign Minister Niaz Naik, US officials threatened the Taliban with military action in October 2001 if they failed to comply with the federalization plan. But the Taliban had no intention of conceding power to its rivals, and rejected the plan. The war on Afghanistan was therefore not really a response to 9/11, but had been planned for at least a year before the terrorist attacks. As a close observer of Afghan politics, I had documented as early as January 2001 an unfolding US operational war plan for Afghanistan, rooted in a wider geostrategy going back several years.[27]

Post-9/11, there remains keen interest in the pipeline project. Regional oil and gas supply agreements have been signed and the Asian Development Bank – of which the US is one of the largest shareholders – is putting up the funds. 'Since the US-led offensive that ousted the Taliban from power', reported *Forbes* in 2005, 'the project has been revived and drawn strong US support' as it would allow the Central Asian republics to export energy to Western markets 'without relying on Russian routes'. Then US Ambassador to Turkmenistan Ann Jacobsen noted: 'We are seriously looking at the project, and it is quite possible that American companies will join it.'[28] The problem remains that the southern section of the proposed pipeline runs through territory that is still *de facto* controlled by Taliban forces – explaining current US and British efforts to use Pakistan as a forward base for the projection of military forces into Afghanistan to rout remaining Taliban fighters.[29]

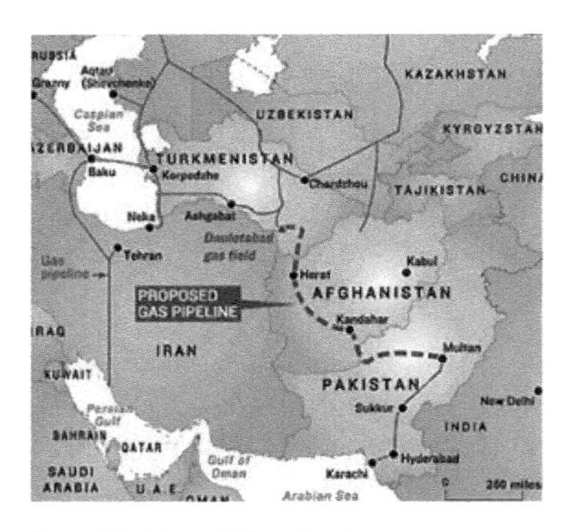

Figure 5.1 Trans-Afghan Pipeline.

Source: BBC News. Available from http://www.ca-c.org/online/2004/journal_eng/cac-05/07.ziyeng.shtml

However, this is not necessarily the only factor. Although British forces have led US, European and local forces in attempts to eradicate Afghanistan's burgeoning drug trade after the Taliban was toppled, as the *Sunday Telegraph* reported, 'to British officials' embarrassment, the level of opium cultivation during their stint at the helm has reached an all-time high of nearly half a million acres'.[30] Before 9/11, the Taliban had in fact banned the cultivation of opium poppies with significant success. Yet after the October invasion, US forces reversed the ban. By 2003, under the reign of US-sponsored Northern Alliance warlords, Afghanistan 'retook its place as the world's leading producer of heroin'.[31] Indeed, the US frequently foiled British counter-narcotics efforts. An early example was CIA pressure on President Hamid Karzai to dismiss the late Mohammed Daud as governor of the Helmand province, which produces over 60 per cent of the world's opium. British military commanders complained that Daud was a central figure in their counter-narcotics campaign: 'The Americans knew Daud was a main British ally, yet they deliberately undermined him and told Karzai to sack him.'[32]

Indeed, both senior British and American officials worked to undermine their own professed anti-opium campaign in different ways. According to Thomas Schweich – former US Counternarcotics Ambassador to Afghanistan – Karzai, the Pentagon, the CIA, and elements of the British military systematically sabotaged counter-narcotics measures: 'They (the Western military) didn't want anything to do with either interdiction or eradication.' Similarly, Robert Charles, Assistant Secretary of State for International Narcotics and Law Enforcement, noted: 'We could have destroyed all the [heroin] labs and warehouses in the three primary provinces – Helmand, Nangarhar and Kandahar – in a week', but the measures were blocked by Defense Secretary Donald Rumsfeld.[33]

This should not be surprising, given that US and British officials had carefully selected ardent veteran drug traffickers to return to power in the new 'post-Taliban' Afghanistan. War correspondent Philip Smucker reports that 'When the Taliban claimed Jalalabad' several years ago, Haji Zaman 'had fled Afghanistan for a leisurely life in Dijon, France'. During his tenure in Jalalabad, Zaman was 'at the top of the heroin trade'. But in late September 2001, 'British and American officials, keen to build up an opposition core to take back the country from the Taliban, met with and persuaded Zaman to return to Afghanistan'.[34]

According to former senior Indian intelligence official Bahukutumbi Raman, who has testified several times as an expert witness before US Congressional Committees, Zaman's ally in the drug trade, Haji Ayub Afridi, was simultaneously released from a Pakistani jail 'reportedly at the request of the CIA', returning to Afghanistan to play his part in US designs. Raman names another major figure in the drug trade, Haji Abdul Qadeer, who 'was the CIA's choice as the Governor of the Nangarhar province [in 2001] in which Jalalabad is located'. During the war against Soviet occupation:

> [Qadeer] played an active role under the control of the CIA and the Directorate-General For External Security (DGES), the French external intelligence agency, in organising the heroin trail to the Soviet troops from the heroin refineries of Pakistan owned by Haji Ayub Afridi, the Pakistani narcotics baron, who was a prized operative of the CIA in the 1980s.

At that time, Qadeer and Afridi were 'close associates in running this drug trade with the blessings of the CIA'. Others associated with the trade were 'Haji Mohammed Zaman and Hazrat Ali', both of whom resurfaced in November 2001 as US proxies against the Taliban. In other words, the 'post-

Taliban' re-narcotization of Afghanistan occurred under the auspices of the Anglo-American intervention, not in spite of it as is conventionally assumed.

Raman cites 'reliable sources in Afghanistan' for his claim that the Anglo-American 'war on drugs' in Afghanistan is a sham:

> The marked lack of success in the heroin front is due to the fact that the Central Intelligence Agency (CIA) of the USA, which encouraged these heroin barons during the Afghan war of the 1980s in order to spread heroin-addiction amongst the Soviet troops, is now using them in its search for bin Laden and other surviving leaders of the Al Qaeda, by taking advantage of their local knowledge and contacts. These Pakistani heroin barons and their Afghan lieutenants are reported to have played an important role in facilitating the induction of Hamid Karzai into the Pashtun areas to counter the Taliban in November, 2001. It is alleged that in return for the services rendered by them, the USA has turned a blind eye to their heroin refineries and reserves.[35]

There is thus growing evidence that the emergence of the Afghan narco-state under US-UK tutelage is not simply an unfortunate byproduct, but part of a wider pattern of Western intelligence liaisons with drug-trafficking networks. As former senior US Drug Enforcement Agency official Dennis Dayle has remarked: 'In my 30-year history in the Drug Enforcement Administration and related agencies, the major targets of my investigations almost invariably turned out to be working for the CIA.'[36]

Fortune magazine thus observes that 'several past and present cabinet ministers, senior law enforcement officials, and even Karzai's own brother are widely suspected of profiting handsomely from the poppy trade, overseeing growing operations or enabling transport of the yield across and out of the country'.[37] The total annual revenue generated from the Afghan heroin trade has been estimated at $400 billion, 95 per cent of which accrues not to Afghan farmers and traders, but to 'business syndicates, organized crime, and banking and financial institutions', largely in the consuming nations (hence only $4 billion accrues to Afghanistan).[38] Leading expert on terrorism financing Dr Loretta Napoleoni reports the existence of a US-backed Islamist narco-trafficking route across Central Asia: 'While the ISI [Pakistan's Inter-Services Intelligence] trained Islamist insurgents and supplied arms, Turkey, Saudi Arabia, several Gulf states and the Taliban funded them...Each month, an estimated 4–6 metric tons of heroin are shipped from Turkey via the Balkans to Western Europe.'[39]

Former FBI translator Sibel Edmonds, in testimony before classified Congressional hearings, has identified senior US officials that played a key role in the international narcotics trade, and received large payments of heroin money. She specifies the role of US, Pakistani and Turkish intelligence in controlling the trade in north Central Asia.[40] Edmonds further reveals direct US and NATO military complicity: 'A lot of the drugs were going to Belgium with NATO planes. After that, they went to the UK, and a lot came to the US via military planes to distribution centers in Chicago and Paterson, New Jersey.'[41] Her allegations are corroborated by local sources. According to Russian Ambassador to Afghanistan Zamir Kabulov, US military transport aviation 'is used for the delivery of drugs from Afghanistan to the American airbases, Ganci in Kyrgyzstan and Incirlik in Turkey'. Anonymous Afghan sources say: '85 per cent of all drugs produced in southern and southeastern provinces are shipped abroad by US aviation.' Officials in Afghanistan's security services also claim that the US military acquires drugs through local officials connected to Afghan field commanders presiding over heroin production.[42]

Indeed, over the last decade, concrete evidence has emerged, verified by British and European authorities, confirming direct CIA complicity in illicit narco-trafficking, including the use of 'extraordinary rendition' planes. In January 2010, the *Independent* reported:

> Evidence points to aircraft – familiarly known as 'torture taxis' – used by the CIA to move captives seized in its kidnapping or 'extraordinary rendition' operations through Gatwick and other airports in the EU being simultaneously used for drug distribution in the Western hemisphere. A Gulfstream II jet aircraft N9875A identified by the British Government and the European Parliament as being involved in this traffic crashed in Mexico in September 2008 while en route from Colombia to the US with a load of more than three tons of cocaine.
>
> In 2004, another torture taxi crashed in a field in Nicaragua with a ton of cocaine aboard. It had been identified by Britain and the European Parliament's temporary committee on the alleged use of European countries by the CIA for the transport and illegal detention of prisoners as a frequent visitor in 2004 and 2005 to British, Cypriot, Czech, German, Greek, Hungarian, Spanish and other European cities with its cargo of captives for secret imprisonment and torture in Iraq, Jordan and Azerbaijan.[43]

This suggests that control of the Afghan heroin trade plays at least an equal role in Anglo-American military strategy to that of dominating the Trans-Afghan pipeline route. But to complicate matters further, the Pakistani ISI continues to be integrally involved in the military, intelligence and financial support of al-Qaeda, including the insurgency in northwest Pakistan and Afghanistan. As US intelligence expert George Friedman observes: 'Pakistan was the key, because it had the closest connections to Al Qaeda and the least cooperative intelligence service, in spite of the apparent cooperation of Pakistan's President Musharraf.'[44] It is partly this situation which has led to the heightened political instability and internal conflict in Pakistan, with the US putting pressure on the Pakistani Army to rout Taliban forces from the Swat valley while continuing to covertly sponsor al-Qaeda affiliates elsewhere.

Just as Musharraf backtracked on public assurances to crack down on the ISI's sponsorship of al-Qaeda, the current Pakistani Army chief, General Ashfaq Pervez Kiani, has also been similarly implicated. General Kiani served as head of the ISI from 2004 to 2007. During this period, the ISI 'presided over the development of a major logistical and training program for the Taliban forces'. The programme was revealed by a NATO report 'of a two-week battle by NATO forces against a determined Taliban offensive in Kandahar province in September 2006'. The NATO account 'described two ISI training camps for the Taliban near Quetta in Pakistan's Balochistan province. It also documented the provision by the ISI of 2,000 rocket-propelled grenades and 400,000 rounds of ammunition – just for that one Taliban campaign.' Further evidence that ISI assistance to Taliban forces continued after Kiani's appointment replacing Musharraf as Pakistan's top army general 'was compiled in an intelligence assessment circulated to the top national security officials of the George W Bush administration in mid-2008'. Indeed, US intelligence intercepted a communication in which Kiani referred to a senior Taliban leader, Maulavi Jalaluddin Haqqani, as a 'strategic asset' – although Haqqani's insurgent network 'has been a key target for the US campaign of Predator drone strikes in Pakistan during 2008 and 2009'. Despite this, 'senior officials of the Barack Obama administration have persuaded the US Congress to extend military assistance to Pakistan for five years without any assurance that the Pakistani assistance to the Taliban had ended'. Rather than demanding that military assistance should be conditional on evidence that the ISI had ended support to al-Qaeda and Taliban forces, the new legislation provided an unconditional $6 billion in military and economic aid over five years, requiring only a certification by the secretary of state that Pakistan is 'making concerted efforts' against the Taliban.[45]

In effect, through the ISI, the US is subsidizing the Afghan insurgency, and thus prolonging the war – raising the question of the war's role in legitimizing both 1) a continued substantive US military presence in the region, and 2) the compliance of key US allies, particularly the UK and Western Europe, with the geopolitical demands of the 'War on Terror'.

THE 'REDIRECTION': AL-QAEDA SPONSORSHIP IN THE MIDDLE EAST AFTER 2003

Iraq: Covert Action and Sectarian Violence

Since the invasion of Iraq in 2003, unconventional warfare strategies linking US military intelligence to al-Qaeda terrorist networks shifted to a new theatre of operations – primarily to the Middle East and Central Asia. Much of the destabilization of Iraq under sectarian conflict was due to this change.

In early 2006, then Senator Joseph Biden – now Obama's vice-president – proposed that the solution to intensified sectarian conflict in Iraq was to partition the country into three ethnically composed autonomous regions, each with their own limited authority for self-rule, but with a central government in charge of 'common interests'. Late the following year, the Senate officially approved a 'soft' partition plan. What is not so well known is that from the very beginning, senior American planners envisaged that in the long term Iraqi territory would be divided to facilitate the Anglo-American military occupation. Fragments emerged in late 2002 of a plan to fracture Iraq along ethnic and religious lines in order to facilitate control of oil reserves and allow population control. Al-Qaeda-affiliated groups financed by Saudi Arabia and Pakistan, and in some cases directly supported by highly classified US military intelligence operations, were mobilized with the effect of initiating and accelerating the dynamic of sectarian conflict in accordance with the carve-up plan.

Richard Perle, who then chaired the Defense Policy Board, the prominent Pentagon advisory group, issued a briefing for Pentagon officials in September that year. *Ha'aretz* reported that, according to a 'top official in the Israeli security services':

> [Perle] showed two slides to the Pentagon officials. The first was a depiction of the three goals in the war on terror and the democratisation of the Middle East: Iraq – a tactical goal, Saudi Arabia – a strategic goal, and Egypt – the great prize. The triangle in the next slide was no less interesting: Palestine is Israel, Jordan is Palestine, and Iraq is the Hashemite Kingdom.[46]

This outrageous vision advocated a total reconfiguration of power across the Middle East, including a greatly expanded Israel fully encompassing the Occupied Territories; the expulsion of the Palestinians to Jordan; and the incorporation of the Sunni areas of Iraq with Jordan to form a wider pro-US Sunni Arab Hashemite Kingdom.

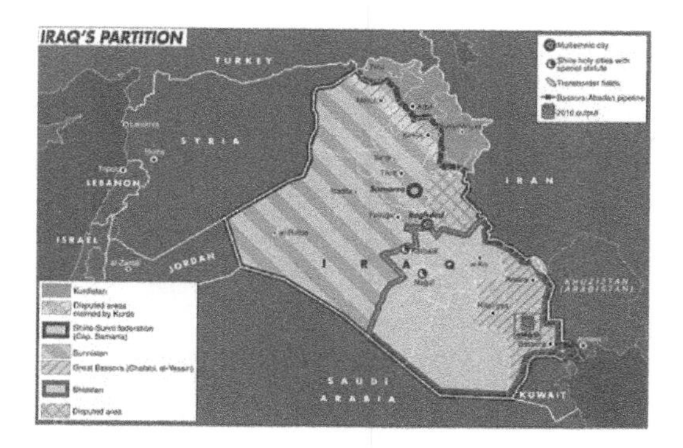

Figure 5.2 US Plan for the Partition of Iraq.

Source: Heartland: Eurasian Journal of Geopolitics (December 2005). Available from http://temi.repubblica.it/limes-heartland/iraqs-partition/1106

According to the private American intelligence firm, Stratfor, this was a highly influential plan. Stratfor reports that the Bush administration was 'working on a plan to merge Iraq and Jordan into a unitary kingdom to be ruled by the Hashemite dynasty headed by King Abdullah of Jordan'. The plan was 'authored by US Vice-President Dick Cheney' as well as 'Deputy Secretary of Defense Paul Wolfowitz', and was first discussed at 'an unusual meeting between Crown Prince Hassan of Jordan and pro-US Iraqi Sunni opposition members in London in July [2002]'. Iraq would be *de facto* ethnically partitioned into three autonomous cantons: The central and largest part of Iraq populated by Sunni Arabs would be joined with Jordan, and would include Baghdad, which would no longer be the capital. The Kurdish region of northern and northwestern Iraq, including Mosul and the vast Kirkuk oilfields, would become its own autonomous state. The Shi'ite region in southwestern Iraq, including Basra, would make up the third state, or more likely it would be joined with Kuwait. Stratfor outlined the perceived advantages for the US as follows (see Figure 5.3):

> First, the creation of a new pro-US kingdom under the half-British Abdullah [king of Jordan] would shift the balance of forces in the region heavily in the US favor. After eliminating Iraq as a sovereign state, there would be no fear that one day an anti-American government would come to power in Baghdad, as the capital would be in Amman [Jordan]. Current and potential US geopolitical foes Iran, Saudi Arabia and Syria would be isolated from each other, with big chunks of land between them under control of the pro-US forces. Equally important, Washington would be able to justify its long-term and heavy military presence in the region as necessary for the defense of a young new state asking for US protection – and to secure the stability of oil markets and supplies. That in turn would help the United States gain direct control of Iraqi oil and replace Saudi oil in case of conflict with Riyadh.[47]

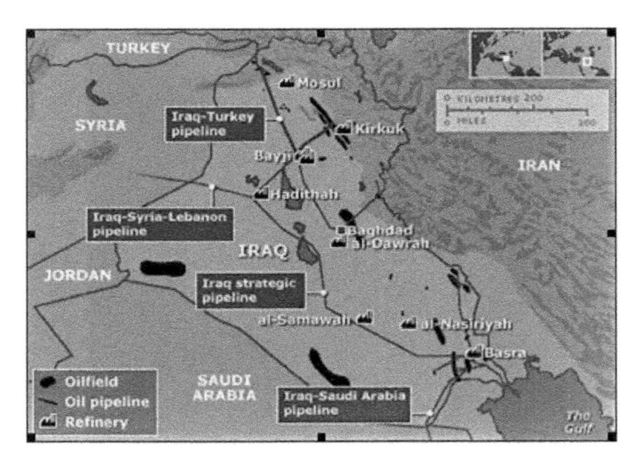

Figure 5.3 US Plans for Iraq Pipeline.

Source: BBC News. Available from http://news.bbc.co.uk/nol/shared/spl/hi/middle_east/02/iraq_key_maps/img/maps/iraq_oil_map485.gif

Such tripartite partitioning of an entire country could never be achieved peacefully. Violent conflict along sectarian lines is inevitable if this were to be achieved. The task of 'eliminating Iraq as a sovereign state' by fracturing the country along ethnic and religious lines, in other words, was *precisely* the postwar strategy being explored by Dick Cheney as the most effective means of securing American control over the country, and the wider region.

It is no surprise then to find US covert action supporting the escalating sectarian violence in Iraq as a tool for consolidating the occupation. The strategy received semi-official acknowledgement in the November 2005 edition of the *US Joint Special Operations University Report*, in which Thomas H. Henriken, a senior fellow at the university and a former member of the US Army Science Board, reported:

> *The post-invasion stage in Iraq also is an interesting case study of fanning discontent among enemies, leading to 'red-against-red' firefights* (this color-coding derives from US training exercises, in which red designates enemy combatants and blue designates friendly forces). Like their SOG predecessors in Vietnam, *US elite forces in Iraq turned to fostering infighting among their Iraqi adversaries on the tactical and operational level...*
>
> Events during fall 2004 within the central Iraqi city of Fallujah showcased the wily machinations required to set insurgents battling insurgents. ... But Fallujah was hardly a unified camp – the city seethed with internecine tensions. [Abu Musab al-]Zarqawi's strict Salafi beliefs clashed with the more moderate Sufi views of the Sunni residents. Additionally, the Zarqawi jihadis and nationalistic Fallujans disagreed over the use of terror tactics. Both wanted the Americans out of Fallujah and out of Iraq, but they differed on the methods.... Evidence of factional fighting between the residents came to light with nightly gun battles not involving coalition forces. *US psychological warfare (PSYOP) specialists took advantage of the internal warring by tapping into Fallujans' revulsion and antagonism to the Zarqawi jihadis. The PSYOP warriors crafted programs to exploit Zarqawi's murderous activities – and to disseminate them* through meetings, radio and television broadcasts, handouts, newspaper stories, political cartoons, and posters – thereby diminishing his folk-hero image. Battles among anti-coalition forces killed enemy combatants and heightened factionalism. Thus, red-on-red battles enhanced the regular blue-on-red engagements by eliminating many insurgents.[48]

The 'dissemination' of the 'murderous activities' of al-Qaeda's Zarqawi as part of US covert operations to escalate sectarian conflict in Iraq appears to have been part of a much wider strategy for influencing the Iraqi population. In November 2004, a joint statement was released on several Islamist websites on behalf of Zarqawi, al-Qaeda's leader in Iraq, and Saddam Hussein's old Ba'ath Party loyalists. Zarqawi's network had 'joined other extremist Islamists and Saddam Hussein's old Baath party to threaten increased attacks on US-led forces'. Zarqawi's group said they signed 'the statement written by the Iraqi Baath party, not because we support the party or Saddam, but because it expresses the demands of resistance groups in Iraq'.[49]

The statement formalized what the *Sunday Times* had already reported, citing post-Saddam Iraqi intelligence and US military officials: 'Al Qaeda terrorists who have infiltrated Iraq from Saudi Arabia and other Arab countries have formed an alliance with former intelligence agents of Saddam Hussein to fight their common enemy, the American forces.' Al Qaeda leaders 'recruit from the pool' of Saddam's former 'security and intelligence officers who are unemployed and embittered by their loss of status'. After vetting, 'they begin Al-Qaeda-style training, such as how to make remote-controlled bombs'. Both Saudi Arabia and Pakistan appear to be integrally involved in the operation. 'The alliance, known as Jaish Muhammad – the army of the prophet Muhammad – is believed to be responsible for increasingly sophisticated attacks on US soldiers.' Jaish Muhammed is smuggling 'millions of dollars, weapons and hundreds of Arab fighters across the desert border with Saudi Arabia'.[50]

Jaish Muhammed is a notorious al-Qaeda affiliated network based in Pakistan with close links to the Pakistani ISI.[51] Indeed, the connection was corroborated in February 2005 by Pakistani military sources who revealed that 'the US has... resolved to arm small militias backed by US troops and entrenched in the population' involved in the Iraqi insurgency. The US had secretly 'procured Pakistan-manufactured weapons, including rifles, rocket-propelled grenade launchers, ammunition, rockets and other light weaponry'. Consignments were bulk loaded onto US military cargo aircraft at Chaklala airbase, arriving from and departing for Iraq. 'The US-armed and supported militias in the south will comprise former members of the Ba'ath Party' – the same people recruited and trained by Zarqawi's al-Qaeda network in Iraq. A Pakistani military analyst noted that the reason US-made arms were not being supplied was to conceal US complicity:

> A similar strategy was adopted in Afghanistan during the initial few years of the anti-USSR resistance [the early 1980s] movement where guerrillas were supplied with Chinese-made AK-47 rifles [which were procured by Pakistan with US money], Egyptian and German-made G-3 rifles.

Military sources added that their destination was not the Iraqi security forces 'because US arms would be given to them'. Rather, the US is playing a double game to 'head off' the threat of a 'Shi'ite clergy-driven religious movement'.[52] Notably, the US appropriation of Pakistani military resources to support elements of the Iraqi insurgency coincides with the ISI's sponsorship of al-Qaeda and Taliban forces in the same period, under Gen. Pervez Kiani's leadership.

The insurgency therefore consisted of two contradictory elements – a genuinely indigenous resistance movement, and a much smaller, insidious, alien element responsible for terrorist violence, among whom were operatives co-opted by US military intelligence in coordination with regional allies. This suggests that destabilization was not merely a consequence of US incompetence – but also part of a plan to use unconventional warfare as a tool to influence the Iraqi population in favour of a permanent US security presence.

Iraq, Lebanon, and Iran: Arc of Destabilization

These covert policies, beginning around 2004, were intensified throughout 2006, and reintensified in 2008, across a large swathe of the Middle East and Central Asia. Seymour Hersh discovered that the Bush administration was actively sponsoring al-Qaeda-affiliated groups across these regions to counter regional Shi'ite Iranian influence. Moreover, much of the finances for these covert operations were being funnelled by Saudi Arabia through Iraq, with US connivance:

> This administration has made a policy change, a decision that they are going to put all of the pressure they can on the Shiites, that is the Shiite regime in Iran... we are interested in recreating what is happening in Iraq in Lebanon, that is Sunni versus Shia... we have been pumping money, a great deal of money, without congressional authority, without any congressional oversight, Prince Bandar of Saudi Arabia is putting up some of this money, for covert operations in many areas of the Middle East where we think that the – we want to stop the Shiite spread or the Shiite influence.
>
> They call it the 'Shiite Crescent.' And a lot of this money... has gotten into the hands – among other places, in Lebanon, into the hands of three – at least three jihadist groups. There are three Sunni jihadist groups whose main claim to fame inside Lebanon right now is that they are very tough. These are people connected to al Qaeda who want to take on Hezbollah...
>
> So America, my country, without telling Congress, using funds not appropriated, I don't know where, by my sources believe much of the money obviously came from Iraq where there is all kinds of piles of loose money... We are simply in a situation where this president is... supporting groups indirectly that are involved with the same people that did 9/11, and we should be arresting these people rather than looking the other way.[53]

Hersh's reporting indicates that al-Qaeda-affiliated networks remain useful as mercenary proxies for Anglo-American regional geostrategy in the Middle East. Even the international structure of state-sponsorship is unchanged, with the US at the helm, Saudi Arabia providing the funds, and Pakistan providing military intelligence support, although the bulk of finances for these operations were being funnelled through Iraq.

In March 2007, Hersh elaborated his findings in *New Yorker* magazine, citing White House insiders and other US government officials, all confirming in perhaps the clearest terms possible that the US was deliberately attempting to control al-Qaeda terrorist activity through Saudi Arabia (among others), in an attempt to re-direct the network against Iran:

> The 'redirection,' as some inside the White House have called the new strategy, has brought the United States closer to an open confrontation with Iran and, in parts of the region, propelled it into a widening sectarian conflict between Shiite and Sunni Muslims.

> To undermine Iran, which is predominantly Shiite, the Bush Administration has decided, in effect, to reconfigure its priorities in the Middle East. In Lebanon, the Administration has cooperated with Saudi Arabia's government, which is Sunni, in clandestine operations that are intended to weaken Hezbollah, the Shiite organization that is backed by Iran. The US has also taken part in clandestine operations aimed at Iran and its ally Syria. A by-product of these activities has been the bolstering of Sunni extremist groups that espouse a militant vision of Islam and are hostile to America and sympathetic to Al Qaeda...
>
> The clandestine operations have been kept secret, in some cases, by leaving the execution or the funding to the Saudis, or by finding other ways to work around the normal congressional appropriations process, current and former officials close to the Administration said...
>
> This time, the US government consultant told me, Bandar and other Saudis have assured the White House that 'they will keep a very close eye on the religious fundamentalists. Their message to us was "We've created this movement, and we can control it." It's not that we don't want the Salafis to throw bombs; it's who they throw them at – Hezbollah, Moqtada al-Sadr, Iran, and at the Syrians, if they continue to work with Hezbollah and Iran.'[54]

Of course, as former CIA official Robert Baer points out, controlling the Saudis is an impossible task.[55] Nevertheless, the policy escalated throughout 2008 with bipartisan support. Early that year, George W. Bush signed a Presidential Finding informing US Congress of his authorization of 'a covert offensive against the Iranian regime... across a huge geographic area – from Lebanon to Afghanistan', to be financed by an additional $300 million. Diverse anti-Shi'ite and anti-Iranian groups were to be sponsored, among them many al-Qaeda affiliates, but also including Kurdish nationalists, Awwaz Arabs in the southwest of Iran, and Mujahideen-e-Khalq – an Iranian terrorist group on the State Department's and EU's lists of proscribed terrorist organizations.[56] Despite the legislative opportunity to challenge the covert programme, neither Republican nor Democrat Congressional representatives did so. Indeed, under Obama's presidency the programme has not ceased. On the contrary, Obama retained Bush's Defense Secretary, Robert Gates – the architect of Bush's Iran policy who strongly favours continued covert action as a more viable mechanism of pressure than direct intervention.[57]

Manipulating Terrorism

Al-Qaeda-affiliated groups have frequently functioned as mercenary proxies mobilized in the service of a distinctive form of Anglo-American imperial expansionism, aimed largely at consolidating control over strategic energy reserves, not just during the Cold War, but also after, into the post-9/11 era. The same addiction to oil that underlies the crises of climate change and peak oil has also led us into a geopolitical dance of death with Osama bin Laden. The same trajectory of covert action has intersected with the centres of international heroin production and trafficking, while also functioning to 'calibrate levels of violence', in the words of Oslo analyst Ola Tunander, to facilitate profiteering from regional war economies and, more importantly, to exert geopolitical pressure on potential US rivals. *Al-Qaeda is not simply an enemy out there. It is an intrinsic feature of the global political economy, like a tumour spawned by its own host.*[58]

There is no doubt that these covert practices, enacted through key states like Saudi Arabia and Pakistan, continue to foster Islamist terrorist activity across the world. Thus, at the heart of the material infrastructure underlying al-Qaeda activity lies a network of different regional states each with their own interests, which in turn are conduits of Western military and financial aid. Not only has the 'War on Terror' failed entirely to dismantle this infrastructure, it continues to strengthen it.

Instead of trying to transform the global political economy and its increasingly dangerous over-dependence on oil, our governments are pointing to Islamist extremism as the principal threat to our security. This is not to deny the reality of this threat, but to point out that al-Qaeda's operational reach is being exacerbated by precisely the same oil addiction and global political economy that are in crisis today. In the process, governments are using the threat of terrorism as a way of deflecting attention from the failures of the system. This, of course, raises profound questions about the legitimacy and integrity of Western national security policy. The truth is that UW strategies are largely not subject to regulation by the public face of the state – at the levels of parliament, cabinet, or even conventional military intelligence. They belong to its deeper, invisible dimension, where we find the nexus between a highly unaccountable, and as such enormously powerful, subsection of the 'security' agencies, private military and defence contractors, energy companies, and entrenched international criminal networks.[59]

This grants us a distinct insight into the way the phenomena of violent extremism and international terrorism will be exacerbated by the escalation of global ecological, energy and economic crises. Given the increasing pressures placed on state militaries in the face of intensifying social upheaval and civil disturbance due to these crises, it is likely that the US will increasingly rely on UW to pursue its strategic aims.

However, this will also entail increased difficulties in efforts to manipulate terrorist groups as proxy forces. The cases reviewed above explore how Islamist terrorist networks have been selectively co-opted by Western military intelligence services in the pursuit of strategic and economic interests throughout the post-Cold War period, and even after 9/11. Yet it is important to recognize the inherent limitations of these covert practices. Clearly, Al-Qaeda networks in strategic regions are not simply front groups consciously operating on behalf of Western intelligence agencies, but are co-opted through the recruitment and mobilization of particular intelligence assets as part of clandestine HUMINT (human intelligence) operation techniques. These assets act either as double or triple agents in senior positions in target groups, attempting to steer these groups in directions favourable to the strategic interests of the agencies on whose behalf the agents operate.[60] This means that the continuity and apparent authenticity of these terrorist groups fundamentally relies on their success within the regional socio-cultural context, in exploiting a local groundswell of political grievances to recruit foot soldiers for their operations. Thus, the majority of the group's members are unlikely to be aware that the group is susceptible to covert manipulation from outside. Ultimately, this means that UW doctrine's advocacy of co-opting terrorist networks remains a highly dangerous activity that constantly threatens to bring about unanticipated consequences.[61]

The problem is dual: 1) As global crises escalate and intensify, they will increasingly aggravate social upheaval, and thus the political grievances that provide recruiting fodder for extremist and terrorist networks. As these grievances escalate, they will magnify the internal pressure on these groups and amplify their ability to act outside of US interests. 2) The escalation of global crises will also create qualitatively *new* grievances, and novel forms of social polarization based on 'identity politics': as social anxieties, fears and frustrations proliferate, political and community leaderships of all kinds will find it increasingly useful and sensible to mobilize on the basis of ethnic, religious and class identities that allow groups to impose order and control.

This means that the escalation of global crises will create space for social groups in general to respond by resorting to ideological extremes that legitimize violence against 'Others.' These trends will lead to new hubs of extremist and terrorist activity operating beyond the original remit of US power, particularly in Central Asia and the Middle East where covert policies of selective co-optation of extremist and terrorist networks persist. Civilian populations of major core powers with a geopolitical stake in these regions – the US, UK, EU, Russia, China, and India – will face the brunt of this impact.

6
The Militarization Tendency

Unconventional Warfare doctrine is only one element in a wider array of state strategies in response to global crises. Although these crises are symptomatic of failures inherent to the very structure of the international system, states are so far attempting to find creative ways to stabilize and strengthen that system. By responding along traditional nation-state lines, they are attempting to maximize national military and economic power so as to: 1) impose order on an international level through intensified geopolitical regulation; 2) impose order on a domestic level through new regimes of surveillance backed by unprecedented police powers. In other words, we are witnessing a comprehensive militarization of our societies, both in their external relations with other less powerful societies and in their internal relations between different social groups. Global crises are driving a Western state tendency toward the maximization of social control, ostensibly justified with reference to only one crisis, international terrorism; and the predominant victims of the measures taken are largely Muslim civilian communities, both in strategic peripheries and at home. In other words, an exclusive focus on the threat posed by Islam and Muslims is increasingly a preoccupation of these broader 'securitization' processes, which are actually driven by the ecological, energy and economic crises.

'Securitization' refers to a 'speech act' – an act of labelling – whereby political authorities identify particular issues or incidents as an *existential threat* which, because of their extreme nature, justify going beyond the normal security measures that are within the rule of law. It thus legitimizes resort to special extra-legal powers. By labelling issues a matter of 'security', therefore, states are able to move them outside the remit of democratic decision-making and into the realm of emergency powers – all in the name of survival itself. These powers are not the prerogative of the parliamentary system, but of the extra-legal dimension of the state – or the 'deep state' – and are therefore beyond public scrutiny and accountability.[1]

This chapter explores the disturbing trends of 'securitization' in the context of the evolution of contemporary international relations as a fundamentally *imperial* system, dominated by the United States, Britain and Western Europe after the Second World War, and thereafter expanding to encompass and integrate peripheral colonies, semi-colonies and imperial protectorates in Africa, South America, Asia and the Middle East. The escalation of global ecological, energy and economic crises has been closely monitored by Western government and security agencies. These crises are recognized not as evidence that the global imperial system is fundamentally unsustainable and therefore requiring urgent transformation, but as vindicating the necessity for Western states to *radicalize* the exertion of their military-political capacities to maintain the existing power structures.[2]

Thus, global crises are being 'securitized' to the extent that they are seen by states as threats with the potential to seriously undermine the integrity of governments, their legitimacy, and their capacity to sustain internal and international order. Indeed, my review of internal assessments by various state agencies of the security impacts of global crises (see below) suggests that these agencies view the crises as *amplifying factors*, in danger of mobilizing the popular will in ways that challenge existing political and economic structures. While this is seen as justifying the state's adoption of extra-legal measures outside the normal sphere of democratic politics, there is a fundamental problem here: The adoption of extraordinary measures and emergency powers cannot be legitimized in Western liberal societies simply with reference to a need to control large-scale civil unrest and popular resistance, stoked by the breakdown of public services due to crises like climate change, energy scarcity, financial collapse, and inadequate food supplies – doing so would underscore the narrow vested interests being served. Indeed, the securitization of global crises is not genuinely designed to ameliorate their persistent causes, but primarily to grant social control powers to the increasingly unaccountable state at a time of heightened uncertainty and civil disorder due to those very crises. As such, securitization processes require an external legitimizing rationale, a narrative framework which can be sold to the public.

The 'War on Terror' provides an ideological framework encouraging the ongoing subliminal construction of an exclusionary Islamist 'Other' as the source of a civilizational threat, against which social control powers can be legitimized in the name of defending the liberal values which those powers themselves are eroding. If continued unchecked, the long-term outlook of these securitization trends is that our societies will sacrifice liberal values in favour of increasingly polarizing and exclusivist conceptions of group identity which normalize political violence: *i.e., militarization*. In this context, the covert activities of the 'deep state', in terms of the selective sponsorship of al-Qaeda terrorist networks, serve a dual purpose: 1) generating geopolitical conditions conducive to strategic dominance; and 2) providing substance to a narrative of 'threat' that can justify a state of permanent global militarization. Here, the concept of *militarization* should be understood as distinct from securitization. The latter refers to the way in which particular issues are labelled or constructed by states (or indeed other actors) as 'security' threats justifying emergency powers; the idea of militarization conveys the *cumulative effect* of these securitization processes in terms of the institutionalization of emergency powers through new state-backed socio-political structures, entrenching the role of state police and military force in the jurisdiction of social life. While securitization processes (the labelling and constructing of particular issues as 'security' problems requiring extra-legal responses) could stop, the resultant apparatuses of militarization that normalize political violence will continue to remain in place.

Yet the very reality of global crises demonstrates that this programme of comprehensive militarization cannot last far beyond the mid twenty-first century, particularly as growing resource scarcities are likely to make conventional forms of industrialized warfare increasingly costly, unsustainable, and ultimately redundant. Such processes will diminish the relevance of the nation-state as the dominant political unit, giving way to more decentralized forms of social organization and political mobilization. This means that the long-term outlook for post-carbon civilization may well be optimistic, creating the possibility of an end of the 'total wars' of the modern period. While this does not mean the end of war *per se* as the growing power of non-state actors will also create new opportunities for inter-communal violence, the prospect of a Long Peace could become a real possibility after mid century, made possible by the obsolescence of industrialized social forms.

EMPIRE

British Legacy, American Destiny

Imperial conceptions of world order were integral to American planning during the Second World War for the postwar reconstruction of the global system. The idea was to create a new political and economic order that would evade the massive social, political and economic crises faced by the old empires, including that of the British. State Department planners had advocated that the US 'must cultivate a mental view toward world settlement after this war that will enable us to *impose our own terms*, amounting perhaps to a *pax-Americana*'.[3]

By this time, Britain's terminal economic decline relative to its newly industrializing competitors in Western Europe was a foregone conclusion. The military and economic impact of the First World War shrank Britain's global economic influence and reduced its bargaining power vis-à-vis its colonies. In response to this crisis, exacerbated by intensifying popular resistance in the colonies, the British empire was compelled to withdraw from its periphery. The imperial experiment of direct rule had failed.[4] Its American successor, in turn, tried a different approach. Already at a unique global geopolitical advantage by the end of the war, the US attempted to transform world order along new 'liberal' principles. Rather than occupying and controlling regions directly, the US promoted the emergence of formally independent nation-states throughout the former colonies in the South, articulating the necessity to *police* these states so that they complied with US interests, and to defend them from interventions – such as from the USSR – that might undermine the new 'liberal' order.[5]

In the postwar period, this included two interrelated strategies designed to subordinate strategic regions to an emerging US-dominated international political and economic order: 1) from 1945 to 1990, the waging of 70-plus military intelligence interventions in multiple peripheries across Africa, South America, Asia and the Middle East to guarantee the emergence of compliant nation-states capable of sustaining minimal regulatory frameworks, ensuring the transnational mobility of American capital; 2) the erection of a US-dominated global governance architecture to regulate this new system of states and maintain the interests of the North, consisting now of the United Nations, the World Bank, the IMF, and the World Trade Organization, among other institutions.[6]

As Birkbeck College empire expert Dr Alex Colas observes, US imperialism is uniquely different from all previous empires, seeking to 'promote self-government, civil liberties and territorial integrity' as integral components of the postwar international system. Yet these same conditions are only deemed indispensable *insofar as they establish a regulatory framework of private property rights and market stability necessary for capitalism to function.* Postwar American preoccupation with imperial 'state-building' cannot be reduced to a drive to 'seek to invade and occupy other territories exclusively to exploit their human and natural resources', but is rather concerned primarily with 'fostering the institutional conditions that will allow capitalist economies to thrive in supervised territories'. As such, the 'politics and ethics' of so-called humanitarian intervention 'cannot be divorced from the political economy of liberal empire'. Consequently, while the US empire is not wholly reducible to capitalism, it is also inexplicable without reference to capitalism. Distinguished by a 'liberal internationalist mission of building viable states and market economies', the US empire is thus 'all the more dependent on local acquiescence and cooperation'.[7]

The Political Economy of Imperial Hubris

Thus, the fostering of formally sovereign nation-states is a geopolitical vehicle for the forcible transformation of social conditions in strategic regions in a manner conducive to the penetration and operation of American and Western capital. One indication of this is an article written on the eve of the Cold War by Lt. Gen. Alfred M. Gray – Commandant of the US Army Marine Corps under President Reagan – conceding that the ultimate aim of 'maintain[ing] military credibility in the next century' was to ensure continued 'unimpeded access' to 'developing economic markets throughout the world' and 'to the resources needed to support our manufacturing requirements'. In an extraordinarily candid paragraph, he continued:

> In fact, the majority of the crises we have responded to since the end of World War II have not directly involved the Soviet Union... The underdeveloped world's growing dissatisfaction over the gap between rich and poor nations will create fertile breeding ground for insurgencies. These insurgencies have the potential to jeopardize regional stability and our access to vital economic and military resources... This situation will become more critical as our Nation and its allies, as well as potential adversaries, become more and more dependent on these strategic resources... If we are to have stability in these regions, maintain access to their resources, protect our citizens abroad, defend our vital installations, and deter conflict, we

> must maintain within our active force structure a credible military power projection capable with the flexibility to respond to conflict across the spectrum of violence throughout the globe.[8]

The extension of US military power is thus a response to growing social crises generated in the context of the capitalist world system's increasing marginalization of the 'underdeveloped' South. How does this fit with the uniquely American vision of 'liberal' empire? While, internally, capitalist markets are designed to work without government interference, the actual creation of such markets in new territories requires a violent transformation of their social relations to take control of productive resources, dispossess large numbers from the land to create wage labourers, and open markets to foreign capital. If such efforts are resisted by local populations, then counter-insurgency measures are required to forcibly establish the 'liberal' conditions of the market – that is, a regulatory private property framework supported by appropriate political, legal and ideological institutions. Hence, military doctrines come hand-in-glove with a potent vision of 'liberal' imperialism, advocating 'the forceful extension of free markets, electoral democracies and human rights', all of which are essential ingredients in the maintenance of 'legitimate states and capitalist markets to secure the expanded reproduction of a liberal world order'.[9]

The duplicity of this liberal vision was alluded to by Lt. Col. (ret.) Ralph Peters in 1997 in the quarterly US Army War College journal *Parameters*. 'Those of us who can sort, digest, synthesize, and apply relevant knowledge soar – professionally, financially, politically, militarily, and socially. We, the winners, are a minority.' This minority will inevitably clash with the vast majority of the world's population. 'For the world masses, devastated by information they cannot manage or effectively interpret, life is "nasty, brutish... and short-circuited."' In 'every country and region', these masses who can neither 'understand the new world', nor 'profit from its uncertainties... will become the violent enemies of their inadequate governments, of their more fortunate neighbors, and ultimately of the United States'. The coming conflict, then, is not really about blood, faith, or ethnicity at all. It is about the gap between the haves and the have-nots. 'We are entering a new American century... we will become still wealthier, culturally more lethal, and increasingly powerful. We will excite hatreds without precedent.' In predicting the future course for the US Army, Lt. Col. Peters argued: 'We will see countries and continents divide between rich and poor in a reversal of 20th-century economic trends.' In this context, he said, 'we in the United States will continue to be perceived as the ultimate haves', and therefore, 'terrorism will be the most common form of violence', along with 'transnational criminality, civil strife, secessions, border conflicts, and conventional wars'. Meanwhile, 'in defense of its interests', the US 'will be required to intervene in some of these contests'. In summary:

> There will be no peace. At any given moment for the rest of our lifetimes, there will be multiple conflicts in mutating forms around the globe. Violent

> conflict will dominate the headlines, but cultural and economic struggles will be steadier and ultimately more decisive. The de facto role of the US armed forces will be to keep the world safe for our economy and open to our cultural assault. To those ends, we will do a fair amount of killing.[10]

Crisis Response, Hegemony Rehabilitation

But the post-9/11 era was not simply a continuation of historic US imperialism. President George W. Bush's new National Security Strategy represented a departure, emphasizing a new doctrine of power projection based on unilateral US power at the expense of multilateralism; the primacy of pre-emptive warfare; the need for long-term US state-building projects in strategic regions; and overall, all this as part of a drive to reconfigure the entire defence system to sustain US pre-eminence in the face of a potentially rapidly changing world order. The Strategy brought together pre-existing US military trends and practices into a coherent policy package. This was a *regressive intensification* of US security policy, a strategic response to the perception of an unprecedented crisis for US hegemony.

Several reports commissioned by then Vice-President Dick Cheney in 2000 and 2001 before 9/11, as well as documents from Cheney's Energy Task Force, reveal that one of the geopolitical crises preoccupying US policymakers at that time was the imminence of a major global oil supply crunch, linked to a danger of world oil production reaching a permanent plateau. Potential unrest across the Middle East, including rising instability in key client states, was flagged up, as were tensions with core powers such as Russia and China over contested energy reserves and transport routes. Iraq was highlighted as a problem due to its volatile behaviour and unwillingness to sell its oil to the US on any terms whatsoever – having signed contracts instead with US geopolitical rivals Russia, China and Europe. Meanwhile, US oil industry executives were integrally involved and often had direct input in US military planning, heavily mapping Iraq's oil fields and infrastructure and discussing their future.[11]

By the dawn of the twenty-first century, the vast majority of the world's productive and financial resources were owned and controlled by a minority of Northern, primarily American, state and financial institutions.[12] But given the dependence of the US-dominated world economy on geopolitical control of oil reserves, the perception of an impending energy crisis signalled the danger of a monumental decline in American power, of which rivals might indeed take advantage.

American and British defence planning documents reviewed below prove clearly that the trajectory of the 'War on Terror' has evolved in response to the convergence of global energy, ecological and economic crises, fixing the focus of geopolitical competition on not only the world's remaining energy reserves in the Middle East, Central Asia, Northwest Africa, South Asia and Latin America, but also on key potential transshipment routes for the transport of energy to the US, Europe, Russia, and China. In the post-9/11 period, the US – with UK support – has sought not merely to establish occupying forces in key regions by which energy reserves can be physically controlled, but also to dominate these transshipment routes precisely to prevent major rivals from challenging its energy hegemony. Russia and China have in response accelerated efforts to outmanoeuvre the US, attempting to force US troops out of key Central Asian republics and secure control over critical pipeline routes and regional energy reserves.[13]

A February 2003 report of the UK Department of Trade and Industry notes: 'our energy supplies will increasingly depend on imported gas and oil from Europe and beyond'. By 2010, the report states, 'we are likely to be importing around three quarters of our primary energy needs. And by that time half the world's gas and oil will be coming from countries that are currently perceived as relatively unstable, either in political or economic terms'. As a result, Britain has moved from being 'self-sufficient to being a net importer of gas and oil', a transformation which 'requires us to take a longer term strategic international approach to energy reliability' as well as 'strategic energy issues in foreign policy'. The solution is diversification of energy supplies, meaning the extension of strategic influence to Russia, the Middle East, North and West Africa, and the Caspian basin.[14]

Further government documents leaked to the *Guardian* later that year reveal that the strategy is to be pursued jointly with the US: 'We have identified a number of key oil and gas producers in the West Africa area on which our two governments and major oil and gas companies could cooperate to improve investment conditions... and thus underpin long term security of supply', reads a US report to President Bush and Prime Minister Blair. Both countries 'have noted the huge energy potential of Russia, Central Asia and the Caspian', and that 'we have similar political, economic, social and energy objectives'.[15]

In December 2003, a British Foreign Office report confirmed that the 'ability to project armed force will be a key instrument of our foreign policy', including 'early action to prevent conflict' – i.e. pre-emptive warfare. The document identified eight 'international strategic priorities', including the 'security of UK and global energy supplies'.[16] Around the same time, the Ministry of Defence produced a white paper outlining the new strategic direction of British military operations. Noting that 'military force exists to serve political or strategic ends', the MoD report observes that 'UK policy aims' derive from the fact that 'the UK has a range of global interests including economic well-being based around trade, overseas and foreign investment and the continuing free flow of natural resources'. British forces now require 'the capability to deliver a military response globally', including 'expeditionary operations' and 'rapidly deployable forces' to be used in 'a range of environments across the world'. The report emphasizes that 'our armed forces will need to be interoperable with US command and control structures'.[17]

In other words, the US and Britain favour military intervention as a primary instrument for the establishment of energy security. Whereas economic competition in world energy markets is a game which can be lost to other rising rivals such as China or India, military intervention potentially guarantees direct control of the world's increasingly scarce hydrocarbon resources. As early as 1999, then Chairman of Halliburton and soon-to-be US Vice-President Dick Cheney spoke about the implications of peak oil at the launch of the London Institute of Petroleum. His comments reveal the extent to which US policymakers and the energy industry have seen peak oil as a defining factor in the underlying strategic objectives of Anglo-American foreign policy:

> From the standpoint of the oil industry obviously – and I'll talk a little later on about gas – for over a hundred years we as an industry have had to deal with the pesky problem that once you find oil and pump it out of the ground you've got to turn around and find more or go out of business. *Producing oil is obviously a self-depleting activity.* Every year you've got to find and develop reserves equal to your output just to stand still, just to stay even. This is as true for companies as well in the broader economic sense it is for the world. A new merged company like Exxon-Mobil will have to secure over a billion and a half barrels of new oil equivalent reserves every year just to replace existing production. It's like making one hundred per cent interest; discovering another major field of some five hundred million barrels equivalent every four months or finding two Hibernias[18] a year. For the world as a whole, oil companies are expected to keep finding and developing enough oil to offset our seventy one million plus barrel a day of oil depletion, but also to meet new demand. By some estimates there will be *an average of two per cent annual growth in global oil demand over the years ahead along with conservatively a three per cent natural decline in production from existing reserves. That means by 2010 we will need on the order of an additional fifty million barrels a day. So where is the oil going to come from?* Governments and the national oil companies are obviously in control of about ninety per cent of the assets. Oil remains fundamentally a government business. While many regions of the world offer great oil opportunities, *the Middle East with two thirds of the world's oil and the lowest cost, is still where the prize ultimately lies*, even though companies are anxious for greater access there, progress continues to be slow... Oil is unique in that it is so strategic in nature. We are not talking about soapflakes or leisurewear here. *Energy is truly fundamental to the world's economy. The Gulf War was a reflection of that reality.*[19]

The magnitude of the problem identified by Cheney can be grasped when one considers that the projected additional 50 million barrels of oil a day (mbd) required in 2010 is more than double what was collectively produced in 2001 by the six major Middle Eastern suppliers bordering the Persian Gulf, namely Saudi Arabia, Iran, Iraq, United Arab Emirates, Kuwait and Qatar (i.e. 22.4 mbd). His reference to the Gulf War illustrates the symbiotic connection between military intervention and access to strategic resources.

This thinking was updated in a 2001 study commissioned by Cheney as US Vice-President, jointly conducted by the Council on Foreign Relations and the James Baker Institute for Public Policy, which noted the 'centrality' of energy policy to 'America's domestic economy and to our nation's security':

> The world is currently precariously close to utilizing all of its available global oil production capacity, raising the chances of an oil-supply crisis with more substantial consequences than seen in three decades. These limits mean that America can no longer assume that oil-producing states will provide more oil... [T]he situation is worse than the oil shocks of the past because in the present energy situation, the tight oil market condition is coupled with shortages of natural gas in the United States, heating fuels for the winter, and electricity supplies in certain localities... with spare capacity scarce and Middle East tensions high, chances are greater than at any point in the last two decades of an oil supply disruption that would even more severely test the nation's security and prosperity.

The impending crisis is increasing 'US and global vulnerability to disruption' and now leaves the US facing 'unprecedented energy price volatility', already leading to electricity blackouts in areas like California. The report warns of 'more Californias' ahead. The 'central dilemma' for the Bush administration was that 'the American people continue to demand plentiful and cheap energy without sacrifice or inconvenience'. But if the global demand for oil continues to rise, world shortages could reduce the status of the US to that of 'a poor developing country'. With the 'energy sector in critical condition, a crisis could erupt at any time [which] could have potentially enormous impact on the US... and would affect US national security and foreign policy in dramatic ways'. The growing energy crisis thus demands 'a reassessment of the role of energy in American foreign policy'. One of the key 'consequences' of the fact that 'the United States remains a prisoner of its energy dilemma' is the 'need for military intervention'. The report thus recommends that energy and security policy be integrated to prevent 'manipulations of markets by any state'. Iraq in particular was pinpointed as a prime threat to US energy security:

> Iraq remains a destabilizing influence to US allies in the Middle East, as well as to regional and global order, and to the flow of oil to international markets from the Middle East. Saddam Hussein has also demonstrated a willingness to threaten to use the oil weapon and to use his own export program to manipulate oil markets... The United States should conduct an immediate policy review toward Iraq, including military, energy, economic, and political/diplomatic assessments.[20]

In the post-invasion period, the Obama administration conceded a full US withdrawal from Iraq by 2011 under the new 'status of forces agreement'. But the agreement's wording carefully concealed the intent to continue the US military occupation. Shortly after he won the November 2008 elections, Barack Obama went out of his way to emphasize his commitment to maintaining a 'residual force' to fight 'terrorism', as well as to train and protect US civilians in Iraq. His security adviser Richard Danzig estimated that the future US force presence could amount to between 30,000 and 55,000 troops. Thus, the proposed withdrawal of troops represented an unofficial continuation of the occupation under a modified legal remit. Briefings by Pentagon officials clarified that this 'residual force' could remain in Iraq long after 2011. Worse, the new security agreement can be vetoed by either side, and the Iraqi government can invite US troops to remain so long as their presence is not officially defined as 'permanent'. Combat troops can also be 're-missioned' as 'support units'. The wording of the agreement prohibits Iraqi government jurisdiction over any crimes committed by US soldiers, unless they can be shown to be 'premeditated' and committed while off-duty.[21]

Finally, the agreement effectively allows the US to continue controlling billions of dollars of proceeds from the sale of exported Iraqi oil. It stipulates that 'the President of the United States has exercised his authority to protect from United States judicial process the Development Fund for Iraq [DFI]' and that the US will 'remain fully and actively engaged with the Government of Iraq with respect to continuation of such protections and with respect to such claims'. The agreement further clarifies that:

> [The US will remain committed to maintaining Iraq's] request that the UN Security Council extend the protections and other arrangements established in Resolution 1483 (2003) and Resolution 1546 (2003) for petroleum, petroleum products, and natural gas originating in Iraq, proceeds and obligations from sale thereof, and the Development Fund for Iraq [DFI].

Resolution 1483 stated that the DFI would be held at the Central Bank of Iraq, and that its funds would 'be disbursed at the discretion of the [Coalition Provisional] Authority' for purposes of Iraqi reconstruction and development. A later resolution (1446) noted that upon the Coalition Provisional Authority's dissolution, this discretion would fall to the Iraqi government. As *Foreign Policy Journal* editor Jeremy Hammond reports, to bypass these terms the US-controlled Coalition Provisional Authority established the DFI

> in an account at the Federal Reserve Bank of New York... the DFI was held on the books of the Central Bank of Iraq and a portion of the fund located in Baghdad. But the US nevertheless remained in control of the money and held most of it in New York.

An executive order issued by President Bush in March 2003 directed 'the transfer of funds controlled by the Iraqi government and its financial and oil institutions to the US Treasury', after which the Federal Reserve Bank created a 'Special Purpose Account' for the funds on the Treasury's behalf. Since then, billions of dollars of funds have gone missing. The problem is that Iraq's approval of the status of forces agreement would 'effectively acquiesce to continued control over these proceeds from the export of Iraqi oil by the US, with merely... a veil of Iraqi control'.[22]

Since then, the world's oil giants – ExxonMobil, Shell, Total, BP, and Chevron – along with several smaller companies, have been 'in talks with Iraq's Oil Ministry for no-bids contracts to service Iraq's largest fields'.

Contracts were drawn up in 2008 by a group of US advisers led by a State Department team.[23] Early on, large oil fields went to American and British companies. Development rights for the huge West Qurna oilfield in southern Iraq were awarded to ExxonMobil and Shell, on an 80 and 20 per cent basis respectively, for the next 20 years. British Petroleum (BP) in partnership with China National Petroleum Corp obtained rights to the Rumaila field, on a 38 and 37 per cent basis respectively. However, overall, the oil companies of the US's major rivals – Russia, China, France, Italy, and even Korea and Malaysia – collectively won the vast majority of bids. The financial crisis had clearly put the US in a weak position, as its Euro-Asian rivals won due to sheer competitiveness – lower labour costs and superior technical know-how – allowing them to offer lower fees per barrel.[24]

In any case, despite the unexpected turn of events, the status of forces agreement had already put the US in a predominant position from a military, geopolitical and legal perspective regarding the 'protection' of Iraqi oil fields and control of oil export revenues. Industry analysts project that Iraq will be able to double existing oil production to about 4 million barrels a day, and perhaps as much as 7 million by 2016 – and even higher thereafter. This may well ameliorate the impact of peak oil by temporarily and partially alleviating exorbitant price rises.[25] A secret February 2003 State Department report, *Moving the Iraqi Economy from Recovery to Sustainable Growth*, established a blueprint for 'a radical makeover of Iraq as a free-market Xanadu including, on page 73, the sell-off of the nation's crown jewels: "privatization... [of] the oil and supporting industries"'. As reported by Greg Palast, the idea – favoured by the neoconservatives – was that

> if Iraq's fields were broken up and sold off, competing operators would crank up production. This extra crude would flood world petroleum markets, OPEC would devolve into mass cheating and overproduction, oil prices would fall over a cliff, and Saudi Arabia, both economically and politically, would fall to its knees.

A second 323-page State Department report in November of that year proposed an alternative scenario favoured by US oil lobbies, in which a state-owned oil company would give the Iraqi government official title to reserves, while granting 'operation and control' to foreign companies – with special favours for US oil corporations.[26] The actual Obama strategy has combined options, most likely in response to economic recession. While maintaining formal Iraqi state title to the reserves – albeit under US 'protection' of both production and profits – the auctioning of development rights to a variety of non-US companies has stoked fears of a shake-up of OPEC and the Persian Gulf oil hierarchy, as Iraqi production capacity might increase to as much as 12 million barrels a day. 'Iraq's emergence as a rival to oil giant Saudi Arabia is causing concern in OPEC and the countries of the Gulf region', reported the German broadcaster Deutsche Welle. 'Some believe a newly oil-rich Iraq could disrupt OPEC's dominance and destabilize the Middle East'.[27] Saudi Arabia and Iran, in particular, could be undermined with lower oil prices denuding them of their profit revenues, weakening their regional clout, and paving the way for the remaking of the Middle East.

This illustrates that the objective was never simply to maximize oil company profits, but rather to rehabilitate the entire Persian Gulf into the global oil supply-system with a view to stabilize US hegemony. This would delay potential tensions with major rivals over energy conflict for some years, undermine Gulf state OPEC hegemony, and flatten oil prices to assuage pressure on the world economy. This link between Middle East interventionism and the global energy crisis was reiterated by Brigadier-General James Ellery CBE, the Foreign Office's Senior Adviser to the Coalition Provisional Authority in Baghdad from 2003 to 2005, who confirmed that Iraqi oil reserves are to play a critical role in alleviating a 'world shortage' of conventional oil. The Iraq War has helped to head off what Brigadier-General Ellery described as 'the tide of Easternisation' – a shift in global political and economic power toward China and India, to whom goes 'two thirds of the Middle East's oil'. 'The reason that oil reached $117 a barrel last week', he said in April 2008, 'was less to do with security of supply... than World shortage'. He went on to emphasize the strategic significance of Iraqi petroleum fields in relation to the danger of production peaks being reached by major oil reserves around the world: 'Russia's production has peaked at 10 million barrels per day; Africa has proved slow to yield affordable extra supplies – from Sudan and Angola for example. Thus the only near-term potential increase will be from Iraq.' Whether Iraq began 'favouring East or West' could therefore be 'destabilizing' not only 'within the region but to nations far beyond which have an interest'. Ellery, now on the Board of Directors of Aegis, one of the largest US defence contractors in Iraq, also pointed to the destabilizing influence of Iran, not only for US plans in Iraq, but for the Middle East more generally.[28]

As conditions in Iraq have improved, Iran has increasingly moved to the centre of unfolding geopolitical rivalry to control oil and gas reserves as well as transshipment routes in the Middle East and Central Asia. As noted by Michael Klare, Five College Professor of Peace and World Security at Hampshire College:

> With a long coastline on the Gulf and a large and growing naval capability, Iran is viewed in Washington as the 'threat after next' – the nation that is most likely to oppose American oil interests once the risk of an Iraqi invasion has been reduced to marginality.[29]

The rationale behind this was elaborated by General James Henry Binford Peay, who explained that:

> [Iran] has a population of over 60 million people, large numbers of highly educated engineers and technicians, abundant mineral deposits, and vast oil and gas reserves. With such resources, Iran retains the means, over

the long-term, to potentially overcome its current economic malaise and endanger other Gulf states and US interests.[30]

Iran is OPEC's second largest oil producer and holds approximately 10 per cent of the world total proven oil reserves – new discoveries in June 2004 put the total at 132 billion barrels. Iran also has approximately 940 trillion cubic feet (Tcf) in proven natural gas reserves – 16 per cent of total world reserves and the second largest after Russia's. Sixty-two per cent of these reserves are located in non-associated fields and have not yet been exploited, implying huge potential for gas development. This puts Iran's combined supply of hydrocarbon energy at equivalent to some 280 billion barrels of oil, just behind Saudi Arabia's equivalent.[31] And as former Bush administration energy adviser Matthew Simmons has documented, though it possesses the world's largest oil reserves Saudi Arabia has most likely already peaked, and production is unlikely to increase significantly any further in the near future.[32]

The critical factor in this equation is not simply quantity of reserves, but the potential for future productive capacity, as Klare points out. With a giant like Saudi Arabia unable to raise output sufficiently to meet swelling global demand due to higher consumption, particularly in the United States, China, and India – expected to rise by 50 per cent over coming decades – Iran still retains 'considerable growth potential'. Currently producing about 4 million barrels per day, Iran is believed 'to be capable of boosting its output by another 3 million barrels or so. Few, if any, other countries possess this potential, so Iran's importance as a producer, already significant, is bound to grow in the years ahead'. The situation is even more promising for gas, of which Iran is presently producing only about 2.7 trillion cubic feet per year:

> This means that Iran is one of the few countries capable of supplying much larger amounts of natural gas in the future. What all this means is that Iran will play a critical role in the world's future energy equation. This is especially true because the global demand for natural gas is growing faster than that for any other source of energy, including oil. While the world currently consumes more oil than gas, the supply of petroleum is expected to contract in the not-too-distant future as global production approaches its peak sustainable level – perhaps as soon as 2010 – and then begins a gradual but irreversible decline. The production of natural gas, on the other hand, is not likely to peak until several decades from now, and so is expected to take up much of the slack when oil supplies become less abundant.[33]

Under Executive Order 12959, US companies are prohibited from trading with Iran, or collaborating in energy exploration and production. Under the Iran-Libya Sanctions Act of 1996, the US government has also threatened to use sanctions to punish foreign firms and nations that trade with Iran. Yet, as the growing global energy crisis has heightened international competition over control of increasingly costly energy resources, countries have flouted US warnings and closed deals with Iran, demonstrating the increasing inability of the US to regulate the behaviour of its most powerful rivals.

China, for example, signed a $70 billion contract with Iran to purchase its oil and gas. Under the deal, China would buy 250 million tonnes of liquefied natural gas over 30 years, and 150,000 barrels per day of crude oil for 25 years at market prices, after commissioning of the Yadavaran field. Already in 2003 China had imported approximately 30 million tonnes of oil from Iran, 14 per cent of its total oil imports.[34]

India and Pakistan have also negotiated with Iran for access to its oil and gas. In January 2005, the Gas Authority of India Ltd (GAIL) signed a 30-year, $50 billion deal with the National Iranian Gas Export Corporation for the transfer of 7.5 million tonnes of liquid natural gas to India per year. In return, India would assist in the development of Iran's gas fields. Furthermore, Indian and Pakistani officials discussed the construction of a $3 billion natural gas pipeline from Iran to India via Pakistan, supplying both with substantial quantities of gas, while granting Pakistan $200–$500 million per year in transit fees. The plan was condemned by then Secretary of State Condoleezza Rice while in India: 'We have communicated to the Indian government our concerns about the gas pipeline cooperation between Iran and India.' At the time, those concerns were ignored. The Iran-India-Pakistan pipeline deal was originally slated for conclusion sometime in 2008, perhaps as early as mid year.[35] If concluded successfully, the negotiations could have heralded an unprecedented era of international cooperation between previously antagonistic states situated along the heart of Eurasia, eliminating incentives for hostility between India and Pakistan, and facilitating new diplomatic and financial ties. Given that both India and Pakistan are nuclear states, and that Iran is pursuing nuclear power, the consolidation of this agreement would have facilitated the emergence of a major Central/South Asian alliance fundamentally challenging Anglo-American ambitions for control of regional resources. The prospects were highlighted in Pakistani President Asif Zardari's optimistic proposals for close India-Pakistan economic and political cooperation as a means to diffuse tensions over Jammu and Kashmir.[36] However, the terrorist attacks in Mumbai in late 2008 permanently sabotaged these negotiations. Although by summer 2008 a timetable had been agreed to see Iranian gas pumping into the Pakistani and Indian markets by 2009/10, the attacks on the Indian financial capital – blamed on Pakistani militants – seemingly confirmed the chronic insecurity problem that would make the pipeline ultimately unviable.[37]

Others have also moved closer to Iran. In early 2003, a Japanese energy consortium purchased a 20 per cent stake in the development of the Soroush-Nowruz offshore field in the Persian Gulf, a reservoir estimated at 1 billion barrels of oil. By 2004, the Iranian Offshore Oil Company awarded a $1.26 billion contract to Japan's JGC Corporation for the recovery of natural gas and natural gas liquids from Soroush-Nowruz and other offshore fields.[38] In the same year, Japan signed a $2 billion deal to exploit Iran's Azadegan oil fields, which has estimated reserves of 26 billion barrels. A State Department spokesman, Richard Boucher, said he was 'disappointed' by the deal.[39]

Even Europe has jumped into the fray. In May 2008, Switzerland signed a 25-year, $22 billion contract with the National Iranian Gas Export Company to deliver 5.5 billion cubic meters of gas annually to Europe through a pipeline scheduled for completion in 2010.[40] According to the *Financial Times*:

> 'The worry is that the Swiss deal will lead others, such as the Austrians, to confirm energy investments in Iran, and that companies like [France's] Total could then follow suit and sign contracts of their own,' said one western diplomat. He pointed out that the EGL agreement ended a period in which European energy companies had largely confined themselves to agreeing only non-binding memoranda of understanding with Iran.[41]

Iran has thus been a central player in a more system-wide shift threatening to undermine US hegemonic control of Middle East and Central Asian energy reserves. While the Bush strategy was particularly direct in its approach to militarization as a mechanism to counter this process, Obama has emphasized the effectiveness of covert operations. These techniques are being extended principally to encircle Iran, debilitate Pakistan, and subordinate India; as well as to outmanoeuvre China, Russia and Europe in competition for Central Asian resources. US policymakers, for instance, exploited the Mumbai attacks to push through a nuclear deal with India; expand the US military presence in both India and Pakistan on the pretext of security; and especially extend and intensify military operations in Pakistan – the net result being the deepening of regional geopolitical divisions, and thus the increasing vulnerability of the region to US influence. In Central Asia, the US has pushed to establish new military bases in Kazakhstan and Uzbekistan; and it signed strategic accords with Ukraine and Georgia, attempting to severely delimit the scope of Russian and Chinese regional influence.[42] Under Obama, such policies will be continued with the overarching objective of destabilizing these strategic regions to prevent the emergence of regional alliances and subordinate regional powers to US influence.

Conquest through Covert Destabilization: Middle East, Central Asia, and Africa

The post-9/11 strategy of destabilization by fostering sectarian conflict and arming breakaway ethnic groups, beginning in Iraq and extending to the greater Middle East, was reconfirmed rather candidly by retired Lt. Col. Ralph Peters, formerly assigned to the Office of the US Army's Deputy Chief of Staff for Intelligence with responsibility for future warfare. In June 2006, Lt. Col. Peters published a detailed paper in the *Armed Forces Journal*, which describes itself as 'the leading joint service monthly magazine for officers and leaders in the United States military community'. The paper outlines a detailed map for how Middle East, Central Asian and Northwest African national borders should be fundamentally redrawn:

> The most arbitrary and distorted borders in the world are in Africa and the Middle East... without such major boundary revisions, we shall never see a more peaceful Middle East...
>
> *Accepting that international statecraft has never developed effective tools – short of war – for readjusting faulty borders*, a mental effort to grasp the Middle East's 'organic' frontiers nonetheless helps us understand the extent of the difficulties we face and will continue to face...
>
> Borders have never been static, and many frontiers, from Congo through Kosovo to the Caucasus, are changing even now... Oh, and one other dirty little secret from 5,000 years of history: Ethnic cleansing works...[43]

Thus, acknowledging that the sweeping reconfiguration of borders he proposes would necessarily involve war and ethnic cleansing, he insists that unless it is implemented, 'we may take it as an article of faith that a portion of the bloodshed in the region will continue to be our own'. Among his proposals is the need to establish 'an independent Kurdish state' to guarantee the long-denied right to Kurdish self-determination. But behind the humanitarian sentiment, Lt. Col. Peters declares: 'A Free Kurdistan, stretching from Diyarbakir through Tabriz, would be the most pro-Western state between Bulgaria and Japan.'[44]

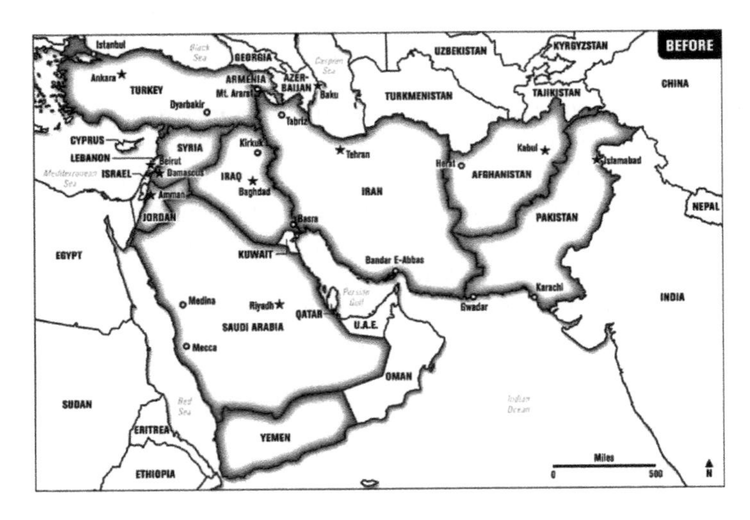

Figure 6.1 Conventional Map of the Middle East.

Source: Armed Forces Journal (2006)

He chastises the United States and its coalition partners for missing 'a glorious chance' to fracture Iraq, which 'should have been divided into three smaller states immediately'. This would have left 'Iraq's three Sunni-majority provinces as a truncated state that might eventually choose to unify with a Syria that loses its littoral to a Mediterranean-oriented Greater Lebanon:

Phoenecia reborn'. Meanwhile, the Shia south of old Iraq 'would form the basis of an Arab Shia State rimming much of the Persian Gulf'. US strategy in establishing a separate Arab Shia state within Iraq is to create 'a state more likely to evolve as a counterbalance to, rather than an ally of, Persian Iran'.[45]

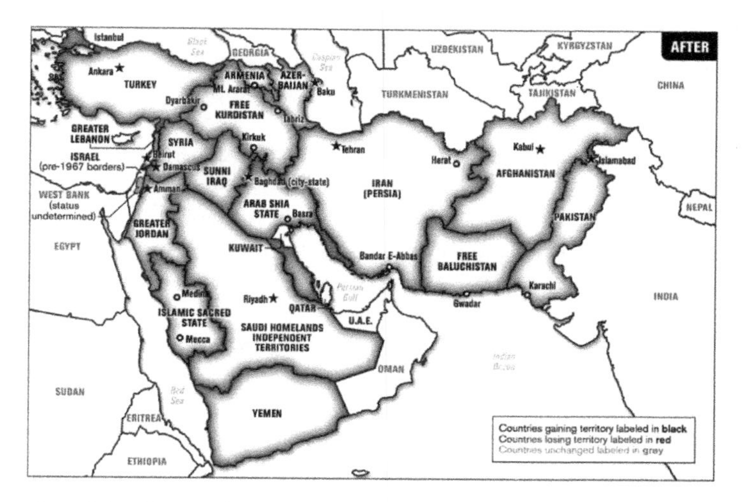

Figure 6.2 US Army Plan for Redrawing Middle East and Central Asian Borders.

Source: Armed Forces Journal (2006). Available from http://proteus.brown.edu/ancientneareast/9089

Jordan, a US-Israeli friend in the region, would 'retain its current territory, with some southward expansion at Saudi expense. For its part, the unnatural state of Saudi Arabia would suffer as great a dismantling as Pakistan'. Saudi Arabia's 'coastal oil fields' would be reassigned 'to the Shia Arabs who populate that subregion', and Saudi's 'southeastern quadrant would go to Yemen'. The House of Saud would then be cut down to size and confined to a rump Saudi Homelands Independent Territory around Riyadh'. Iran too would 'lose a great deal of territory to Unified Azerbaijan, Free Kurdistan, the Arab Shia State and Free Baluchistan, but would gain the provinces around Herat in today's Afghanistan'.[46]

Some parts of western Afghanistan would return to Persia, but overall its eastern territories would be extended 'as Pakistan's Northwest Frontier tribes would be reunited with their Afghan brethren'. Peters then turns to Pakistan, which he describes as 'another unnatural state', which will 'lose its Baluch territory to Free Baluchistan', leaving the remaining Pakistani territory to 'lie entirely east of the Indus, except for a westward spur near Karachi'. The rest of the Middle East would have a 'mixed fate'. Some of the 'city-states' of the United Arab Emirates might 'be incorporated in the Arab Shia State ringing much of the Persian Gulf'. However, Dubai, Kuwait and Oman would 'remain within' their 'current borders', permitted to retain their 'playground status for rich debauchees'.[47]

Although this vast imperial programme might be difficult to implement now, with time 'new and natural borders will emerge', driven by 'the inevitable attendant bloodshed'. As for the goals of this plan, Lt. Col. Peters is equally candid. While including the necessary caveats about fighting 'for security from terrorism, for the prospect of democracy', he also mentions the third all-important issue – 'and *for access to oil supplies* in a region that is *destined to fight itself*'. In summary, the Middle East represents a 'worsening, not an improving situation', characterized by an 'oversupply of terrorists' and a growing 'paucity of energy supplies'.[48]

The Peters plan is reminiscent of a grand strategy for the Balkanization of the Middle East set out in 1982 by Israeli Foreign Ministry official Oded Yinon, which advocated that Arab states should be broken down into smaller ethnically-divided units, including the 'dissolution of Iraq into a Shi'ite state, a Sunni state and the separation of the Kurdish part'.[49] The Israeli strategy was integral to the formulation of neoconservative foreign policy doctrines as eventually adopted by the Bush (II) administration. In 1996, five years before they took up posts in George W Bush's White House, David Wurmser, later Dick Cheney's Middle East adviser; Douglas Feith, Bush's Undersecretary of Defense for Policy; and Richard Perle, who under Bush became adviser to the Pentagon's Defense Policy Board, co-authored a report for then Israeli Prime Minister Benjamin Netanyahu (who re-took this position in 2009). The report called for a plan to 'contain, destabilize, and roll-back' Israel's rivals. Among its recommendations were 'removing Saddam Hussein from power in Iraq' along with 'striking Syrian military targets in Lebanon, and should that prove insufficient, striking at select targets in Syria proper'.[50]

As remarked by former CIA official Robert Baer, who worked in the CIA's Directorate of Operations for over two decades, spending most of his career in the Middle East:

> [The Bush administration wanted] to divide up Syria, give part of Iraq to Turkey, overthrow the monarchy in Saudi Arabia, restore the Hashemites to the Hijaz... The underlying motivation... is Israel. They think the demographics are going badly for Israel, for the US.[51]

Other veteran CIA officials confirm that this programme of regional destabilization is not just about Israel, but simultaneously about energy. According to former CIA political analyst Kathleen Christison and former Director of the CIA's Office of Regional and Political Analysis Bill Christison:

> [T]wo strains of Jewish and Christian fundamentalism have dovetailed into an agenda for a vast imperial project to restructure the Middle East, all further reinforced by the happy coincidence of great oil resources up for grabs and a president and vice president heavily invested in oil. All of these factors – the dual loyalties of an extensive network of policymakers allied with Israel, the influence of a fanatical wing of Christian fundamentalists,

and oil – probably factor in more or less equally to the administration's calculations on the Palestinian-Israeli situation and on war with Iraq.[52]

There is a crucial distinction, though, between current plans and those articulated in the 1980s. The Peters plan advocates that Israel return to its 1967 borders in exchange for the redivision of the region under American auspices, including the dismantling of key regimes constituting potential threats, especially Iran, Lebanon and Syria, and even Saudi Arabia and Pakistan. This indicates the potential for both significant strategic convergence, as well as *underlying tensions*, regarding regional plans between the US and Israel. Clearly, the Balkanization strategy outlined here is not just a neoconservative policy. The neoconservative foreign policy vision called for a much more concerted form of direct military intervention and long-term occupation throughout the Middle East.[53] The Peters plan, in contrast, envisages a permanent US role in the region, but one that uses the occupation of Iraq and geopolitical pressure on states like Saudi Arabia and Pakistan as strategic pivots by which to extend an arc of destabilization using multiple armed groups and ethnic factions, including al-Qaeda affiliates. This instability would allow the US to more easily manipulate a wider diversity of territorially debilitated, politically disunited, and militarily weakened entities in order to dominate regional resources. The important aspect of this plan is not the consolidation of new borders as such, but the heightened capacity for politico-economic manipulation and penetration in the context of engineered discord. It is likely that actual policy will evolve according to developing geopolitical conditions.

Hence, in late December 2008, signifying its contempt for the idea of a withdrawal to its 1967 borders, Israel launched a major attack on Gaza, purportedly in response to Hamas rocket attacks along the border. The operation was the beginning of a well-planned strategy to create a security pretext for Israel to reinsert itself into the Occupied Territories. Having agreed to disengage from Gaza in 2005, Israel began war preparations by blockading Gaza off from the entire world. The siege deprived Gaza, a discontinuous and divided strip of land carrying a population of 1.5 million Palestinians, of electricity, fuel, food imports, medical supplies, and vital maintenance goods and spare parts. As water and sanitation services deteriorated, hunger and ill health intensified, and mortality rates increased, international aid agencies warned of a major public health crisis.[54] On 19 September 2007, Israel's security cabinet unanimously declared the entire Gaza Strip an 'enemy entity' due to ongoing Hamas rocket fire. Yet that rocket fire was and is a response to continued indiscriminate Israeli military bombardments. In January 2007, Israeli Defence Forces (IDF) staged three days of air strikes, killing 30 Palestinians, and on 17 January the Gaza Strip was placed under total closure. In response, over 150 rockets and mortars were fired by Hamas into Israel between 15 and 18 January. Yet while these caused *four* fatalities to Israelis, in that same period nearly *700 Palestinians* (including 224 civilians, of whom 78 were children) were killed by Israeli extra-judicial executions.[55]

Richard Falk, the UN Special Rapporteur for Human Rights in the Occupied Territories, observed:

> Is it an irresponsible overstatement to associate the treatment of Palestinians with this criminalized Nazi record of collective atrocity? I think not. The recent developments in Gaza are especially disturbing because they express so vividly a deliberate intention on the part of Israel and its allies to subject an entire human community to life-endangering conditions of utmost cruelty. The suggestion that this pattern of conduct is a holocaust-in-the-making represents a rather desperate appeal to the governments of the world and to international public opinion to act urgently to prevent these current genocidal tendencies from culminating in a collective tragedy... But it would be unrealistic to expect the UN to do anything in the face of this crisis, given the pattern of US support for Israel and taking into account the extent to which European governments have lent their weight to recent illicit efforts to crush Hamas as a Palestinian political force.[56]

The siege was a strategy to prepare the ground for a protracted military operation, known as 'Cast Lead'. Although justified on the grounds of stopping Hamas rocket fire, the operation was planned over six months earlier at the end of 2008.[57] Hints that the scope of the operation would be far broader than hitherto admitted came when Deputy Defence Minister Matan Vilnai told Israeli Army Radio that the Palestinians would 'bring upon themselves a bigger holocaust (shoah) because we will use all our might to defend ourselves'. By the late January 2009 ceasefire, the Israeli operation had killed 1300 Palestinians and injured 5000.[58]

Increasingly, the Israel–Palestine conflict has become radicalized in the context of the discovery about a decade prior by the British firm BG International of substantial natural gas reserves just off the Gaza coast. The reserves contain 1.2 trillion cubic feet of gas valued at just over $4 billion, unlike Israel's indigenous gas fields north of the Gaza marine field which are projected to be depleted in a few years. The British Foreign Office described the reserves as 'by far the most valuable Palestinian natural resource'. Tel Aviv journalist Arthur Neslen cites an informed British source saying, 'The UK and US, who are the major players in this deal, see it as a possible tool to improve relations between the PA [Palestinian Authority] and Israel. It is part of the bargaining baggage.' The project could provide up to 10 per cent of Israel's energy needs, at around half the price the same gas would cost if purchased from Egypt. The Gaza Strip would be effectively circumvented, as the gas would be piped directly ashore to Ashkelon in Israel. Neslen reports another source noting 'an obvious linkage' between the BG-Israel deal and 'attempts to bolster the Olmert-Abbas political process'. Yet this process was designed precisely to marginalize the Palestinian people, as Neslen reports that 'up to three-quarters of the $4bn of revenue raised might not even end up in Palestinian hands at all'. The 'preferred option' of the US and UK is that the gas revenues would be held in 'an international bank account over which

[PA President Mahmoud] Abbas would hold sway'. No wonder then, that Ziad Thatha, the Hamas economic minister, had denounced the deal as 'an act of theft' that 'sells Palestinian gas to the Zionist occupation'. But before any deal could be finalized, Hamas won the 2006 elections to the Palestinian Legislative Council, provoking a bitter power struggle between Hamas and the pro-West Fatah, fuelled by the supply of US and Israeli arms to the latter. This culminated in the Palestinian Authority split in 2007, with Hamas taking control of Gaza and Fatah taking control of the West Bank. Having been excluded from the US-UK brokered gas deal between Israel and the PA, one of the first things Hamas did after being elected was to declare the natural gas deal void and in need of renegotiation. There is therefore no doubt that the IDF operation in late December 2008 and January 2009 was the first effort to head off imminent Palestinian democratic elections that threatened to consolidate Hamas's power and legitimacy and undermine the Israeli gas deal.[59] Coming years will see new fault lines of regional resource conflict emerging. The further discovery in early 2009 by Israeli and US companies of another natural gas deposit off the Haifa coast, of over 3 trillion cubic feet and valued at up to $15 million, has been contested by Lebanon, which says part of the field may be in Lebanese territorial waters.[60]

The inauguration of the Obama administration in early 2009 thus saw the acceleration of Anglo-American military engagements in multiple regions: the successful subordination of the Iraqi state to US demands for an ongoing military presence, guaranteeing privileged domination of Iraqi oil reserves; tacit support for expanded Israeli offensives in the Occupied Territories (certainly in the supply of military and financial aid by the US and UK); the debilitation of Pakistan and subordination of India; the commitment to an expanded military occupation of Afghanistan; and the entrenchment of a network of military-financial alliances with major Central Asian republics. The cumulative goal of these efforts is a process of encircling and marginalizing the regional influence of America's major geopolitical competitors – such as China, Russia, Europe and Iran – in order to dominate strategic peripheries. Under Obama's presidency, resource control will increasingly be a defining feature of such geopolitical contestations.

Obama's National Security Advisor is US Marine General James Jones – previously NATO's Supreme Allied Commander Europe (SACEUR) and Commander of the US European Command (COMUSEUCOM) under the Bush administration. Toward the eve of the Bush administration's departure from office, on 1 October 2008, the US Army officially launched its Africa Command (AFRICOM) – a massive military structure whose area of operations covers fifty-three nations – whose chief architect was none other than Gen. Jones himself. At the heart of AFRICOM's rationale is energy security. In February 2006, Jones announced that 'the Pentagon was seeking to acquire access to two kinds of bases in Senegal, Ghana, Mali and Kenya and other African countries'. The new US strategy was reportedly 'based on the conclusions of a May 2001 report of the President's National Energy Policy Development group chaired by Vice-President Richard Cheney and known as the Cheney report'. Indeed, Jones's interests in energy extended far beyond his NATO position. Through 2008, he was President and CEO of the US Chamber of Commerce's Institute for 21st Century Energy, and a board member of Chevron. After his formal selection as nominee for National Security Advisor in November 2008, Jones himself stated that 'as commander of NATO, I worried early in the mornings about how to protect energy facilities and supply chain routes as far away as Africa, the Persian Gulf and Caspian Sea'. This was the remit of Jones's work under NATO since shortly after 9/11, when then Defense Secretary Donald Rumsfeld 'sent marching orders to Marine Gen. James L. Jones, telling him that the US European Command needed an overhaul to meet the unique challenges of the 21st century'. The first stages of his plan were executed in 2002, with NATO 'moving troops into Eastern Europe and setting up forward operating sites in Africa'. In his own words: 'Our strategic goal is to expand... to Eastern Europe and Africa... The United States is not unchallenged in its quest to gain influence in and access to Africa' – a less-than-subtle allusion to Chinese influence. He later described seeing a 'potential role for the alliance in protecting key shipping lanes such as those around the Black Sea and oil supply routes from Africa to Europe'.[61]

The thrust of Jones's vision for US force projection in Eastern Europe and Africa was reported in 2005 by UPI:

> NATO's top military commander is seeking an important new security role for private industry and business leaders as part of a new security strategy that will focus on the economic vulnerabilities of the 26-country alliance. Two immediate and priority projects for NATO officials to develop with private industry are to secure the pipelines bringing Russian oil and gas to Europe... to secure ports and merchant shipping, the alliance Supreme Commander, Gen. James Jones of the US Marine Corps said Wednesday... A further area of NATO interest to secure energy supplies could be the Gulf of Guinea off the West African coast, Jones noted. Oil companies were already spending more than a billion dollars a year on security in the region... pointing to the need for NATO and business to confer on the common security concern.[62]

In early 2006, Jones 'raised the prospect of NATO taking a role to counter piracy off the coast of the Horn of Africa and the Gulf of Guinea, especially when it threatens energy supply routes to Western nations'. NATO's plan was to 'safeguard oil and gas fields in the region', and 'to protect the routes transporting oil to Western Europe'. In summary, Jones's national security strategy for the Obama administration privileges US military control over regions containing substantial underexploited oil and natural gas reserves in Africa's Gulf of Guinea, the Black and Caspian Seas, and the Persian Gulf, as well as the Arctic Circle and the northern parts of South America and the Caribbean. The US will also consolidate Europe's dependence on NATO for its energy security, thus solidifying EU support of the US strategy to control global energy resources and transportation routes.[63]

This type of geopolitical jockeying for the control of energy reserves will proliferate across key regions that hold substantive quantities of petroleum. As oil-exporting countries in the Middle East, Central Asia, Central and North Africa, and South America continue to peak in production, leading to exponential declines in their exporting capacity due to the demands of internal consumption, outbreaks of increasingly violent international conflict will become more frequent, as states seek to guarantee their energy security through military means. Already, as of 2005, the world's top five net oil exporters – Saudi Arabia, Russia, Norway, Iran and the UAE – together accounting for about half of world net oil exports, have experienced a collective decline in output.[64] Major military contestations, be they direct or by proxy, may well break out within the next decade, depending on the dates and regional significance of future oil production peaks. The US, UK, EU, Russia and China will continue to be principal players in these efforts to sustain privileged access to strategic areas – and the US and UK in particular are actively developing military doctrines designed to maintain and increase their force presence in these positions.

The general trend in new security thinking about the future of interstate conflict is evident from two military planning documents, American and British respectively. A US Department of Defense and Department of the Army report, *2008 Army Modernization Strategy*, sets out the future of international conflict up to 2050. Early on, the report states that:

> We have entered an era of persistent conflict... a security environment much more ambiguous and unpredictable than that faced during the Cold War... We face a potential return to traditional security threats posed by emerging near-peers as we compete globally for depleting natural resources and overseas markets.

This thinly veiled reference to increasing resource-competition with Russia and China is supplemented by a focus on the impact of global inequalities and radical ideologies of resistance. The report notes that the 'expanding, inter-connected global economy' will continue to 'drive prosperity' – for the North – exacerbating 'wealth and power disparities between populations'. This will fuel 'extremist ideologies and separatist movements'. The rise of radicalization will be further fuelled in the context of 'population growth – especially in less-developed countries – [which] will expose a resulting "youth bulge"'. This rapid growth of a younger demographic, particularly in the poorer South, will lead to 'anti-government and radical ideologies that potentially threaten government stability'. The report then highlights the danger of 'resource competition induced by growing populations and expanding economies', particularly in the South, which 'will consume ever increasing amounts of food, water and energy'. Even the danger of climate change is factored in, primarily in its capacity to 'compound' the destabilization of the South through humanitarian crises, population migrations and other complex emergencies. This is, of course, a direct reference to an escalating convergence of global crises and the constraints these will place on world production and consumption in the context of massive new demand from a larger world population. The rest of the document attempts to recommend strategies by which a downsized US military 'might ensure survival of the fittest for the US and its allies in future resource wars over water, food and energy', and states explicitly that the US military is preparing to fight continuous resource wars 'for the long haul'.[65]

A British Ministry of Defence report is in some ways even more explicit. Drawn up by the MoD's Development, Concepts and Doctrines Centre (DCDC), the report attempts to identify the 'key risks and shocks' that will define the 'future strategic context' facing the armed forces over the next 30 years. The major trends are as follows: By 2035, world population is likely to grow to 8.5 billion, with less developed countries accounting for 98 per cent of this growth. The report warns of a 'youth bulge', with some 87 per cent of people under the age of 25 inhabiting the South. In particular, the population of the Middle East will increase by 132 per cent, and of sub-Saharan Africa by 81 per cent – predominantly in Muslim regions. This will make these regions highly unstable: 'The expectations of growing numbers of young people [in these regions] many of whom will be confronted by the prospect of endemic unemployment... are unlikely to be met.'

This is likely to result in growing resentment among the rising numbers of young people in these regions toward their undemocratic regimes, which will be channelled through 'political militancy, including radical political Islam whose concept of Umma, the global Islamic community, and resistance to capitalism may lie uneasily in an international system based on nation-states and global market forces'. Yet the report articulates an exception to this prognosis. While noting that Iran will grow in economic and demographic strength due to the strategic leverage granted by its energy reserves and geographic location, the report sees no need for a protracted military intervention in that country, predicting that its government could be transformed: 'From the middle of the period, the country, especially its high proportion of younger people, will want to benefit from increased access to globalisation and diversity, and it may be that Iran progressively, but unevenly, transforms... into a vibrant democracy.' This draws attention to an ongoing disagreement between the US, Britain and Israel over the mechanisms used to discipline Iran. Elements of the British security establishment clearly favour covert methods to weaken the current Iranian government and increase the probability of its overthrow by a more compliant regime, but pressure from American and Israeli neoconservatives could still lead to an escalation that would culminate in direct conflict.

In most cases, however, the 'youth bulge' centred around the less developed countries, and in Muslim regions especially, will lead to the persistence not only of Islamist terrorism, but of a 'terrorist coalition' consisting of a wide range of ultra-nationalists, religious extremists and even militant environmentalists, who could together conduct a 'global campaign of greater intensity'. Potential recruits for this coalition might be drawn from the more than 60 per cent of the world population who are likely, by 2035, to live in urban rather than rural areas, in conditions of widespread social deprivation, increased homelessness and shanty towns, generating 'new instability risks'. In this context, the report rules out interstate conflict after the next two decades, emphasizing instead the increased prevalence of 'inter-communal conflict', much of which will occur across national boundaries along class, rather than purely ethnic, lines.

Indeed, the report warns: 'The middle classes could become a revolutionary class, taking the role envisaged for the proletariat by Marx.' This could occur on a transnational scale, due to an increasing global divide between a super-rich elite and the middle classes, as well as the rise of an urban underclass, in which case: 'The world's middle classes might unite, using access to knowledge, resources and skills to shape transnational processes in their own class interest.' The persistence and escalation of global inequality amidst prevailing moral relativism could lead people to seek the 'sanctuary provided by more rigid belief systems, including religious orthodoxy and doctrinaire political ideologies, such as popularism and Marxism'.[66]

American and British security agencies are preparing for the Long War, based on their anticipation of intensifying resource scarcities, economic inequalities and global warming in the coming decades. The imperative emphasized in the new security doctrines is to view the potentially catastrophic consequences of these converging global crises as largely inevitable, generating a whole new range of threats demanding expanded and creative state-led militarization strategies. The perceived outcome of these strategies is not systemic transformation, but *systemic stabilization* – that is, the preservation of extant socio-political and economic structures; the defence of deep transnational class divides; and the perpetuation of an inherently unequalizing dynamic – regardless of the social, environmental and human costs.

DEFENDING THE HOMELAND

Security Agencies, the Muslim Problem, and Population Politics

The imperial trajectory of the 'War on Terror' can therefore be explained as a strategic response to the convergence of global crises, designed to develop a plausible ideological-security framework by which to consolidate state power, establish domination over hydrocarbon energy reserves, and control domestic populations. This complex of crises, as we have seen, springs not only from resistance in the peripheries of the global imperial order, but also from the capitalist world system's own internal contradictions. The scale of this Crisis of Civilization may be so huge that it will threaten the viability and integrity of the entire system in the coming decades. It is in this context that Western governments and social institutions have responded militarily, socio-politically, economically, and indeed ideologically to this crisis through the 'War on Terror'. To reiterate, the ideological relations of the 'War on Terror' consist of the construction of bifurcated civilizational group identities on a global scale, in which an 'outside' uncivilized 'Other' is considered as a subhuman aggressor and labelled as the fundamental cause of Western insecurity.[67]

It is no accident that this tendency focuses overwhelmingly on the demonization of Islam both inside and outside the West. Within the US, Britain and Western Europe – and indeed even in Russia and China – Islam and Muslims are increasingly constructed and represented as a dangerous fifth column, inclined to a terrorism that aims to destroy Western civilization. Yet officially, most Western governments are at pains to clarify that the 'War on Terror' does not target Islam and Muslims in general, and to recognize that the latter are not supportive of terrorism.

Yet despite this official position, the actual framing of the threat of international terrorism as a specific form of Islamist extremism – by the core centres of social power in the West including government security agencies and the corporate media – has tended to generate the opposite perception of Islam and Muslims as a generalized security threat. In the early 1990s Willy Claes, then NATO Secretary-General, identified 'Islamic fundamentalism' as a new threat to Western Europe, replacing the defunct USSR.[68] By the late 1990s, a number of hearings had been held in Congress and the Senate on the Islamist threat from the Middle East and Central Asia.[69] The publication of Samuel Huntington's influential thesis on the clash of civilizations was a decisive turning point in the crystallization of this thinking.[70] And throughout this period, as noted in chapter five, Islamist extremist networks associated with al-Qaeda were being covertly mobilized by the security agencies of the very same governments now declaring 'Islamism' to be the next great threat.

After 9/11, security agencies increasingly characterized the threat of Islamist terrorism as being, despite its marginality, nevertheless *widely dispersed throughout Muslim communities*, necessitating comprehensive regimes of surveillance, policing, and in some regions counterinsurgency. A briefing paper published by the Pentagon agency Counterintelligence Field Activity (operational from 2002 to 2008, after which time its activities were subsumed by the Defense Intelligence Agency) argued that 'political Islam wages an ideological battle against the non-Islamic world at the tactical, operational and strategic level. The West's response is focused at the tactical and operation level, leaving the strategic level – Islam – unaddressed.' The paper concludes that 'Islam is an ideological engine of war (Jihad)', and 'no one is looking for its off switch' due to political 'indecision [over] whether Islam is radical or being radicalized'. Attempting to review the Qur'an and prophetic traditions, the paper infers that 'Strategic themes suggest Islam is radical by nature... Muhammad's behaviors today would be defined as radical.' Western policymakers can no longer afford to overlook the 'cult characteristics of Islam'. Indeed, even Islam's advocacy of charity – the principle known as Zakat and considered an obligatory 'pillar' of Islam – is described as 'an asymmetrical war-fighting funding mechanism'. The fact that the US has not suffered sporadic insurgent terrorist attacks – as opposed to the single catastrophic attack of 9/11 – is primarily due to its relatively small Muslim population. Accordingly, the threat of such insurgency will increase as the Muslim minority grows and gains more influence. The Pentagon cites attempted and successful terrorist attacks in Britain, along with the predominantly Muslim riots in France, as examples.[71]

Similarly, a leaked classified operational briefing note by MI5's Behavioural Science Unit on the phenomenon of Islamist radicalization in the UK concludes that the 'British terrorist' is 'demographically unremarkable'. Purportedly based on hundreds of case studies (although the number of terrorist convictions in the UK to date is about a dozen), terrorists are described as usually British nationals who do not practise their faith regularly and either lack religious expertise or are novices with regard to Islam, rather than being obvious fundamentalists. Few were raised in religious households, most are converts, and some even consume drugs and alcohol. Many have steady relationships, and among those most have children. Ultimately, the security service concludes that British Muslim terrorists are 'a diverse collection of individuals, fitting no single demographic profile, nor do they all follow a typical pathway to violent extremism'.[72] The implication is that Islamist extremists are deeply embedded amongst normal British Muslims, and cannot be easily marked out as different from moderates, bolstering the legitimacy of comprehensive surveillance programmes designed to proactively police Muslim communities as a whole.

This tendency to generalize the threat of Islamist extremism as amorphously entrenched in seemingly peaceful and moderate Muslim communities is increasingly common across Western Europe, where there is 'a significant shift in the balance between the logic of intelligence, based on suspicion, and the logic of the legal system, which requires proof. Suspicion is becoming more significant than established guilt.'[73] In this context Tim Savage, division chief at the State Department's Office of European Analysis, argues that Europe's Muslim population is expected to double while its non-Muslim population is projected to fall by at least 3.5 per cent. At worst, he speculates that by mid century Muslims might outnumber non-Muslims not only in France, but throughout Western Europe. European intelligence analysts already estimate that up to 2 per cent of the continent's Muslims – half a million people – are involved in extremist activity. This number, for which no corroborating evidence exists, is so huge according to Savage not because of the role of Islamic fundamentalism *per se*, but rather due to the inevitable 'chemistry resulting from Muslims' encounter with Europe [which] seems to make certain individuals more susceptible to recruitment into terrorist activities'. Therefore, he implies, terrorists are supposedly born simply from the identity crisis generated by Muslim immigration to Europe – 'A larger group of terrorists by far is recruited from the masses of young men, many of them middle-class, who experience a sort of culture shock in Europe and become radicalized "born again" Islamists.'[74]

Security agencies are preoccupied with population politics not only in the context of alleged dangers posed by the rising number of Muslims in the US, Britain and Western Europe, but also in terms of growing populations in the South in general, and intensifying migration toward the West as people attempt to escape the calamities created by climate change, food insecurity and resource scarcity. In this sense, security agencies predict a direct connection between the phenomena of civil unrest, global crises, and rising populations of 'Others', particularly Muslims.

According to then CIA Director Michael V. Hayden, rising world population and immigration 'could undermine the stability of some of the world's most fragile states, especially in Africa, while in the West, governments will be forced to grapple with ever larger immigrant communities and deepening divisions over ethnicity and race'. Noting the projected 33 per cent growth in global population over the next 40 years, he warned that regional friendly oil-exporting regimes 'like Niger and Libya will be forced to rapidly find food, shelter and jobs for millions, or deal with restive populations that "could be easily attracted to violence, civil unrest, or extremism"'. Adding to the fears outlined above, he added that the rising world population would also have a dangerous impact within the West, due to the ethnic differences between comparatively shrinking and expanding population groups:

> European countries, many of which already have large immigrant communities, will see particular growth in their Muslim populations while the number of non-Muslims will shrink as birthrates fall. 'Social integration of immigrants will pose a significant challenge to many host nations – again boosting the potential for unrest and extremism,' Hayden said.[75]

These sentiments have been echoed not only by increasingly vocal and popular right-wing political parties and movements across the EU, but also by mainstream leaders and in popular referendums. In October 2007, for instance, the French Prime Minister Nicolas Sarkozy shocked listeners when he complained that there were far 'too many Muslims in Europe' and that there is an intrinsic 'Muslim difficulty towards integration'.[76] The populist Swiss minaret ban in December 2009, in this context, 'represents a potentially important legitimation of anti-Islamic views' according to Daniel Pipes, who noted that online polls by mainstream media in France, Germany and Spain found 73 to 93 per cent endorsing the Swiss ban, a 'signal that Swiss voters represent growing anti-Islamic sentiments throughout Europe'.[77]

Information Warfare: Creating the Enemy Within

The growing propensity among security agencies and publics to view Islam and Muslims in general as a monolithic potential threat coincides with the corporate media's increasingly consistent demonization of Islam and Muslims. This is no accident. On the one hand, the corporate media relies relatively uncritically on government and security agencies for its information on foreign policy and intelligence matters, including terrorism.[78] On the other hand, there have also been direct efforts from security agencies to actively influence the media. Since 1990, for instance, the Pentagon has bribed, pressured, and censored Hollywood filmmakers to adapt storylines in support of military propaganda.[79] Reviewing over a thousand Hollywood movies,

Jack Shaheen, Professor Emeritus of Mass Communication at Southern Illinois University, found that:

> Today's image makers regularly link the Islamic faith with male supremacy, holy war, and acts of terror, depicting Arab Muslims as hostile alien intruders, and as lecherous, oily sheikhs, intent on using nuclear weapons. When mosques are displayed onscreen, the camera inevitably cuts to Arabs praying, and then gunning down civilians.[80]

Similarly, other studies of the representation of Islam and Muslims by news media consistently show that the latter are portrayed in an overwhelmingly negative way. A study commissioned by the Mayor of London found that in a single week in 2006, 91 per cent of newspaper articles published nationwide about Muslims were negative 'in tone and content'. Muslims in Britain 'are depicted as a threat to traditional British customs, values and ways of life', and most media reporting implies 'there is no common ground between the West and Islam, and that conflict between them is accordingly inevitable'.[81] Another study by Cardiff University's School of Journalism, analysing UK press coverage of British Muslims from 2000 to 2008, found that 'the bulk of coverage of British Muslims – around two thirds – focuses on Muslims as a threat (in relation to terrorism), a problem (in terms of differences in values) or both (Muslim extremism in general)'. Further, it concluded: 'Four of the five most common discourses used about Muslims in the British press associate Islam/Muslims with threats, problems or in opposition to dominant British values.'[82] A similar study of Islamophobia in the American media concluded that 'media stereotyping' after 9/11 'primed Americans to understand the 9/11 attacks as representative of Arab political culture and Islamic devotion', and was 'an important factor in the backlash that afflicted these communities post-9/11'. The study further documents an 'alarming deterioration in Islamophobic hate speech in the media' up to 2006.[83] A 2008 World Economic Forum study of the way 'Muslim-West' relations are covered across global media in a total of 24 Muslim-majority and non-Muslim-majority countries found that 'negative coverage was 10 times more frequent than positive coverage' with Muslims being 'associated with fundamentalist and extremist activities more than six times as often as other religious protagonists'.[84]

The growing ideological problematization of Islam and Muslims is linked to multiple practices of violence justified as counterterrorism policy, and which tend to target innocent Muslim civilians. After the US Department of Justice passed a regulation on 20 September 2001 allowing for indefinite detention, nearly 1200 Arabs and Muslims were secretly arrested and detained without charge.[85] The US National Security Entry-Exit Registration System (NSEERS) 'call-in' program required male visitors from twenty-four Arab and Muslim countries (and North Korea) to register with Immigration and Naturalization Service (INS) offices. No terrorists were found, yet over 13,000 of the 80,000 men who registered were threatened with deportation, and many were 'detained in harsh conditions'.[86] In the UK, more than a thousand Muslims have been detained without charge under anti-terror laws, and only a handful of these have been convicted of terrorist offences. More than 100,000 Muslim men worldwide – victims of the CIA's 'extraordinary rendition' programme – are being held without charges 'in secretive American-run jails and interrogation centres similar to the notorious Abu Ghraib Prison' under conditions which violate the Universal Declaration of Human Rights, the Geneva Conventions on the Treatment of Prisoners, and UN Standard Minimum Rules for the Treatment of Prisoners.[87]

Such practices are a counterpart to Anglo-American military engagements in predominantly Muslim theatres of war, regions often described as dangerous failed zones harbouring potential Islamist terrorists who plan to inflict apocalyptic forms of mass destruction on Western civilization.[88] These operations also tend to result in the indiscriminate killings of Muslim civilians, and invariably are carried out in strategic locations vis-à-vis contested energy reserves in the Middle East, Central Asia and Northwest Africa. A recent US Army War College study makes reference to Huntington's clash of civilizations thesis, arguing that, while it 'captured the possibilities' already emerging in the 1990s:

> [T]he future and its implications are even darker than what Professor Huntington suggested.... The confluence between the world's greatest reserves of petroleum and the extraordinary difficulties that the Islamic world is having, and will continue to have, in confronting a civilization that has taken the West 900 years to develop will create challenges that strategists are only now beginning to grasp.[89]

It is thus clear that the massive military violence that has been inflicted on the civilian populations of these regions is only possible because they have been 'Islamophobically' constructed as having lives that are of less value than those of Western citizens.[90]

Architecture of Control: Toward the Police State

Across the West, the 'War on Terror' has granted governments the moral and political capital to justify the adoption of increasingly draconian anti-terror laws. These encroach on longstanding civil liberties and human rights, on the pretext of allowing the police and security services to more easily protect civilian populations from the threat of Islamist terrorism, which, it is claimed, has amorphously infiltrated Western societies. Yet although the justification is the fight against terrorism, in practice governments have applied these laws indiscriminately as measures of population control, not merely against Muslims and immigrants, but also against political dissidents and protestors against Western domestic and foreign policies.

US documents obtained in July 2008 through a Public Information Act lawsuit against the Maryland State Police confirmed that police forces had engaged in covert surveillance of local peace and anti-death-penalty groups for over a year, from 2005 to 2006. In 43 pages of summaries and computer logs, there were no references to criminal or even potentially criminal acts, 'other than a few isolated references to plans for completely nonviolent civil disobedience' and 'lawful political activities'. Although police reports consistently depicted activists as having 'acted lawfully at all times', the spying continued with detailed surveillance reports 'sent to at least seven federal, state, and local law enforcement agencies, including the National Security Agency, Baltimore City, Baltimore County, and Anne Arundel County police departments, and the state General Services police'. The American Civil Liberties Union (ACLU) described the revelations as 'only the tip of the iceberg'.[91]

By October, legislative hearings revealed that the police had classified as terrorists at least 53 US citizens engaged in nonviolent activism, entering their names and personal information into state and federal databases that track terrorism suspects. More broadly, entire protest groups were also entered into the databases as terrorist organizations, despite admissions that the state police have 'no evidence whatsoever of any involvement in violent crime' by those groups. Subsequently, undercover troopers used aliases to infiltrate organizational meetings, rallies and group email lists.[92]

The British government has similarly criminalized legitimate political dissent and protest. In 2004, for instance, protestors at the Docklands arms fair were stopped and searched under the Terrorism Act 2000, although they were not committing or threatening any violent act. In 2003, anti-terrorist powers were used to prevent a coach full of people at Fairford from attending a demonstration altogether. Walter Wolfgang, an elderly gentleman and long-time Labour party member, was forcibly removed from the Labour Party conference under anti-terror laws for heckling then Cabinet Minister Jack Straw, who was justifying Britain's involvement in Iraq. In other cases, members of the public have been searched and detained on anti-terror grounds simply for wearing anti-government slogans on their T-shirts, or engaging in legitimate campaign activities. One of the most notorious incidents occurred in August 2007, when anti-terrorism powers were invoked for the questioning of anyone approaching the Climate Camp near Heathrow airport, and even of residents in a nearby village who were preparing to march against the impending loss of their homes to airport expansion.[93] The scope of these powers' application is potentially limitless, and their utility for the state, particularly in times of major social and global crisis, is obvious. A clear example of this came to light when the British government used anti-terror powers to take control of assets held in Britain by Icelandic bank Landsbanki, when it collapsed in the wake of the October 2008 global banking crisis.[94]

This phenomenon is certainly not confined to the West, and with Western support it is increasingly spreading to governments of less developed countries, encouraged to develop anti-terror powers to deal with their own populations. Across the South 'many governments were committing human rights abuses and imposing unlawful restrictions on the movement of activists'. These measures were conducted under the rubric of 'new security laws' purportedly for the protection of citizens, but in practice used 'to create a climate of fear'.[95]

As part of their contingency planning for the domestic impact of global crises, Western security agencies are preparing to impose unprecedented measures of population control, permitted under little-known clauses of anti-terror and emergency power legislation. According to US Representative Henry Waxman (Los Angeles)and former congressman Dan Hamburg, since 1999 the US government has contracted Halliburton subsidiary Kellogg Brown and Root 'to build detention camps at undisclosed locations within the United States', and several other companies 'to build thousands of railcars, some reportedly equipped with shackles, ostensibly to transport detainees'. The contract is part of a Homeland Security plan titled ENDGAME, whose goal is the detention of 'all removable aliens' and 'potential terrorists'. As Hamburg asks in the *San Francisco Chronicle*, 'What kind of "new programs" require the construction and refurbishment of detention facilities in nearly every state of the union with the capacity to house perhaps millions of people?'[96]

Indeed, the US Military Commissions Act of 2006 allows for the indefinite detention of anyone who speaks out against government policies or donates money to a 'terrorist' organization. Worse, the law calls for secret trials of American citizens and noncitizens alike. Section 1042 of the 2007 National Defense Authorization Act (NDAA), 'Use of the Armed Forces in Major Public Emergencies', authorizes the president to activate the military in response to 'a natural disaster, a disease outbreak, a terrorist attack or any other condition in which the President determines that domestic violence has occurred to the extent that state officials cannot maintain public order'. In the same year, the White House issued National Security Presidential Directive 51 (NSPD-51) to ensure 'continuity of government' in the event of a 'catastrophic emergency', in which case the president is individually empowered to do whatever he deems necessary – including everything from cancelling elections, to suspending the constitution, to launching a nuclear attack. Congress has yet to hold a single hearing on NSPD-51, and President Obama has shown no interest in repealing these laws.

In September 2008, as the global financial crisis was unfolding, US Army combat units previously operating in Iraq were reassigned for operations inside US territory. The 3rd Infantry's 1st Brigade Combat Team, reported the *Army Times*, was returning from Iraq to defend America as an 'on-call federal response for natural or manmade emergencies and disasters, including terrorist attacks', and would be operational from the beginning of October for at least twelve months:

> [T]his new mission marks the first time an active unit has been given a dedicated assignment to NorthCom, a joint command established in 2002 to provide command and control for federal homeland defense efforts and *coordinate defense support of civil authorities*... After 1st BCT finishes its dwell-time mission, expectations are that another, as yet unnamed, active-duty brigade will take over and that *the mission will be a permanent one*... In the meantime, they'll learn new skills, *use some of the ones they acquired in the war zone* and more than likely will not be

shot at while doing any of it... They may be called upon to *help with civil unrest and crowd control* or to deal with potentially horrific scenarios such as massive poisoning and chaos in response to a chemical, biological, radiological, nuclear or high-yield explosive, or CBRNE, attack... *Training for homeland scenarios has already begun at Fort Stewart* and includes specialty tasks... *The 1st BCT's soldiers also will learn how to use 'the first ever nonlethal package that the Army has fielded,' 1st BCT commander Col. Roger Cloutier said, referring to crowd and traffic control equipment and nonlethal weapons designed to subdue unruly or dangerous individuals without killing them.*[97]

Confirmation of the growing danger of civil unrest in Western states, and particularly in the US, came from an internal client memo from US bank and Federal Reserve member Citigroup, authored by the bank's chief technical strategist Tom Fitzpatrick. The memo warned of:

> [C]ontinued financial deterioration, causing further economic deterioration, with the risk of a feedback loop... This will lead to political instability. We are already seeing countries on the periphery of Europe under severe stress. Some leaders are now at record levels of unpopularity. There is a risk of domestic unrest, starting with strikes because people are feeling disenfranchised... We're already seeing doubts emerge about the sovereign debts of developed AAA-rated countries, which is not something you can ignore.[98]

A US Army Strategic Studies Institute report released in December 2008 further revealed that security agencies were actively preparing for heightened civil unrest due to economic and other crises. The report warned that the US must prepare for a 'violent, strategic dislocation inside the United States' provoked by 'unforeseen economic collapse' or 'loss of functioning political and legal order'. If such crises lead to organized violence that exceeds the capacity of local, state and national authorities to 'restore public order and protect vulnerable populations, DoD [Department of Defense] would be required to fill the gap'. In this case, circumstances would force DoD to 'put its broad resources at the disposal of civil authorities to contain and reverse violent threats to domestic tranquility', including the 'use of military force' in extreme circumstances 'against hostile groups inside the United States'. This could in turn require a deeper direct role for the DoD in government – that is, the full-fledged militarization of the state: 'Further, DoD would be, by necessity, an essential enabling hub for the continuity of political authority in a multi-state or nationwide civil conflict or disturbance.' Thus, the DoD plans to deploy 20,000 troops nationwide by 2011 for emergency response purposes.[99]

Indeed, what took the Americans several pieces of legislation to accomplish, the British achieved in one fell swoop with the UK Civil Contingencies Act 2004, drafted to grant the government extraordinary powers to deal with social crises. According to Tony Bunyan of the London-based EU legislation monitoring organization, Statewatch, the Act 'paves the road to an authoritarian state'. Among the powers enabled by the act, the government can declare a state of emergency at its discretion without a parliamentary vote, and without publicly declaring a state of emergency; ministers can introduce 'emergency regulations' under the Royal Prerogative without recourse to parliament; such regulations can 'give directions or orders' of virtually unlimited scope, including the destruction of property, prohibiting assemblies, banning travel and outlawing 'other specified activities'. Failure to comply with the regulations would become a criminal offence punishable by up to three months in jail. The armed services can be deployed without parliamentary notification or approval, and emergency regulations may be passed 'protecting or restoring activities of Her Majesty's Government'. In Bunyan's words:

> At a stroke democracy could be replaced by totalitarianism... The powers available to the government and state agencies would be truly draconian. Cities could be sealed off, travel bans introduced, all phones cut off, and web sites shut down. Demonstrations could be banned and the news media be made subject to censorship. New offences against the state could be 'created' by government decree.[100]

While the Civil Contingencies Act thus empowers the central government to act unilaterally and dictatorially in any situation it defines as an 'emergency', the act does not sufficiently address the development of appropriate resources and procedures that would help local authorities respond effectively to plausible emergency scenarios. *BAPCO*, a specialist journal for civil contingencies responders, is especially critical:

> While the Government's Civil Contingencies Secretariat may be making positive noises about Britain's preparedness for a variety of disasters – from an influenza pandemic to terrorist attack – the reality is somewhat different. With no clear direction or dedicated budget and a complete lack of Act-specific assessment, the majority of the local authorities and NHS Trusts in England & Wales have achieved little more than an effective 'passing the buck' exercise. If the Government is truly committed to protecting the nation, why are Ministers not using the powers provided by the Civil Contingencies Act to proactively monitor the true state of preparedness across the country?[101]

There is, in other words, *significant scope for reasonable doubt as to whether emergency and anti-terror powers are genuinely concerned with protecting public safety, as opposed to instituting unchallengeable state social control powers in times of unprecedented crisis.*

Similar processes are also broadly under way under the umbrella of the European Union. In 2005, for instance, a landmark ruling by the EU Court of Justice in Luxembourg gave the EU the power to compel the national courts of its 25 member-states to fine or imprison people for breaking EU laws, even if their own home government or parliament remained opposed to those laws. The ruling gave Brussels the power to harmonize and enforce criminal law across the EU.[102] Yet this is only the tip of the iceberg. The EU has implemented wide-ranging new laws granting state authorities unprecedented powers of comprehensive surveillance of citizens including their communications, financial transactions, and travel, and is driving efforts to tighten anti-terrorism legislation. According to civil liberties groups, these measures 'go well beyond what the US Government finds acceptable and palatable' – and are being pushed through with no parliamentary consultation or approval at the level of individual member-states.[103]

Unfortunately, the weight of the evidence on record from the last few decades points to a clear trend of intensifying social control powers, with the state-intelligence apparatus encroaching on the rights of due process and *habeas corpus* as well as judicial checks and balances, all of which are necessary for the retention of a semblance of accountability and democracy in liberal societies. With the erosion of these fundamental rights and liberties, Western states have quietly erected a comprehensive (and expanding) security-surveillance architecture that makes possible new forms of control over their own citizens and others. It seems that they are at least partially attempting to pre-empt large-scale civil unrest that might be provoked as global crises unfold and increasingly converge, with potentially devastating consequences for human life and disastrous costs for the earth.

The Obama administration will not stray from this fundamental course, the origins of which do not lie solely in neoconservative ideology. While Obama's presidency may offer new opportunities for progressive forces to limit the damage, their space for movement will ultimately be constrained by deep-rooted structural pressures that will push Obama to rehabilitate American imperial hegemony rather than transform it. Indeed, the radicalization of Anglo-American political ideology represented by the rise of neoconservative principles and their support by Anglo-American business classes was itself a strategic response to global systemic crises. As the crises intensify, this response is likely to undergo further radicalization, rather than change course in a meaningful way.

This is clear from the political and ideological backgrounds of Barack Obama's appointees within the administration, largely sympathetic to neoconservative ideals particularly on security matters, and whose social and intellectual connections link them to neocon think tanks and policymakers.[104] A scathing *New York Times* editorial noted that President Obama's economic team, put together to tackle the economic and financial meltdown, consisted of the very same people who had 'played central roles in policies that helped provoke today's financial crisis'. These include Tim Geithner, who as president of the Federal Reserve Bank in New York 'helped shape the Bush administration's erratic and often inscrutable responses to the current financial meltdown, up to and including this past weekend's multibillion-dollar bailout of Citigroup'; and former World Bank chief Larry Summers, who 'championed the law that deregulated derivatives, the financial instruments – aka toxic assets – that have spread the financial losses from reckless lending around the globe'.[105] Similarly, analysis of Federal Election Commission data on the largest financial donors to both the McCain and Obama presidential campaigns reveals that the opposing candidates were sponsored largely by the same banks, financial institutions, and corporations – except that Obama received significantly more corporate financing. This suggests that the Obama administration will have little choice but to broadly represent the interests of American capital.[106]

Already, the Obama administration has pursued policies of *hegemony rehabilitation* and *systemic stabilization* to perpetuate existing structures in the global political economy, rather than attempting to transform them. This has involved a far more ostensibly multilateralist approach than that of his predecessor, avowing respect for international law and institutions, thus allowing the US to regain the moral high ground so completely squandered by the Bush administration's open policies of unilateralism, endorsement of torture, and unabashed violations of international law. Yet in effect, this has involved relocating, relabelling, or simply concealing the practices that have served to undermine US authority in the eyes of its allies and the world – as is evident from Obama's immediate measures to close down Guantanamo Bay, delegitimize torture, and challenge CIA practices of extraordinary rendition.

The distinction here is clear. CIA training manuals from the 1960s, '70s, '80s, and '90s prove that the CIA consistently practiced torture long before the Bush administration attempted to legitimize the practice publicly.[107] The Obama administration thus represented a return to recognizing the obvious, that open acknowledgements of covert practices of torture as official policy are detrimental, not conducive, to US hegemony. The new president's first executive orders addressing these issues were met with much celebration worldwide. They implied an end to the official legitimization of torture, yet they did not necessarily mean the full cessation of such practices, but rather a return to the pre-Bush period when those practices were carried out in deep secrecy – practised widely in covert operations but never *officially* acknowledged as part of US intelligence policy. Close scrutiny of the executive orders makes clear that this was a shift in *public relations* and *legal* strategy, that they were designed to allow illegal US military intelligence practices to continue in secret without legal obstruction by redefining the character of those practices (while retaining their substance).

While Obama demanded that interrogation policy be harmonized with a purportedly Geneva Conventions-compliant US Army Field Manual, unacknowledged revisions made to the manual in 2006 – 'in particular, a ten-page appendix known as Appendix M' – 'go beyond the Geneva-based restrictions of the original field manual'. Indeed, the manual accepted 19 forms of interrogation and the practice of extraordinary rendition. Retired Admiral Dennis Blair, Obama's Director of National Intelligence, told a Senate confirmation hearing that the Army Field Manual would itself be changed – potentially allowing new forms of harsh interrogation – but that any such changes would remain classified. Obama's supposed banning of the CIA's secret rendition programmes did not prevent the CIA from extra-judicially apprehending and detaining innocent civilians without evidence or due process, but it was emphasized, in the words of one White House official, that 'There is not going to be rendition to any country that engages in torture.' The problem here is that rendered detainees had *already* been sent to countries across the EU that do not officially sanction torture – where they *were nevertheless tortured*. Secret CIA detention facilities have been hosted in Poland, for instance, and were previously justified by the Bush administration on exactly the same grounds – that Poland does not engage in torture. Even Obama's own counterterrorism adviser John Brennan insisted that rendition is 'absolutely vital'. Finally, while purportedly banning the CIA's use of secret prisons, the prohibitions 'do not refer to facilities used only to hold people on a short-term, transitory basis'. Yet by not specifying an actual *time limit*, Obama's injunctions effectively *continued* to permit indefinite detention, simply by officially *reclassifying* the period of detention as 'short-term' and 'transitory'.[108]

The end result was a successful reconfiguration of the *public presentation* of US military intelligence practices, coupled with legal caveats permitting them to continue relatively unimpeded – essentially a giant PR exercise. Meanwhile, the vast post-9/11 domestic national security apparatus – denying *habeas corpus*, undermining due process, and facilitating mass surveillance, as well as intrusive social control powers brought in by the Bush administration – was not repudiated, but in fact retained.

This pattern of responses by Western states and institutions to global crises raises disturbing questions, brought to light by sociological studies of violence and genocide such as those of Professor Jacques Semelin, research director of the Centre for International Studies and Research in Paris and founder of the *Online Encyclopedia of Mass Violence*. Semelin, wanting to understand how ordinary people can become not just participants but leaders of genocidal movements, underlines the critical role of a prior context of entrenched social, political and economic crises in generating deep-seated anxieties throughout an entire society. Such anxieties are not unfounded, but are rooted in very real conditions that undermine the social fabric.[109] This is *stage 1*. Such crises – currently unfolding on a massive and unprecedented global scale with profound local impacts – mean that conventional points of reference begin to crumble, both for political leaders increasingly challenged by circumstances beyond their control and a public afflicted by an incurable sense of impotence.

In response to these reality-driven anxieties, *stage 2* begins, a process of ideological radicalization, an attempt to impose a new kind of order and influence on what is otherwise beyond control by constructing new group identities based on the stigmatization of differences.[110] Here we can recognize the significance of the ideological construction and increasing problematization of Islam, Muslims, minorities and immigrants, among other social groups within the West, as Western leaders search vainly for certainty in traditional military-oriented security solutions and politico-economic structures, while publics project their anxieties onto 'dangerous' excluded groups.

This then leads to *stage 3*, in a process which is neither automatic nor haphazard, but is largely driven by specific political actors who may already be rooted in particular ideological traditions, and who become increasingly radicalized in response to intensifying social crisis. They exploit this crisis (consciously or otherwise) to ensure self-consolidation. Large-scale public anxieties are deflected by identifying a newly constructed 'Other' as the purported cause of social crisis, and therefore suggesting that the destruction of this 'Other' might be society's salvation. Through this process, violence against a specific group becomes legitimated as a rational strategy to attain social stability.[111]

The evidence discussed in this study suggests that we are already at stage 3 of this process. This applies not only to the escalation of anti-Western terrorist activity, but also to systematic Western state violence against Muslim groups in multiple theatres of war in the Middle East and Central Asia, as well as at home in the form of indiscriminate surveillance, incarceration, ill-treatment, and even torture.

Semelin shows that historically these three stages can easily form a positive feedback loop, intensifying one another, and dragging ordinary people into a vicious cycle of radicalization and violence. Ultimately a 'final solution' is arrived at wherein violence reaches genocidal proportions, and there results a concerted attempt to physically and culturally liquidate the demonized 'Other'. In this way, ordinary people with no history of, or inclination to, murderous violence can be absorbed into a vortex of extremism from which there is no easy escape.[112]

This model of violence illustrates that the intensification of global crises will create social conditions conducive to the increasing polarization of communities along a variety of identity markers, including ethnic, religious, and even class lines. In particular, as the economic crisis deepens, aggravated by the onset of ecological and energy crises, this polarization will increasingly appear as a deepening *class* divide between powerful elites who dominate the world's wealth and resources, and the majority who are denied access to the world's resources and suffer varying forms of dispossession. How these dynamics play out will depend on the actions of elites, who may wish to polarize communities along ethnic and religious lines to deflect from class antagonisms; and on the decisions of communities themselves, who will experience powerful pressures to push for a meaningful transformation of social relations toward more sustainable and just forms of living.

In summary, over the coming decades, as global crises converge and intensify, these political trends – the maximization of state power, the legitimization of imperial and inter-communal violence, and the normalization of the surveillance society – will intensify in concert. Without remedying the *underlying systemic causes* of these processes as essentially hegemonic responses to converging global crises, these tendencies can only be slowed in the short term and not stopped completely.

As resource scarcity becomes more widely recognized and social crises more entrenched, states may respond to the rising danger of civil disorder by resorting to more intrusive measures of social control, such as enforced carbon rationing and the labelling of dissident groups as terrorists. There is no sign yet (beyond political rhetoric) that present or future US administrations will make any effort to obviate this process. The danger is that, unchecked, the strategy of militarization could increasingly damage democratic structures, polarize communities along ethnic and religious lines, and escalate into a form of genocidal fascism.

The future contours of intra- and international conflict will therefore principally concern geopolitical competition for the dominance of regional and local resources, including oil, gas, coal, raw materials, food and water. The current structure of global governance in the form of the United Nations and other related institutions will become increasingly ineffective, both as a means to resolve international disputes and to legitimize the imposition of Anglo-American policy. Yet even as conflict intensifies, the discourses mobilized to justify violence will continue to reinforce ethnic, religious and ideological markers of 'difference'. However, although the threat of major (and potentially nuclear) interstate conflict will drastically increase over the next 10 to 15 years if these problems are not addressed, the long-term outlook after mid century is quite different.

Given the forecasts for peak oil, it is likely that by the mid twenty-first century the limits to oil production will constrain the ability of states to sustain the technologies of modern industrial warfare, implying a corresponding reduction in their ability to monopolize the means of violence. This will not mean the end of war, but rather the increasing inability on the part of Western states to decisively win wars, the reversion to forms of military power less reliant on hydrocarbon energies, the general contraction of state power and, hence, the diffusion of political and military power. It also suggests that while there will be a heightened danger of interstate friction, war, and nuclear proliferation over the immediate decades, by mid century major interstate wars will be increasingly unlikely simply due to overwhelming practical and technological constraints. Thus, while states, banks and corporations are likely to be the principal actors in ongoing militarization processes over the next decade, these processes and their long-term futility will also generate increasing numbers of detractors in the form of popular grassroots movements opposing war and calling for social justice. Over the long term, large state structures like the US, Russia, China and the EU will find it increasingly difficult to sustain their internal territorial integrity, facing rising demands for autonomy by different political and national groups within their borders. While the short-term outlook sees a potential escalation of conflict, the long-term outlook is for increasing military-political decentralization. The dangers should not be underestimated, but we should also recognize that such a development will create a new and unprecedented opportunity for grassroots communities to reclaim political space based on the values of peace and cooperation.

7
Diagnosis – Interrogating the Global Political Economy

We have now developed a body of empirical data unequivocally demonstrating the potentially catastrophic convergence of six global crises over the coming decades. It is essential to now develop a conceptual apparatus that can make sense of these crises by examining their systemic-causal origins in the structure of the global political economy.

To this end, the current chapter begins by conducting an overview and summary of our empirical analysis. It then critiques orthodox International Relations approaches in the forms of Realism and Liberalism, before moving on to a critical review of the most significant interdisciplinary studies of global crises based on historical analysis of the rise and fall of civilizations. While deeply compelling, these studies often lack the sufficient theoretical tools to accurately diagnose contemporary systemic crises, largely because of inattention to the concrete *social relations* by which modern industrial civilization is constituted.

Here, I attempt to identify those social relations in the context of the global political economy and its neoliberal capitalist form, using a holistic methodology that accounts for the interdependence of military, political, economic, and ideological structures. While conceding that a theory of capitalism is the best way of understanding the *social form* of the world system, the analysis must be broadened to understand how this form transfigures the military, political, cultural and even ethical constitution of that system.

On this basis, I also identify the specific structural features that can be causally implicated in the acceleration of global crises, and which thus require urgent transformation. These are delineated in the text as *Key Structural Problems*, providing the foundation for our concluding recommendations for social structural 'reform'. The analysis, then, is not simply a negative one – rather it demonstrates that an alternative is not only possible but *necessary* if civilization is to survive the twenty-first-century transition to the post-carbon era.

THE CONTINUUM OF CRISIS, AND CIVILIZATIONS AS COMPLEX ADAPTIVE SYSTEMS

The world is on the verge of a potentially cataclysmic convergence of crises, which fundamentally threaten the viability of modern industrial civilization. Seemingly diverse phenomena such as international terrorism, climate change and resource depletion are in fact intensifying manifestations of the same structural dynamic generated by the inherently dysfunctional character of the global political economy, its ideology, its value system, and the interrelation of these with state policies and individual action.

There is now an overwhelming scientific consensus that climate change is a consequence of human-induced carbon dioxide (CO_2) emissions, which are produced by the burning of hydrocarbon fossil fuels, namely oil, gas and coal. Over-dependence on these sources of energy, and the concomitant structurally induced resistance to a rapid transition to alternative, renewable and sustainable energy sources, means that the introduction of multiple 'green' economic mechanisms cannot reduce CO_2 emissions sufficiently to prevent or slow down global warming. On its own terms, over-dependence on hydrocarbon energy sources is also likely to reach a critical point early in this century. Leading oil industry experts (and even some elements within the US Army) now agree that the reaching of an irreversible peak in world oil production is either imminent or has already passed. Yet even according to the most 'optimistic' predictions, such as that of BP, peak oil production is about 30 years away.[1] Either way, there is very little time to respond in a way that will forestall or prevent a crisis that will have devastating import for the general public, governments and investors alike. The inertia of policymakers over these issues, despite their potentially dire consequences for industrial civilization over the next few decades, is not due to a bureaucratic failure of imagination, but ultimately explainable by the ontological and ethical limitations of the ideology of neoliberal capitalism.

Issues such as climate change and resource depletion illustrate that neoliberal ideology fails to fully reflect the real conditions of human life as fundamentally embedded in the natural environment.[2] Neoliberal ideology, due to its equation of human well-being with unlimited economic growth driven by short-term profits, is unable to recognize the limitations to that growth. It therefore cannot respond effectively to global crises of climate change and resource scarcity, even though these threaten to undermine its premises. This is because it remains irrevocably preoccupied with short-term concerns to sustain unlimited growth by maximizing profits for the next quarter's turnover, with little or no regard for external risks or costs.[3]

Similarly, because neoliberal ideology rests on the twin principles of the Washington Consensus – deregulation and financial liberalization – it lacks the intellectual resources and the financial mechanisms to respond effectively to the growing instability of the global economy. The increasingly crisis-prone nature of financial markets is precisely due to the generalized and unlimited application of these Washington Consensus principles – a consensus unable to offer any long-term macroeconomic solution to the problems of its own making.

The global crises we face are not separate but fundamentally interlinked: the excessive exploitation of hydrocarbon resources is tied to the rampant escalation of CO_2 emissions, fuelling global warming and the acceleration of climate change, devastating ecosystems, killing millions of people and bringing the extinction of thousands of species. Yet despite the clearly annihilatory impact of these processes, the exploitation of the earth's resources and their overconsumption by a minority of the world's population continues with no regard for limits or boundaries, revealing a deeply irrational dynamic, and the ultimately *unnatural* character of the world economy.

The logic of 'growth' at all social and environmental costs is driving the depletion of hydrocarbon and other natural resources at unprecedented, and unsustainable, rates – such that oil, gas and hydrocarbon resources in general are, for all intents and purposes, running dry. Both climate change and energy crises are impacting detrimentally on our ability to sustain global food production. Water shortages and hotter weather are destroying the viability of agriculture, while predicted fuel shortages are set to undermine agribusiness which is heavily dependent on oil and gas. The increasing inability to meet consumer demand for food is also linked to the destructive 'growth'-driven technologies of a hierarchical agribusiness industry monopolized by short-sighted corporate conglomerates, and a skewed international system of distribution that marginalizes two thirds of the world's population.

Finally, the world economy, even judged on its own terms, is on the verge of collapse. Geared to serve the interests of corporate profit maximization, the world economy systematically generates widening inequalities that result in the deprivation of a majority of the world's population, causing deaths on an increasing scale. But in doing so the economic system ignores its own internal contradictions and systemic fragilities, fuelling unsustainable 'virtual' growth trajectories in Northern centres of power at the expense of the South, and creating bubble economies just waiting to burst.

So we are approaching critical junctures on four fronts simultaneously – the climate, energy dependence, the economy, and the food supply. These are amplifying in turn two further mutually reinforcing dangers, in the form of international terrorism and the militarization of Western societies; in other words, the normalization of political violence for both state and communities, fundamentally undermining the democratic values and ideals we associate with liberal society. The scale of these crises has been vastly underestimated by officials, and even by some experts, because their cumulative impact is not properly understood. Experts tend to analyse these crises as separate processes, and thus to offer isolated solutions for each. But these crises are not distinct at all – they are fundamentally linked to the functioning of the global political economic system, and as they accelerate they will feed into and exacerbate each other. We should therefore recognize them as local manifestations of a *failed global system* that institutionalizes norms and values which are out of tune with life and nature.

Contemporary state responses to global crises focus largely on climate change and financial instability, and can be broadly categorized into two predominant types: 1) strategies of militarization to boost an individual state's resilience to crisis through intensification of social control mechanisms; 2) strategies of international cooperation to establish new global governance regimes by which states can develop treaties and agreements to encourage mitigating action. Unfortunately, however, as the previous chapters have shown, while the first set of strategies proceeds apace with little or no public or parliamentary consultation, the second set of strategies continues to result in dismal failure, with states unable to agree on the scale of the crises concerned, let alone the policies required to address them. Both these strategies follow logically from the two predominant orthodox approaches to International Relations (IR) theory, namely neorealism and neoliberalism.

Neorealism understands interstate competition, rivalry and warfare as inevitable functions of states' uncertainty about their own survival, arising from the anarchic structure of the international system. Gains for one state are losses for another, and each state's attempt to maximize its power relative to all other states simply reflects its rational pursuit of its own security. The upshot, of course, is the normalization of political violence in the international system, including practices such as over-exploitation of energy and the environment, as a 'rational' strategy – even though this ultimately amplifies systemic insecurity. Inability to cooperate internationally and for mutual benefit is thus seen as an inevitable outcome of the simple, axiomatic existence of multiple states. The problem with this perspective is that it cannot explain in the first place the complex interdependence or worsening of global crises. Unable to situate these crises in the context of an international system that is not simply a set of states, but a transnational global structure based on a specific exploitative relationship with the natural world, neorealism can only theorize global crises as 'new issue areas' appended to existing security agendas.[4] Thus, the US Army tends to see climate change as a 'stress-multiplier' that will 'exacerbate tensions' and 'complicate American foreign policy'; while the EU perceives it as a 'threat-multiplier which exacerbates existing trends, tensions and instability'.[5] Yet, by the very act of 'securitizing' global crises, neorealism renders itself impotent to prevent or mitigate them by theorizing their root structural causes. In effect, despite its emphasis on the reasons why states seek security, neorealism's approach to issues like climate change actually *guarantees greater insecurity* by promoting policies which frame these issues purely as amplifiers of threats. Neorealism thus entirely negates its own theoretical and normative value. For if 'security' is the fundamental driver of state foreign policies, then why are states chronically incapable of effectively ameliorating the global systemic amplifiers of 'insecurity'?

Although neoliberalism shares neorealism's assumptions about the centrality of the state as a rational actor in the international system, it differs fundamentally in the notion that gains for one state do not automatically imply losses for another; therefore states are able to form cooperative, interdependent relationships conducive to mutual power gains, which do not necessarily generate tensions or conflict. While neoliberalism therefore encourages international negotiations and global governance mechanisms for the resolution of global crises, it implicitly accepts the anarchic states-system and world capitalist economy as unquestionable 'givens', which themselves are not subject to debate or to change. The focus, then, is on developing the most optimal ways of exploiting the natural world to the maximal extent, and neglected is the very role of global political economic structures (such as capitalist markets) in both generating global crises and inhibiting effective means for their amelioration. Neoliberalism is unable to view the natural world in anything other than a rationalist, instrumentalist fashion, and legitimizes the 'unlimited growth' imperative; by focusing on market mechanisms it continues to subordinate environmental and ecological issues to the competitive pressures of private sector profit-maximization.[6]

Both theoretical approaches focus on trying to understand different aspects of *interstate* behaviour, conflictual and cooperative respectively, but each lacks the capacity to address *the relationship of the interstate system itself to the natural world* as a key analytical category for understanding the acceleration of global crises. In doing so, they are unable to acknowledge the profound *irrationality* of collective state behaviour, which systematically erodes this relationship, generating insecurity and poverty on a massive scale – in the very process of seeking security and prosperity. *Indeed, by reducing this destructive state behaviour to a function of instrumental reason, both approaches rationalize the deeply irrational collective human actions that are destroying the very conditions of our existence.*

To effectively theorize the convergence of global crises therefore requires a move away from orthodox IR theories, toward a far broader interdisciplinary systems-oriented social scientific approach which can integrate the role of the world capitalist economy, the political strategies of militarization, and their relationship to the environmental and ecological integrity of the natural world. An emerging body of literature that engages with the historical rise and fall of civilizations as discrete social systems is perhaps most pertinent here. The seminal book is arguably Joseph Tainter's *The Collapse of Complex Societies*. Tainter, an archaeologist from the University of Utah, argued that *complexity* is the key to understanding the vulnerabilities inherent to any society or civilization. Because survival involves cumulative problem-solving, each new solution brings a degree of success, a new level of social organization and complexity, which also brings its own set of new problems. As new solutions are found, new layers of complexity are required. But each extra organizational layer exacts a price, whether it involves building canals or roads, or educating scribes, or innovating agricultural techniques. Each layer imposes a cost in terms of energy, a cost that increases as social organization grows more complex in its attempt to support more people, resources, information and management to solve a wider diversity of problems. Thus, as civilization evolves, not only does it use up more energy, it becomes more complex in its organization and corresponding problems and becomes less efficient. Tainter calls this a 'law of diminishing returns', meaning that all civilizations arrive at a point when all available energy and resources are devoted solely to maintaining the current level of complexity. At that critical point, resources are insufficient to deal with stresses to the system, be they environmental problems, war, or rebellion, leading to the breakdown of overstretched institutions and the collapse of civil order. From the remains might emerge a new, less complex society.

In sum, civilizational survival implies a need for constant change, through a continual process of problem-solving, where each solution brings a new level of social organization fraught with greater complexity, and hence, a new layer of problems. This inherently evolving complexity leads civilization to its own downfall, bringing it to an 'omega-point', a stage when there is insufficient energy to sustain continued growth. The civilization then 'collapses' – that is, undergoes a process of socio-political *simplification* to replace an unsustainably high level of complexity with a more sustainable lower level. This process occurs on a timescale of 'no more than a few decades'.[7]

While Tainter's analysis highlights a key issue – the relationship between a civilization's levels of consumption and its actual energy resource base – his model of civilizational collapse is contradicted by many of his own examples. The collapse of the Western Roman empire, for instance, did not occur over decades through a single protracted collapse-process, but rather consisted of a series of crises over a period of centuries. Each crisis led to the establishment of temporary stability at a less complex level of social organization. Each such level in turn proved to be unsustainable, and was followed by a further crisis and reduction of complexity. The first major breakdown in the Roman imperial system came in 166 CE, and further crises followed until the empire ceased to exist in 476 CE.[8]

University of Toronto political scientist Thomas Homer-Dixon has attempted to build on Tainter's work, in an effort to understand the crises facing modern industrial civilization. Following Tainter, and bringing in the work of ecologist Crawford Holling, he argues that societies should be understood as *complex adaptive systems* which pass through cycles of growth, collapse, regeneration, and growth again. He also focuses on the example of Rome's fall to illustrate the dynamics of civilizational collapse. The expansion of the Roman empire, he argues, depended on its ability to extract agrarian energy surpluses from its peripheries, enabling the development of increasing organizational complexity in the core. However, as Rome's imperial bureaucracy grew bigger in order to extract more taxes and maintain military power, the lack of significant new conquests after the first century AD meant that energy inputs declined over time. Eventually, the complexity of the centre could not be maintained without instituting an even more draconian taxation regime, requiring more intrusive military policing. Ultimately, this provided only a temporary and increasingly ineffectual solution as the empire continued to face increasing 'stresses' in the form of 'barbarian' invasions, a crumbling central state, and civil disorder. In Homer-Dixon's words, 'the empire could no longer afford the problem of its own existence'. This underscores his concept of Energy Return On Investment (EROI) as a measure of the real sustainability of a social system based on a level of energy inputs that exceeds its outputs. If the energy consumed by a social system exceeds that which is available to it, it cannot survive.[9]

In applying the lessons of Rome to modern industrial civilization, Homer-Dixon applies Tainter's model of collapse to argue that the current level of organizational complexity is becoming increasingly unsustainable, and identifies five 'tectonic stresses': population growth, energy depletion, environmental degradation, climate change and financial instability. These stresses are multiplied by two factors – the increased connectivity of the global economy and the heightened capacity of small groups to carry out acts of destruction. He then utilizes the concept of 'negative synergy' to identify the potential for these different stresses to compound one another's negative effects. The cumulative intensification of these stresses, and the growing intricacy of the global system's distributed networks for the moving of money, information, materials and energy across borders, adds up to an interconnectedness which globalizes crisis, amplifying and transmitting shocks across the world. For this reason modern industrial civilization is uniquely vulnerable to what Homer-Dixon calls 'synchronous failure' – a simultaneous peak of multiple stresses that would overwhelm the adaptive capacity of even rich and powerful societies, potentially precipitating a speedy breakdown of world order.

The good news is that the danger of collapse also heralds the hope for 'catagenesis' – renewal through reversion to a simpler state, followed by the emergence of a novel form of society. Using Holling's work, Homer-Dixon demonstrates that catagenesis is an integral and routine stage in the life cycles of natural ecosystems, rather than an anomaly. Homer-Dixon is thus deeply positive about the prospects for civilizational renewal. In this context, global crises are not simply symptoms of *global system failure* – they are simultaneously symptoms of *civilizational transition*.

Yet this approach is not free from theoretical and empirical problems. By relying more or less uncritically on Tainter's model of civilizational collapse, Homer-Dixon likewise makes the erroneous assumption that the process of socio-political simplification involved in the collapse-process is necessarily a relatively speedy event transpiring over mere decades. In contrast, John Michael Greer's theory of 'catabolic collapse' critically reformulates Tainter's model against the historical record of multiple empires, finding that civilizations rarely collapse so swiftly, but rather descend through a protracted process of 'succession' whereby increasing levels of socio-political simplification are reached, stabilized, then breached as they become unsustainable. While Greer's approach does not necessarily undermine Homer-Dixon's warning about the danger of synchronous failure, it emphasizes that even a major collapse event might be more protracted than sudden. This lends weight to Homer-Dixon's optimism about the prospects for civilizational renewal, and to an understanding of the current period as a *transitional* one.

The reliance on Tainter, however, also creates a tension in Homer-Dixon's discussion of five key stresses exerting pressure on industrial civilization. While he clearly recognizes that these stresses are generated by the global system itself, he does not systematically explore how and why this is the case. Therefore his diagnosis of the problem is exceedingly thin – ultimately, it is a matter of too much 'organizational complexity' which cannot be maintained by industrial civilization's resource base, a problem which in itself creates more stresses as new 'solutions' introduce increasingly costly and more complex layers. His prognosis is therefore correspondingly vague –people and institutions should work together to attenuate the chance of synchronous failure, and thus create opportunities for a more mitigated transition to lower levels of sociopolitical complexity based on self-sufficiency. Yet this approach leaves out the specific military, political, economic, and ideological features of the global system, and the manner in which these interlocking structural features create intensifying world-scale crises. Thus, Homer-Dixon is unable to identify the specific social-structural transformations required to create a system capable of overcoming such crisis tendencies. We are left with the impression that *complexity* itself is the problem – that a simplification of our societies is the only solution.

Most critically, Homer-Dixon misunderstands the historically specific dynamics of the neoliberal capitalist global political economy, which are fundamentally different from the feudal tributary dynamics of growth and decline exhibited by past empires such as that of Rome. In trying to identify the uniquely unstable and crisis-prone dynamic of modern capitalist economies, he refers to only two problems: 1) the 'growth imperative'; 2) positive feedback loops due to modern communications. The origin of the growth imperative in relation to capitalism is never convincingly explained beyond some cursory discussion of the chronic need to counter 'inadequate demand' since the 1930s. Yet he makes no attempt to identify the distinctive socio-political relations – especially class, property and production relations – by which contemporary capitalism is constituted. Indeed, he never convincingly defines capitalism, and is thus unable to grasp whether or how it differs from pre-capitalist social systems, for example Rome's. As for the second point, he fundamentally misrepresents the actual nature of the contemporary financial system:

> [L]ike any complex system a capitalist economy can sometimes exhibit unbalanced and capricious behaviour. Instead of acting like a smoothly functioning and predictable machine... it can act more like the planet's climate with its synergies, feedbacks, multiple equilibriums and threshold effects. This is what happened in East Asia in mid-1997, when a self-reinforcing feedback of investment, profit, consumption and more investment flipped overnight to a vicious circle of falling investment, failing banks and crashing consumer demand.[10]

By arguing that the dynamic of the financial system under capitalism operates 'like any complex system' displaying inherently unpredictable 'capricious behaviour', Homer-Dixon effectively erases the role of capitalist actors from the picture. But his account of the 1997 Asian financial crisis is precisely what did *not* happen. Rather than a case of 'capricious' overnight system-wide flipping from investment to disinvestment, the crisis was the direct consequence of an oft-used financial strategy deployed by international capitalist investors. The latter acted strategically to deliberately hype local shares, purchasing them in bulk and thus escalating their prices, then collectively selling them to local investors at their highest prices, thus reaping massive profits while leaving local share prices to plummet. The ability to conduct this sort of financial transaction as a way of extracting unprecedented profits is unique to contemporary capitalism, but cannot be conflated with the unpredictable dynamics of 'any complex system'. Capitalism is not like *any* complex system. It is a very specific social form – originating in England in the sixteenth century and evolving since then – with its own peculiar dynamic, discontinuous from previous socio-economic systems which were largely tributary, sedentary, or nomadic.[11] It is not enough to talk generically about 'complex adaptive systems', but to specify the *constitutive social relations* of these systems in terms of their political, military, economic and ideological forms. Consequently, Homer-Dixon's overall argument is hampered by an inability to convey the unique dynamics that are distinctive to the contemporary global political economy, as opposed to being generically applicable to all societies in world history.

The real problem, then, is not civilizational complexity *per se*, but the lack of parity between socio-political complexity and the available resource base of a given civilization. High civilizational complexity is therefore theoretically sustainable as long as there is: 1) a balance between its level of complexity and its relationship to the environment, such that the former never outgrows the latter; and 2) a balance in the internal distribution of resources throughout the system so that the benefits of advanced complexity are not concentrated among a few at the expense of the majority. But to grasp what this implies in today's context, we need to go beyond generalizations about 'complex adaptive systems' and delve into the *socio-political constitution* of modern industrial civilization.

The central problem with civilizational complexity theory is its tendency to conceptualize 'civilization' itself as the homogenous locus of historical agency. Yet it is not 'civilizations' which adapt or fail, but rather the complex nexus of different social groups within a particular civilization, whose own respective interests, conflicts and relationships configure a civilization's structure. The central role of *class* in delineating the conflicting interests of different social groups in relation to the distribution of resources is underplayed. Indeed, all the civilizations studied by Tainter and Homer-Dixon, among others, were *deeply unequal class societies*. Their over-exploitation of the natural environment was intimately linked with the *unequal relationships of different classes to the materials and technologies of production*. Consequently, they ignore the role of privileged classes in accelerating collapse by their unwillingness or inability to pursue wider social goods in lieu of their own narrow class interest.

It is therefore necessary to recognize the pivotal role of the world capitalist economy in configuring the complex interrelationships between different global crises, and the policy decisions that lie behind them. This requires the development of a theoretical framework that can account for the structure and direction of the system in terms of its impact on human societies and the natural environment. Traditionally, such issues are the concern of the theoretical tradition of Historical Materialism, due to its axiomatic focus on *humanity's relationship to nature* through the process and organization of productive labour. In the next chapter, we will critically explore the salience of a Historical Materialist approach to theorizing global crises, and thereby develop a holistic framework for understanding the contemporary world system, its historical evolution, its *key structural features*, and on this basis the current civilizational conjuncture's *Key Structural Problems* requiring urgent reform.

STRUCTURAL AND SOCIO-SYSTEMIC DYNAMICS OF CAPITALISM

Human Metabolism with Nature

Historical Materialism should not be confused with the wider theory of *philosophical materialism*, which is essentially an ontological theory (a theory concerning the nature of existence) assuming that physical matter is the only reality, and that consequently everything can be explained purely in terms of it. Rather, Historical Materialism is a *social theory* specifically concerned with explaining the dynamics of human societies as a function of their relationship with the natural world. Associated with Karl Marx, it is often viewed as an outmoded, economistic, and deterministic blueprint for the establishment of Soviet-style social engineering, whose viability was permanently refuted with the collapse of the socialist economy of the USSR in 1989. Yet this view, partially based on the more traditional interpretations of Marxism pursued by dissident historians, sociologists, and development economists in the 1960s and '70s, has given way to a much more flexible and holistic understanding of Marx's work. More than ever, the skewed evolution of neoliberal capitalism actually vindicates, rather than discredits, Marx's seminal critique of capitalism as a socio-political system.

That is not to suggest that Marx's work should be taken at face value and drawn upon uncritically – but rather that a critical analysis of the global political economy cannot afford to ignore Marx's contribution. Indeed, the classical Marxist literature on capitalism and imperialism, despite its own internal inconsistencies, consistently postulated five key theses which accurately forecasted the actual development of the world capitalist economy. Fred Halliday, Professor Emeritus of International Relations at the London School of Economics, identifies these as follows:

1. The expansion of capitalism on a world scale.
2. The competitive, expansionist and warlike nature of developed capitalist states.
3. The reproduction of socio-economic inequalities on a world scale deriving from the unequal nature of capitalist expansion.
4. The entrenchment of structures of unequal power and wealth in social, economic, political, legal and cultural sectors.
5. Capitalist expansion's generation of resistance and anti-imperialism.

As all these predictions turned out to be valid, notes Halliday, this indicates that Historical Materialist theories of capitalist imperialism are 'not just a possible, but a necessary, part of any comprehension of the contemporary world'.[12] Indeed, political economist Hugo Radice of Leeds University's School of Business Studies points out that despite the 'analytical weakness and political posturing' of classical Marxist theories, 'they can be seen in retrospect to have accurately identified and predicted almost every feature of present-day peripheral capitalism'.[13] Hence, Marx's critique continues to provide the theoretical groundwork for any attempt to get to grips with the way capitalism works. This is very different, however, from assuming that everything Marx said about capitalism was true, or that his vision for progress – that workers should control their collective lives in terms of production, distribution and consumption – translates easily into a viable prescription for solutions to our contemporary predicament. As such, here we approach Marx's work critically and creatively with a view to identifying the key systemic failures of capitalism that might explain the contemporary convergence of global crises.

Perhaps Marx's most central – and underrated – insight was that at the core of any social system is its constitutive *relationship to the natural world*. Ultimately, the sustainability of human communities and societies is predicated on the means by which they extract resources from the natural environment to produce food, clothes, energy and other products for consumption. As the late anthropologist Eric Wolf noted, Marx's concept of the 'mode of production' is not merely a reductionist category of economic determinism, but rather represents the dynamic and transformative relationship between human beings and nature. The human being, itself an 'outgrowth of natural processes', is not merely a 'passive product' of those processes, but has the ability to 'transform nature to human use'. This transformation is effected by the exploitation of nature through labour to produce goods for social consumption. But this transformative exploitation of nature in turn correspondingly transforms human life. In Marx's words, by 'changing it, [man] at the same time changes his own nature'. There is, in other words, a *dynamic interrelationship* between the structures of society and the way in which society exploits nature through labour and production. This active relation is facilitated through 'technology, organization and ideas'. It feeds back on language and even on human biology, with major transformations in way of life accompanied by changes in physical stature, diseases, immunology, rate of sexual maturation, demography and sensory perception.[14] The mode of production, therefore, is not merely an abstract economic relation, but a way of appropriating nature through social labour, which is inherently related to prevailing forms of social organization, technological mobilization, as well as culture and ideology.[15]

This means that rather than positing an economistic mode of production which determines a social and political superstructure, as do variants of 'vulgar' Marxism that are now largely rejected, the relationship between what we call the 'economy' and other social arenas such as politics and ideology is seen as far more interdependent. At the core of human beings' relationship with nature is energy, denoting the manner in which natural resources are converted through labour into the capacity for production. In this respect, *the means of production signifies the manner in which a society derives and makes use of energy*, thus defining the parameters of its productive relationship to nature. Yet the conditions of this relationship are, in turn, mutually implicated in the way a society *conceptualizes its relationship to nature*. In other words, the way a society relates to nature through the exploitation of its resources is inseparable from that society's overall *self-conception* (its understanding of human nature and social relations) and that society's *conception of nature* (its understanding of the natural world). This relationship will therefore also have specific implications for the way society is politically and culturally organized, and technologically constituted. Put another way, a society's predominant means and technologies of production are *co-extensive* with its self-conception, worldview, and its political and cultural organization.

This can be illustrated with a specific example. For dramatic effect, we may take the specific form of capitalism adopted in Nazi Germany. Although 1930s Germany retained capitalist relations of wage labour, it also engaged in heavy state intervention into all aspects of the economy, particularly in the development of the military-industrial complex. The latter goal took precedence due to Nazism's desire to advance Germany's ability to wage war within Europe in accordance with the objective of securing sufficient '*lebensraum*' or 'living space' – more territory – for an expanding German population, purportedly beleaguered by parasitical enemies (among them communists, the mentally and physically disabled, and different ethnic groups, principally of course the Jewish people). This very specific direction of capitalism, in turn, generated a militarized climate which favoured capital investment in specific industries, marginalized dissent, gave free reign to particular anti-Jewish paramilitary groups, and thus facilitated the anti-Semitic radicalization of large sectors of German society.

Thus, Hitler mobilized German capitalism in a specific direction on the basis of his particular ideological, religious, political and military perceptions and objectives. Simultaneously, the militarization of the German economy facilitated the emergence and proliferation of anti-Jewish radicalism while marginalizing and criminalizing dissenters – the economy thus had a major impact on German ideology. In turn, an increasingly radicalized ideology and culture of fascism continued to manifest in a particular militarized variant of state-led German capitalism. It is therefore clear that the state-oriented capitalist relations of production adopted in Nazi Germany were linked indelibly to, in fact were *mutually constituted* by, the evolution of Nazi ideology. This indicates that strategic actors may adopt different relations of production on the basis of a variety of subjective considerations and interests, not all of them reducible to a purely economic rationale. Yet it is also clear that specific relations of production, once adopted, will have a very specific impact, in turn, on those ideological, religious and military considerations and interests; and that those considerations and interests simultaneously impact on the way actors configure their relations of production.[16]

Social systems as a whole, therefore, are not adaptive. Rather, their different constitutive classes adapt to their social and environmental circumstances in accordance with how they understand those circumstances, their position within society relative to other groups, and their opportunities and interests (material and otherwise) within this socio-political and economic context. We can therefore identify the following as key features of social systems:

1. Relations of production
2. Economic and financial structure
3. Mechanisms of political administration
4. Socio-cultural ethical norms and values
5. Ideology: worldview of life, nature, and their relationship

It is critical to emphasize, therefore, the *organic interdependence* between relations of production and a society's political, cultural and ideological forms. Thus, *relations of production* encompass issues such as: relations with the microbial world and enviro-physical conditions (e.g. seasonality); as well as relations with nature in production specifically including the type of goods produced, from which inputs (energy, raw materials and natural resource), by which producers, under whose decision, for which markets, and under which conditions.[17] Production is, therefore, not a separate sphere from the political, precisely because the organization of productive relations is ultimately *governed by relations of power*. As York University political scientist Ellen Meiksins Wood has argued through the concept of *social-property relations*, 'relations of production' should be 'presented in their *political* aspect, that aspect in which they are actually *contested*: as relations of domination, as rights of property, as the power to organize and govern production and appropriation'. A mode of production is not simply a 'technology but a social organization of productive activity', a 'mode of exploitation' which is therefore 'a relationship of power' constituted of political organization within and between contending classes that in turn conditions the 'nature and extent of exploitation'. Therefore, relations of production 'are historically constituted by the configuration of political power that determines the outcome of class conflict'. The configuration of political power, moreover, is itself constituted by myriad structures, particularly the capacity for organized violence (military power) and access to ideological or cultural resources for the legitimization of power, among others.[18]

In this context, relevant changes in culture and ideology, if they significantly alter a society's self-understanding and conceptions of nature, can impact on a society's concrete relationship with nature, that is, its relations of production, which would also necessitate changes in the political institutions that regulate this relationship. Simultaneously, alterations in a society's relations of production necessarily also involve corresponding alterations in its political regulatory structures, its conceptualization of the world, and its cultural and ethical norms. Thus, *specific productive relations are structured by and in turn affect the structure of different forms of power*, because they are inherently, interdependently, and organically conjoined to them. In summary, the means of production do not monocausally determine social organization, but are rather *organically conjoined with social organization.*[19]

Therefore, a social system cannot be fully understood by examining only its core relations of production, nor by understanding only its dominant culture or ideology, or even its defining political institutions. Rather, a social system needs to be understood holistically as an interdependent and complex whole, in which its constituent relations of production are organically related to its social, political and cultural practices. For this reason, the global political economy should be seen not only in terms of its core productive relations, but simultaneously in terms of *how* those relations are organically conjoined with and mutually constituted by its socio-political structures, its ideological principles, and its cultural and ethical norms.

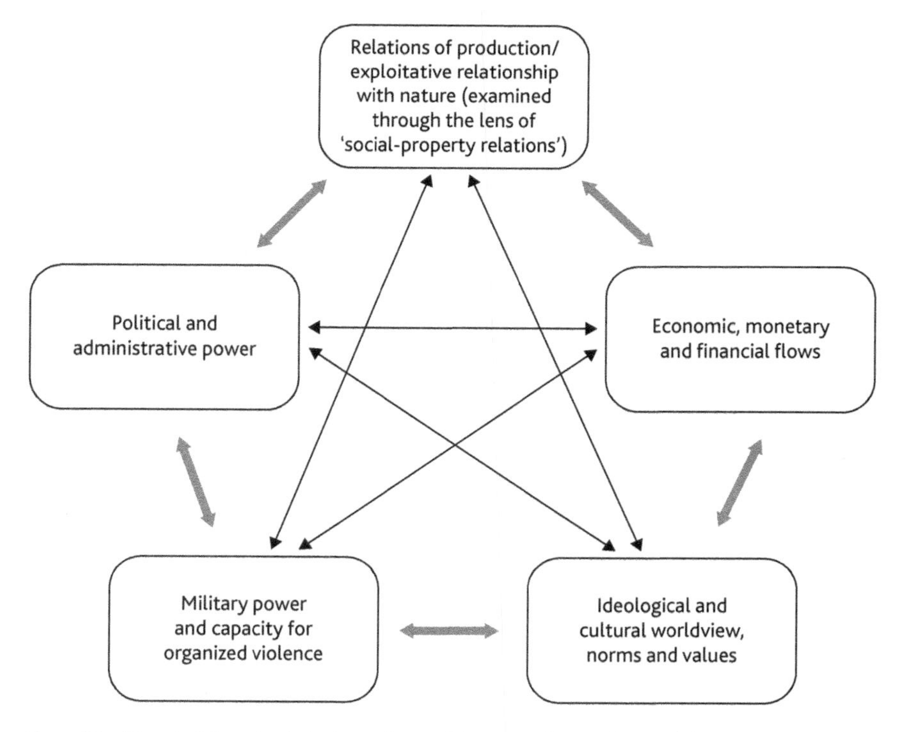

Figure 7.1 Model of Complex Interdependence of Sources of Power in a Social System

In understanding the contemporary world, this model is further complicated by the historical reality that the globalization of neoliberal capitalism in a world of modern industrial nation-states represents only the latest socio-political development in a history of evolving and interlocking social relations. Kees van der Pijl, Professor of International Relations at Sussex University, points out that prior to the emergence of the nation-state some centuries ago, peoples were organized according to quite different 'modes of foreign relations' which, although they have been successively superseded, have not been eradicated. The chronology is as follows:

1. Tribal relations, dating back to pre-agricultural society
2. Empire/nomad relations, following the invention of settled agricultural society and agricultural empires
3. Relations of 'sovereign equality', the traditional focus of International Relations
4. Relations of 'global governance', emerging with neoliberal capitalist globalization

Despite this chronological progression, van der Pijl illustrates that the emergence of new modes of foreign relations does not render the previous modes obsolete, but rather subordinates and transforms the old social relations without destroying them – generating a new social form that evidences both a new, irreducible dynamic as well as a continuing susceptibility to pressures and tendencies (often regressive) linked to older forms. Thus, in the era of 'global governance', the re-emergence of tribal modes of foreign relations is not a surprise, but the very consequence of the imposition of neoliberal capitalist globalization upon older tribal forms in an attempt to subject them to market discipline without eradicating the historical patterns of allegiance by which they are constituted. This suggests that the contemporary capitalist global political economy exhibits a complex amalgamation of dynamics, including structural pressures related uniquely to the capitalist organization of production, as well as other mutually constitutive political, ideological and cultural dynamics that cannot be reduced to the logic of 'capital', but which nevertheless indelibly impact the evolution and development of capitalist practices.[20]

The focus of the current study is on the former issue – examining the structural pressures unique to capitalism, and how these relate to global crises. It is clear, however, that these crises, undermining industrial civilization's own productive capacities, are increasingly chipping away at the contemporary facade of international order and unleashing all manner of historical allegiances, affiliations and latent social bonds in ways that will further amplify the impact of those crises.

Capitalism and the Social Relationship with Nature

The global political economy is a world-scale social system currently functioning in the context of the ideology of neoliberal finance capitalism, and adopting specific assumptions about the natural world and human nature. These assumptions underpin a framework of policy and behavioural principles concerning the production, distribution and consumption of resources by social actors – identified as beneficial, overall, to society. These principles are regulated, protected and enforced not only by the establishment of stable sovereign states that provide appropriate socio-political, cultural and legal backdrops for their functioning, but also through an array of interstate agreements and network of intergovernmental institutions that together constitute a global governance architecture.

The global political economy therefore can be seen as composed of a variety of interpenetrating and interdependent structural features, namely:

1. An implicit spectrum of worldviews suitable for different classes based on (often unstated but nonetheless prevalent) assumptions about the natural world and human nature;
2. Ensuing policy prescriptions for how human societies should relate productively to nature – that is, how natural resources should be exploited, distributed and consumed by human societies;
3. An implicit set of values and ethical axioms which establish broad guidelines for human behaviour in accord with these prescriptions, and which rationalize the latter as 'good';
4. A social, cultural and political architecture at state and global levels, in the form of institutions of government as well as associated institutions of media, education and so on, serving to regulate and reproduce this worldview, these behaviours, and these values.

These features are not simply suspended in mid air within 'the system', but are actively promulgated as the dominant paradigm by key agents of the most powerful classes which dominate the world's productive resources at the system's centre. Thus, they are never set in stone, but are subject to continual recalibration and development in the context of the trajectory of conflict between different social groups and classes. Marginalized groups are able to develop their own responses to the dominant paradigm across all these features (and through forms of political activity), which in turn elicit reactionary responses from elites. The outcomes of this continual contestation rest with the overall 'balance of class forces' – the variety of sources of power available to these different groups and the manner in which they are mobilized. This balance, however, can be tilted in many different ways, and operates not simply in terms of quantitative resources, but also in terms of the efficacy and dynamic of their mobilization relative to different populations, their geographies, and their cultural inclinations, among other issues.

A viable post-carbon civilization will need to adopt a political economy that fundamentally transforms these four key features. To explore the kinds of transformations that will be necessary, we will first need to explore the characters of these structural features and their specific roles in sustaining the system. This will pave the way for recognition of the kinds of changes required to surpass the problems that these structures have created. This analysis does not assume that any of these features are ontologically or causally prior to any other, but rather that they are organically interdependent – changes in one will effect corresponding changes in the others.

The current form of neoliberal finance capitalism is not just capitalism – it is a historically specific form of capitalism that emerged in the late twentieth century, with structural properties that have built upon and in fact worsened the destructive tendencies of previous forms of capitalism.

At its core, capitalism is a specific relation of production based on the *dispossession of labour*, a politically constituted social relationship whose historical origins and structural dynamic have been compellingly dissected by Ellen Wood and social historian Robert Brenner, head of the University of California's Center for Social Theory and Comparative History. The first capitalist system, they show, was formed in England in the 1600s. Previously, under the feudal and later absolutist systems, exploitation was premised on 'tributary' forms of rule. In other words, pre-capitalist social systems were largely defined by the *access of peasant producers to their means of subsistence*, in this case, productive land for agricultural or pastoral use. There were, of course, wide and significant variations in the shapes that pre-capitalist systems took in different regions of the world. But their common feature was essentially that producers lived off the land, and ruling classes used force to compel them to pay tribute. In England, peasants were forced by local lords (under feudalism), and later by the Crown (under absolutism), to pay a percentage of their produce in the form of a tax or rent. Lords did not exercise a direct claim to the land as such, which formally belonged to the peasant producers who inhabited and worked it. But they exploited their control over the means of violence to extract their tribute from peasants in their localities. Peasants' ties to the land also meant they were politically unfree, serfs who were the subjects of their lords – whose authority derived ultimately from military power. Under pre-capitalist systems, there was little incentive to invest in improving the means of production. To increase surplus revenues required escalating the means of force at one's disposal, and deploying it to dominate larger territories, populations of which could be subjected to increasingly harsh regimes of tribute-payment.

This situation changed with the mid-sixteenth-century rural enclosure movement. Faced with a variety of eco-demographic social crises due to the plague, lords faced increasing difficulties in extracting tribute from peasant producers, particularly due to lower rents and escalating peasant rebellions. They responded to these crises by adjusting their claims, with Crown approval. Rather than demanding tribute, they began to claim legal ownership of the land itself, and exercised violence to force the peasants off the land. As this process intensified, greater numbers of peasants were expelled from the countryside. They no longer had access to their own means of subsistence and were homeless, leading them to seek out shelter in the towns and cities. As rural populations thus declined, the urban population swelled.

Enclosure created a historically novel social condition. *Peasants were now dispossessed from the land, and had no means of subsistence* – except to sell their labour power to those who now exerted control over that land. In other words, a quite unpredictable transformation in the relations of production was effected by the response of lords to crises in the tributary system. Thus, a new class of wage labourer began to emerge in England, where dispossessed peasants were compelled by their changed circumstances to work for others who now owned the means of production. While the new wage labourers were politically free and equal, they were also economically subjugated by their condition of dispossession.

The effect was a transformation in England's very relations of production. *These new social relations were the basis of capitalism*, and they generated an unprecedented dynamic. Those who owned the means of production, in this case capitalist landowners, could now maximize their profits by lowering wages. The lower the wages paid to the new wage labourers who worked their land, the greater the profits they could acquire. Profits could also be enhanced by producing larger quantities of better quality (or apparently better quality) goods. The better the produce, the larger the quantity, and the better value for money, the more likely it would outsell competitors on the market. Capitalists whose workers produced poor quality goods at higher cost would inevitably have difficulty in the market compared to those whose workers produced more efficiently, at lower cost, at greater speeds, and with higher quality results.

So capitalist social relations also transformed the very logic of the market, which was now driven by competition between rival capitalists. If they wanted to out-perform their rivals, capitalists had to do far more than just cut wages. They had to *reinvest a portion of their profits in improving the means of production*, finding cheaper sources of raw materials, developing better technologies to harvest them and convert them into goods or manufactures, locating faster and cheaper channels for transporting them to markets, and so on. Failure to do so would result in lower sales, being pushed out of the market, and eventually business failure. Thus capitalism created new conditions of competition and unlimited growth. In fact, *unlimited growth itself became a condition not simply for maximizing profits, but for capitalist survival.* Without continually expanding profits, businesses would be unable to invest larger surpluses in improving the means of production. Failing to develop new, cheaper and more efficient methods and technologies of production would mean a less competitive business, declining sales, and ultimately the collapse of the enterprise. For the first time, then, under capitalism, the unrestricted maximization of profits, and thus unlimited growth, became an imperative for capitalist actors *as a matter of sheer survival* within the capitalist market. In the words of Sussex University social scientist Benno Teschke:

> On this basis, capitalism does not mean simply production for the market, but competitive reproduction in the market based on a social-property regime in which propertyless direct producers are forced to sell their labour-power to property-owners. This separation of direct producers from their means of reproduction and their subjection to the capital relation entails the compulsion of reproduction in the market by selling labour-power in return for wages. This social system is uniquely dynamic, driven by competition, exploitation and accumulation.[21]

Yet capitalism does not render pre-capitalist techniques of exploitation irrelevant or impossible. Capitalists are still perfectly capable of utilizing force directly in order to maximize surplus. The fundamental difference is that while under pre-capitalist social systems such techniques were sufficient, under the pressures of the capitalist market, direct force is unlikely to maximize surpluses alone without investment in developing the means of production. However, this also depends on the social context – if circumstances prevail in which direct force is more effective than investment in production, capitalists are at liberty to utilize it. Capitalism thus represents a specific kind of relationship between human society and the natural world, generating a pressure for unlimited growth based on unconstrained exploitation of natural resources, requiring the continual technological advancement of the means of production.

Key Structural Problem: *The dispossession of labour – the separation of the majority of the population from access to the means of subsistence; the ownership of the means of production – productive resources and the technologies to extract and exploit them – by an elite minority, with 'ownership' itself constituting a construct allowing the 'owner' to use, modify or alienate a resource as seen fit.*

The Expansionary Logic of Capitalism

Capitalism's general expansionary logic derives from the pressures generated by the capitalist relation of production, whereby labour, separated from the means of its own subsistence, is commodified and sold to the owners of the means of production, who in turn are driven to maximize their surplus value (profits) in competition with other capitalists. Depending on the historically specific and regional conditions, capitalists may well remain free to resort to traditional 'pre-capitalist' forms of exploitation to obtain profits, including slavery, extortion, and war outside the sphere of capitalism. But the new social relation of capitalism itself – the dispossession of labour – opens up new methods of maximizing surplus value, including 1) driving down the cost of labour (i.e. wages), and 2) driving down the cost of production. These fundamental methods can in turn be achieved by different secondary methods, such as 3) increasing productivity by employing more workers at low wages; 4) reinvesting in the means of production to generate new labour-saving technologies to increase productivity; 5) monopolizing markets or raw materials; and 6) marketing and propaganda techniques.

All of these methods come up against their own inherent limits, potentially engendering a crisis of capital accumulation, and a subsequent reconfiguration, refinement and thus qualitative expansion of the capitalist system and its policies in order to continue and expand accumulation.

Driving down labour costs within a closed system could impoverish workers to such a degree that they are increasingly unproductive due to deprivation, resulting in either the increasing debilitation of the labour force or the possibility of intensifying working-class resistance and revolt. Such a process can not continue indefinitely, and can only be surpassed by capitalist expansion that seeks out low-cost labour outside the local territory, permitting labour costs at home to rise to levels more acceptable to the domestic working class. Similarly, attempting to sustain higher numbers of low-cost labourers to maximize production would ultimately serve only to shrink the supply of available labour, thus increasing the demand for workers and pushing wages back up. Capital therefore seeks not only to expand in search of cheaper labour reserves beyond its home territory; it also reinvests a portion of its surplus in the labour-saving technologies which, by increasing output per worker, drives down the cost of production and maximizes productivity while decreasing the demand for labour and permitting lower wages – or perhaps higher wages for the few.

There are two further methods of maximizing surplus value that follow from this: 7) maximizing productivity by seeking out a) new means of production, and b) cheaper means of production – including for instance new and cheaper agricultural land, or reserves of raw materials and minerals (as inputs into both the product itself and the expanding and diversifying technologies of production); 8) competing with other capitalists by monopolizing as much as possible not only the available means of production, but also the market itself by diversifying the goods produced for sale.

These methods, which capitalists must for the most part deploy at home if they wish to sustain capital growth, drive the expansion of capitalism, but do not automatically appear – they ultimately derive from the strategies of capitalist actors, as subjects with particular interests that can only be discerned in their historically specific context. The use of these devices by capitalists permits a process of extra-territorial capitalist expansion in the search for cheaper and more lucrative reserves of labour and raw materials. Thus Marx argued that under the competitive pressures of capitalist production, a form of economic growth and expansion ensues in which increasing numbers of the world's population are appropriated by capitalists as sources of labour for the creation of surplus value, while the cycle of production and consumption is perpetually expanded both internally (by penetrating and commodifying ever more aspects of human activity) and externally (by absorbing ever greater numbers of people into the market sytem).[22]

Whereas pre-capitalist systems necessitated the union of military, political and economic structures – with political power premised on the direct application of violence for the forcible extraction of surplus from peasant producers, capitalism allowed the role of force, though no less important in reality, to seemingly recede into the background. Here force was used firstly to establish exclusive capitalist domination of productive resources at the expense of labour; and then was applied in order to regulate this condition of dispossession through private property rights, laws of contract, and other legal mechanisms. This permitted the differentiation of military-political and economic forms of power into separate spheres that were nevertheless interdependent. Capitalism is thus 'eminently compatible' with the separation of political and economic spheres engendered by the territorial-complex of state sovereignty; at the same time it is unhindered in its mobility across sovereign borders because its constituent relations of exchange and accumulation, by their fundamentally 'economic' nature, can cross national politically-constituted boundaries. As such, while a specific constellation of capital originates and operates from within a given demarcated territory, it does not practically recognize that territory's borders as a boundary of its operations.[23]

'Capital as representing the universal form of wealth is the limitless, measureless drive to transgress its limits', notes Marx. 'The quantitative frontier of surplus value appears to it only as a natural limit, as necessity, which it perpetually seeks to overcome, perpetually to transgress.' From its inception, capitalism's logic has been global. In the pursuit of the maximization of surplus value, in the search to find lower-cost reserves of labour and raw material, as well as larger markets for profitable investment, capitalists do not recognize territorial limits. Hence capitalists will eventually seek to open the entire world to capital investment and accumulation. In Marx's words, capital accumulation is 'an end in itself, for the valorization of value only within this constantly renewed movement. The movement of capital is therefore limitless.'[24]

Yet while capitalism thus underpins the drive for 'globalization', this is clearly not a natural or spontaneous process – for capitalist expansion requires the forcible subordination of far-flung productive resources and labour reserves to the original centres of accumulation; distant societies to be politically reorganized in order to open their markets to capitalist penetration; and, if possible, the imposition of new governance mechanisms by which to regulate the integration of these territories into the capitalist system, and subordinate them to the compulsion for unlimited growth.

Key Structural Problem: *The globalization of capitalist social relations – the total domination of the world's productive resources by a minority, and the worldwide intensification and acceleration of the dispossession of labour.*

Cycles of Crisis and the Generation of Inequality

Another feature of capitalism which tends to exacerbate its expansionary tendencies is the cycle of growth and contraction. Marx's exploration of the underlying logic of capitalism's cyclical crises should not be understood as delineating the specific nature of every such crisis, but rather as a theoretical exercise that illustrates the general tendency of the system toward crisis. He argued that the very compulsion to maximize surplus value by maximizing production will eventually saturate the market, creating a crisis of overproduction or overaccumulation that manifests as a fall in the rate of profits – an overabundance of unsellable stock such that capital piles up without avenues for productive investment. The capitalist class responds to this crisis by curtailing production and shedding workers in an attempt to minimize its losses, increasing unemployment. When the store of products is depleted to the point where there are too few products to meet demand, overaccumulated capital is creatively destroyed through mechanisms such as devaluation – leading to a crisis of underproduction. Thus, the cycle begins anew with the drive to expand the employed labour force, invest in technological innovations, and so on, in order to maximize production. Through this process of cyclical crises, less efficient capitals are wiped out in bankruptcies caused by lower prices, permitting only the most competitive and efficient to survive and hence improving the technical quality of capital. Thus Marx theorized what we now know as 'business cycles', identifying this tendency toward cyclical crises as an inherent dynamic of the capitalist social relation.[25]

Although capitalism thus grows increasingly refined and powerful through these cyclical 'boom' and 'bust' crises, Marx argued further that this nevertheless demonstrates a long-term structural tendency for the rate of profits to fall. To accumulate capital, the capitalist must invest in raw materials, plant and machinery (constant capital) and in labour (variable capital). The drive to increase the productivity of labour fuels accelerating technological change, which means that the ratio of constant capital to variable capital varies across different units of capital. The 'technical composition of capital' denotes the productivity of labour, measured in terms of raw material used and goods produced. The relationship between this and the value it embodies (i.e. between the raw materials/goods produced, and their value) is for Marx the 'organic composition of capital'. The problem is that as individual capitalists seek to use less than the socially necessary labour time when producing commodities in their sector, they invest increasingly in technology. This means that the organic composition of capital – *in other words, the value of raw materials used, machinery deployed and goods produced* – continually rises. This also means, in effect, that the more that labour is squeezed out by more advanced and expensive technologies, the greater the necessary investment in such technologies, and thus the greater the degree to which profits need to be reinvested to further improve productivity.[26] Additionally, the value of new technologies/techniques is reduced over time. This means that after several production cycles, not only are current technologies of production devalued, but further technological improvement now requires costly investment – *not* in existing cheapened technologies (which would be pointless in terms of productivity), but in developing new, innovative techniques. As the late Marxist theoretician Chris Harman notes:

> The fact that undertaking investment would cost less after the second, third or fourth round of production does not alter the [high] cost of undertaking it before the first round [at the point of innovation]. The decline in the value of their already invested capital certainly does not make life any easier for the capitalists. To survive in business they have to recoup, with a profit, the full cost of their past investments, and if technological advance meant these investments are now worth, say, half as much as they were previously, they have to pay out of their gross profits to write off that sum. What they have gained on the swings they have lost on the roundabouts, with 'depreciation' of capital due to obsolescence causing them as big a headache as a straightforward fall in the rate of profit.

The implications of Marx's argument are far reaching. The very success of capitalism at accumulating leads to problems for further accumulation. Crisis is the inevitable outcome, as capitalists in key sections of the economy no longer have a rate of profit sufficient to cover their investments. And the greater the scale of past accumulation, the deeper the crises will be.[27]

These theoretical observations have been borne out empirically. Several economic studies show that world capitalism has indeed suffered from a 'long downturn' since the early 1970s, although they emphasize different causes. Rather than being a crisis of overproduction as such, this has been a crisis of 'overaccumulation' – surplus capital with too few outlets for productive investment and continued growth. There is a wide range of evidence that the rate of profits in the US and the West has declined since the 1970s, evident in various phenomena including a productivity slowdown (compared to historical standards), a striking decline in the growth rate of real wages tied to a decline in the profit rate, and a slowing accumulation of capital.[28]

Sociologist Immanuel Wallerstein, a Senior Research Scholar at Yale University, attributes the contemporary tendency for the rate of profit to fall to rising costs from wages, material inputs for production, and taxes. Even if industries are shifted to locales where labour is cheap, wages there too gradually rise as workers in the new zones of production make gains through class struggle. Also, costs of material inputs for production have risen. Wallerstein highlights examples such as detoxification – the waste byproducts of production are traditionally dumped at negligible cost. But as the world is increasingly running out of prospective dumps, the cost of detoxification is rising, and governments are under pressure to make users pay the costs of disposal. Similarly, as primary resources are exploited for production at increasingly faster rates, natural processes of renewal are unable to keep pace (an issue we will return to later). This creates new pressure to restrain resource use or invest in resource renewal, generating new costs. Finally, Wallerstein points to rising costs from taxation due to increasing government expenditures on security and welfare. While technological advances bring increasingly expensive weapons and security products, demands for public welfare in terms of education, health and lifelong income have also risen. This is due to the political dynamic of the world capitalist economy, in which the dominant strata is pushed to make concessions to potentially dangerous lower classes, and is also due to rising populations, lower mortality rates, and the adoption of more advanced (but more expensive) technologies. Thus, traditional methods of combating the tendency of the falling rate of profit are failing, leading to an intensifying squeeze in profits. Wallerstein thus concludes that the capitalist 'world-system' has been in terminal structural crisis since the 1970s, and further that it will make a transition over the next 50 years to another kind of system altogether – elites are attempting to make this new formation even more hierarchical and exploitative than the last.[29]

On a different note, Robert Brenner argues that this 'long downturn' is ultimately the consequence of a decline in prices caused by excess international competition within manufacturing industries.[30] Professor David Harvey, the most cited academic geographer in the world, further argues that there is a crisis of overaccumulation because surplus capital increasingly has difficulty locating opportunities for productive reinvestment. There are a variety of ways to solve this problem systemically, including devaluing the overaccumulated capital. This can involve the development of 'a means to visit the costs of devaluation of surplus capitals upon the weakest and most vulnerable territories and populations'. Thus, the costs and burdens of devaluation are transferred to other regions and nations in territorially bound speculative ventures – in Harvey's words, a 'spatio-temporal fix'.[31]

This profit squeeze over the last thirty years of the twentieth century partly explains the rising predominance of finance in the world economy. As the rate of profit fell relative to rising productivity, and the *sphere of production* in the real economy could offer no solution, the only viable outlet appeared to be outside the real economy, in the realm of financial services. Thus, increasingly, capitalists turned to ingenious financial instruments and mechanisms – exemplified in Harvey's spatio-temporal fixes – to generate profits. Indeed, the sphere of 'structured finance' permitted the generation of massive new profits purely through speculative investment and financial transactions, irrespective of manufacturing capability and production more generally, resulting in asset and real estate booms. Other observers point out that even as the structural crisis of capitalism has continued, capitalists have always adapted, finding new ways to consolidate their power and bolster the accumulation of profits.[32] This emphasis on financial and information services, and the degree to which the real economy became increasingly dependent on them to sustain growth, led to the mistaken belief that Western core powers had become 'post-industrial' societies – that is, somehow beyond the traditional constraints of industrial production. Yet ultimately the resort to 'structured finance' as a means to generate profits in place of production did not overcome capitalism's structural crisis tendencies, but only offset them, setting up the world economy for an even more devastating crisis.

These cyclical crises thus fuel the process of capital's qualitative and geographical expansion within a world system, a process which increasingly transforms and appropriates all of humanity and nature according to capital's logic. In Marx's words:

> Exploration of all of nature to discover new, useful qualities in things; universal exchange of the products of all alien climates and lands; new (artificial) preparation of natural objects, by which they are given new use values. The exploration of the earth in all directions, to discover new things of use as well as new useful quantities of the old;... the development, hence, of the natural sciences to their highest point; likewise rediscovery, creation and satisfaction of new needs arising from society itself; the cultivation of all the qualities of the social human being as the most total and universal possible social product... [are all] conditions of production founded on capital.[33]

Although in his earlier writings Marx saw a progressive element to capitalism's immediate impact, his later writings show a considerable shift in his understanding. He eventually argued firmly that capitalist expansion is an uneven process that simultaneously generates development in the capitalist core and underdevelopment in the imperial periphery.[34] This was not an antecedent of the *dependency* and *world system* paradigms, whose proponents argue that underdevelopment is largely a function of unequal exchange (buying cheap and selling dear);[35] rather it can only be explained in the context of the internationalization of capitalist relations of class and social property (i.e. capital's increasing control of means of production across national boundaries, and the corresponding dispossession of labour from those means), and the increasing capacity to appropriate wealth from pre-capitalist peripheries for conversion into capital at home.

The key factor in the inherently unequal character of capitalism's expansion is *productivity*. For as soon as the capitalist out-innovates her rivals, other rivals may well out-innovate her in turn. Because machines of the same type and function are being produced elsewhere for less and less, the machine's value is reduced and the capitalist's competitive advantage is innovated away. Ultimately, intercapitalist competition is competition over the innovation of productive techniques. As pointed out by economist Alan Freeman:

> When, at any given moment, a capitalist introduces a new cost-saving technique, it is true that this brings an increase in profits without any increase in labour time – for *this* capitalist. For the other capitalists (or other users of labour) the effect of an innovation elsewhere is exactly the opposite. If we consider an Indian producer of cotton cloth faced with competition from Lancashire, what she or he immediately faces is the 'heavy artillery of prices' – a steady, and... dramatic – drop in the value and market price of the cotton cloth.

In this example, the nineteenth-century Indian manufacturer cannot respond by duplicating Lancashire techniques of which he has no knowledge, nor sufficient capital to acquire, leaving only three options: force employees to work harder, cut wages, or, in the last instance, go bankrupt. This degeneration of 'general conditions in India' under the impact of capitalist innovation in Britain 'alters the moral and historical value of the Indian wage and Indian labour', depressing the wage, lengthening working hours, and degrading the conditions of labour. In other words, under capitalism, the 'introduction of superior, more modern technical conditions in one place, simultaneously degradates them elsewhere. Technical backwardness is an inevitable product of technical advance.' Rather than capitalist innovation generating 'an equalisation of profit rates, a convergence of technique and a homogenisation of social conditions', its impact is inherently variegated, leading to 'a dispersion of profit rates, a divergence of technique, and a heterogenisation of social conditions'.[36]

Key Structural Problem: *Cyclical generation of socio-economic crises of over-accumulation and a long-term structural tendency for the rate of profit to fall;*

which exacerbate inequalities between historically advantaged core capitalist states and peripheral states relegated to being suppliers of cheap raw materials and labour, and markets for core agricultural (and other) exports.

Capitalism and the Exhaustion of the Earth

The unevenness of capitalism is not related solely to this tendency to marginalize peripheries in the course of its expansion, nor simply to the crisis tendency tied to the cycle of overaccumulation and underemployment rooted in the property relations of human labour. Implicit in Marx's theory of labour value is an ecologically sensitive energy-centred theory of value, emphasizing capitalism's tendency toward *crisis through self-exhaustion*. According to the late Teresa Brennan, Schmidt Distinguished Professor of Humanities at Florida Atlantic University, Marx focused on human labour power (variable capital) as the primary determinant in production of surplus value (profits), rather than natural resources and technology (constant/fixed capital) – mirroring the mistake made by neoclassical economists.[37]

But Brennan points out that, although he did not elaborate it himself, Marx sowed the seeds of a much broader understanding of production when he noted that all human communities appropriate 'the natural conditions of labour, of the earth as the original instrument of labour as well as its workshop and repository of raw materials'. Brennan thus argues that 'all natural sources of energy entering production should be treated as variable capital and sources of surplus-value'.[38]

Extending Marx's explanation of labour power as the energy transferred to a person by means of nourishment, she utilizes the first law of thermodynamics on the conservation of energy to point out that energy derives not solely from human labour power, but also from the non-living means and technologies of production themselves – whatever form they take, from minerals to oil:

> For a particular capital to stay in the race, it must speed up the materialisation of energy, the value added by natural substances in production, and so speed up the rate of surplus-value extraction. While this speeding-up diminishes overall use-value in the long term, it works compellingly in the short term. It leads to speedier profit via speedier production. But this speedier production is out of joint with the time of natural production.[39]

The generational time required for the natural reproduction of the earth's resources, in other words, is outpaced by the speed of capitalist exploitation – which by its profit-maximizing growth imperative accelerates the scope and speed of production using technological methods which are incapable of paying back to nature the real value of the energy extracted. According to Brennan, a historical limitation of Marx's theory is 'the assumption that the reproductive time of natural substances cannot be speeded up; that is, the tomato plant's inner workings cannot be regulated or controlled'. But as she points out, 'capital has indeed found social ways of speeding up various forms of natural reproduction... by regulating their conditions of production'.[40] This creates an irresolvable problem:

> [E]nergy cut off from its natural source of origin cannot continue to regenerate the natural energy that enabled it to come into being. However, as more and more natural substances assume this form, and as more and more natural substances are bound in fixed capital, they require more and more supplies of external energy to enable them to keep producing, due to two obvious facts: the binding and fixing of natural substances in the form of commodities diminishes the overall quantitative supply of nature; and these bound forms have no regenerative energy of their own.
>
> Yet as capitalism cuts back the supply of natural substances, it not only diminishes the energy or 'nourishing matter' labour-power relies upon, it also diminishes the conditions for other natural sources of energy to regenerate as well. Natural reproduction time is seriously out of joint. Capital, to cope, must speed up agricultural production, which of course makes things worse. Speeding up agriculture in turn feeds into the speed of acquisition through consumption, which depends on wide networks of distribution, as does industrial production. In fact, this is where the two forms of speeding things up are tied together. To acquire more at a faster speed for production means distributing more, and consuming more natural substances. This puts a pressure on agriculture to produce at a rate comparable with other aspects of production and distribution. As available sources of energy in either agriculture or industrial production are diminished, capital has to create routes for the old sources of energy to come from farther away, or create new sources of energy altogether, ranging from chemicals to 'nuclear power.'
>
> The reproduction of natural substances and sources of energy goes according to a natural cycle taking a certain amount of time. The hyperactive rhythm of capitalism, on the other hand, means that the conversion of substances into energy leads to the further conversion of already converted substances into other energetic forms, which are more and more coming to be the sources of energy overall, with lunatic environmental consequences. As by now should be abundantly plain, the reality is this: short-term profitability depends on an increasing debt to nature, a debt that must always be deferred, even at the price of survival.[41]

The inherent contradiction in capitalism, for Brennan, is 'between the substantial energy of nature – the source of value – and the artificial speed of production'. As capital binds more and more living energy in forms which cannot reproduce themselves (without relying on further intensive extraction of energy), 'nature is not paid for in terms of its reproduction time'. Instead, maintaining large-scale profit requires that natural substances must be exploited faster than they can reproduce themselves.[42] In summary, capitalism tends inherently toward the exhaustion of natural resources with no recognition of limits or boundaries, and thus accelerates environmental degradation. But Brennan's seminal reconceptualization of the source of value as energy in nature also has critical implications for capitalism's relationship to geopolitical expansion:

> Yet the contradiction between energy and speed is not simple. Two forms of time are at issue: the generational time of natural reproduction, and speed, the artificial time of short-term profit. Speed, as I have already indicated, is about space as much as time as such. It is about space because it is about centralisation and distance. Speed, measured by distance as well as time, involves a linear axis, time, and the lateral axis of space... I want to emphasise how, in the capitalist or consumptive mode of production, the artificial space-time of speed (space for short) takes the place of generational time. For to the extent that capital's continued profit must be based more and more on the speed of acquisition, it must centralise more, command more distance, and in this respect space must take the place of generational time.[43]

The disjointedness between capitalism's artificial speed of production for maximum short-term profit, and the longer generational time of natural reproduction, translates into a tendency for capital to *absorb and subordinate growing tracts of geographical territory containing natural resources.* Obviously, this tendency does not act in itself, but functions as a direct pressure on capitalist actors who then develop strategies to navigate and resolve these pressures.

Key Structural Problem: *The capitalist social relation's pressure to exploit ever larger quantities of the world's natural resources for production for profit, on an ever expanding geographical scale that will soon be beyond the natural world's ability to self-replenish.*

NEOLIBERAL COMPUTATIONAL FINANCE CAPITALISM

The Imperial Globalization of Industrial Capitalism

Three historical developments served to transform the character of capitalism as it expanded across the world: the rise of industry, the evolution of the banking system, and the proliferation of financial instruments by information technologies. The industrial revolution in England switched the focus of natural resource exploitation from organic to inorganic materials, and thus to hydrocarbon resources. This in turn permitted the extraction of vast sources of energy which could be injected into the circuits of technological development and economic growth with diverse applications. Industrialization thus paved the way for the creation of technologies that allowed the unprecedented integration of world markets, even as colonial powers were increasingly competing for control over territories.

All countries including the most isolated were affected by global changes, largely driven by the activities and rivalries of the advanced industrial states.

There is no doubt that the capitalist economy was going global, increasingly so as it was extended to remote parts of the world, characterized by falling transport costs and trade barriers, unprecedented integration of world markets, rising international capital flows as a share of national output, and increasing cross-border migration.[44] As British historian Eric Hobsbawm notes, the very essence of capitalism 'recognized no frontiers':

> [I]t functioned best where nothing interfered with the free movement of the factors of production. Capitalism was thus not merely international in practice, but internationalist in theory. The ideal of its theorists was an international division of labour which ensured the maximum growth of the economy. Its criteria were global: it was senseless to try and grow bananas in Norway, because they could be produced much more cheaply in Honduras.[45]

The developed capitalist core as a whole constituted 80 per cent of the international market, and its dominant position meant that it determined the development of the peripheral countries which effectively became satellite economies, and whose growth was contingent on their assigned role in supplying the needs of the core. Furthermore, such growth also meant new destinations for the export of goods and capital from the core.[46]

The dividing up of the imperial periphery – or the South – between the rival capitalist empires was accompanied by the establishment of regional specialization in the production of raw materials, food crops and stimulants. Although some of this specialization had already existed since the earlier era of mercantilism, the remainder occurred in response to capitalist development. This contributed to the creation of an international division of labour controlled largely by the competing industrial empires: 'Regional emphasis on a monocrop or single raw material product demanded, in turn, that other areas raise crops to feed the primary producers, or furnish labour power to the new plantations, farms, mines, processing plants, and transport systems.'[47]

While some regions specialized in foodstuffs or raw materials for industry, others processed and consumed the latter, returning manufactured goods. In this way, the South was forcibly drawn into the capitalist world market. Regional specialization included the conversion of entire regions into plantations for the production of one or two cash crops – including sugar, tea, rubber and coffee – and the dedication of other areas to agriculture and industry. Consequently, regional concentration on particular means of production, to produce particular commodities, implied their dependence on other regions for the production of other commodities.[48]

'Malaya increasingly meant rubber and tin, Brazil coffee, Chile nitrates, Uruguay meat, Cuba sugar and cigars', observes Hobsbawm. The technological development that was so crucial to the advancement of industrialization relied on the import of raw materials from areas outside of Europe. The internal combustion engine, for instance, required oil and rubber. Although oil was then largely acquired from America and Europe, competition for access to

Middle East oilfields was intensifying. Rubber came from the rainforests of the Congo and the Amazon, and later from Malaya. Tin could be extracted from Asia and South America. Major reserves of copper, needed by the electrical and motor industries, could be located in Chile, Peru, Zaire and Zambia.

Consequently, the colonial and semi-colonial territories, as specialized producers of one or two products, became dependent on the sale of their produce on the world market. At that time, Southern industrial development was not welcomed by the already industrialized metropoles, since their function 'was to complement metropolitan economies and not to compete with them'.[49] This situation changed more and more in the twentieth century as firms and corporations attempted to outsource manufacturing and labour to Southern peripheries in the search for ever lower costs and greater efficiency. Yet even then the establishment of industries in the South was primarily for the production of inexpensive exports to Western markets, and Southern industry remains dominated by Western ownership.

Key Structural Problem: *The systemic over-dependence on hydrocarbon resources for industrial production, sustained by an international division of labour designed not to meet the needs of local populations, but purely to maximize profits for primarily Northern banks, corporations and governments.*

Fractional Reserve Banking

Capitalism's inherent tendency to generate widening social inequalities, and intrinsic vulnerability to cyclical crises, was exacerbated by the adoption of fractional reserve banking as the fundamental basis of the monetary system. Its application in the context of capitalism endowed this new social system unprecedented financial power across borders, magnifying exploitative opportunities and associated inequities.

The origins of fractional reserve banking lie in seventeenth-century Europe, in which trade was conducted primarily in gold and silver coins. As coins were difficult to transport in bulk and required secure storage facilities, people began depositing them with goldsmiths, who owned the most impenetrable safes. Goldsmiths issued convenient paper receipts which were often traded in place of the coins they represented. These receipts were also issued when people requested loans – instead of receiving actual coins, they received paper receipts representing a quantity of gold. At any one time, only between 10 and 20 per cent of a goldsmith's receipts were actually returned for redemption. This allowed them to accelerate their lending. They exploited these circumstances and issued paper receipts as loans for quantities of gold several times larger than what was actually stored in their depositories. By charging interest, they ensured that loan repayments generated a profit and were able to avoid liquidity problems by maintaining an equivalent to the value of 10 to 20 per cent of outstanding loans in order to meet the demand for gold. In effect, they created 'paper money', in the form of receipts for loans of gold worth several times the quantity of gold they actually held. By charging interest, and as long as they were careful not to overextend their 'credit', the goldsmiths could make substantial profits without producing anything of value. This generated a cycle of debt and deprivation from which the goldsmiths benefited:

> Since only the principal was lent into the money supply, more money was eventually owed back in principal and interest than the townspeople as a whole possessed. They had to continually take out loans of new paper money to cover the shortfall, causing the wealth of the town and eventually of the country to be siphoned into the vaults of the goldsmiths-turned-bankers, while the people fell progressively into their debt.[50]

By the nineteenth century, this model had developed into a system called fractional reserve banking, whereby private banks issued their own banknotes in sums of up to ten times their actual reserves in gold. Under this system, only a fraction of the total deposits managed by a bank were kept in reserve to meet the demands of depositors. In 1913, in the United States, the private banknote system became consolidated into a national system under the Federal Reserve – a corporation owned by several of the largest American private banks, and granted legal authorization to issue Federal Reserve Notes and lend them to the US government. These notes came to form the basis of the national money supply. At the time, dollars were still backed by gold by 40 per cent. By 1933, in the context of the Great Depression, which saw the flight of American gold to Europe and thus a depreciation in the value of the dollar, the currency was delinked from the gold standard. This created a new system whereby bank reserves consisted of nothing but government bonds, that is, IOUs or debts.[51]

In other words, banks were literally creating fiat money – money out of nothing – simply by entering a record into the books, as a loan to governments charged with interest. As noted by British economist John Gilbody, the commercial banking system, as a byproduct of its normal business operations, 'can create bank deposits to an extent which far exceeds the amount of notes and coin being explicitly deposited in the banking system by the general public in any period'.[52] The role of fractional reserve banking in the monetary system is also well described by American economists Arleen and John Hoag in their *Introductory Economics*:

> A fractional reserve system requires that banks hold a percentage of their deposits as reserves... Banks are permitted to make loans as long as a portion of the total deposits are held in reserve. Anything not held in reserve can be loaned out. Because of a fractional reserve system, banks can create money. They do not create paper money or coins, they create checking accounts. Since checking accounts are a part of the money supply, the banks create money...
>
> Who creates and destroys money? Not the government as is commonly thought, but the commercial banking system. Any bank that accepts checkable deposits is a commercial bank. Commercial banks are privately owned and profit-seeking businesses. *Money is created when banks make*

> *loans. Money is destroyed when loans are paid back...* If you go to a bank and ask to see the money, you might be shown the money in the bank vault. This could be a substantial sum, but it is only a small part of the total money held by the bank. Banks have money that *exists only in their books*. This money is the list of accounts that they owe their depositors and must pay out when the depositor demands it by writing a check.
>
> Can these amounts really be money? When a depositor writes a check, the check serves as a medium of exchange and is therefore money. These checkable deposits are the major form of money that the bank holds. When someone takes a loan from a bank, the bank records the borrower's name in its list of accounts and credits the individual with a checkable deposit of the amount of the loan. What did it cost the bank? Nothing. Well, maybe only a few bytes of memory. Are we certain that the checkable deposit of this newly created loan is money? Certainly, since checkable deposits are money and now checkable deposits and hence the money supply have increased. The borrower may now make a purchase with a check. Or the borrower may cash the check and take currency out of the vault instead. Now there is more currency in circulation than before. *Whether the purchase is made by check or currency, money is created when banks make loans.*[53]

In the US Federal Reserve's own words:

> Of course, they [banks] do not really pay out loans from the money they receive as deposits. If they did this no additional money would be created. What they do when they make loans is to accept promissory notes in exchange for credits to the borrower's transaction accounts.[54]

In an earlier document, the Federal Reserve concedes that the expansion of money supply is ultimately premised on the creation of government and public debt. It observes that:

> [The] depositor's balance... rises when the depository institution extends credit – either by granting a loan to or by buying securities from the depositor. In exchange for the note or security, the lending or investing institution credits the depositor's account or gives a check that can be deposited at yet another depository institution. In this case, no one else loses a deposit; the money supply is increased. New money has been brought into existence.[55]

Thus, it is not governments that issue money, but rather governments (and the public at large) that borrow money from central banks, money which is created by fiat simply by book or computer entry, the repayment of which is charged at compound interest. But this generates an elementary contradiction which is the basis of further crisis tendencies in the financial system. Professor Rodney Shakespeare, of the postgraduate economics and finance programme at Trisakti University, Jakarta, points out:

> Consequently, the bank has created enough money with which to pay the principal (the original sum) of the loan – which on repayment is cancelled – but it has not created enough money with which to repay the interest... The banking system as a whole, therefore, has to pump more money (bearing interest) into the economy thereby occasioning more and more interest-bearing debt which occasions an increasing need to create more and more interest-bearing debt... Very obviously, if there is to be enough money in the economy as a whole, there is always an impelling need to create more and more debt. And because the debt is interest-bearing, the need gets greater and greater. Thus the level of debt in the world is certain to increase, and to greatly, even exponentially, increase, so that it will never be capable of being repaid... This is preposterous and, at some stage, must result in the collapse of the whole economic system.[56]

Fractional reserve banking *at compound interest* thus dramatically amplifies the unequal structures of the global financial system under capitalism. As the economy continues to grow, it simultaneously becomes more indebted. This acts as a magnifier on the capitalist pressure to maximize profits, by intensifying the demand for creating more debt-money with which to settle enlarging and unrepayable debts. The need for more debt-money thus intensifies pressure to increase production for profit, and thus *exponentially escalates the limitless exploitation of natural resources*. The corresponding expansion of the money supply also results in a long-term structural tendency toward inflation along with the devaluation of currency, which together amplify the impact of 'boom and bust' crises.

Key Structural Problem: *The world monetary system is based on fractional reserve banking – that is, the creation of fiat money as credit at interest – which serves to subjugate the population to a growing and unrepayable debt that is the basis of self-reproducing profits for banks; and which compounds the imperative for unlimited growth through unconstrained exploitation of natural resources.*

Computational Finance – The New Capital Accord

Under *neoliberal* capitalism, which came to the fore by the late 1970s, these tendencies were exacerbated yet again. While capitalism is consistent with a variety of state-political forms, neoliberalism in theory advocates the abolition of government intervention in the economy, including the lifting of all restrictions on commerce and trade, the removal of barriers to manufacturing, and the elimination of tariffs. In practice, as seen in chapter four, these principles are applied selectively in order to open Southern markets to Northern capital, whose own markets – far from 'free' – remain protected behind trade and tariff barriers. Nevertheless, even here, government intervention to maintain economic stability and ensure the common good is officially considered anathema, paving the way for policies of deregulation and liberalization.

With the rise of information technology and the growing sophistication of global telecommunications and computer modelling techniques, the global banking system has bridged the distance between the financial/monetary system and the real economy, generating immeasurable profits at the cost of the unprecedented globalization and augmentation of national debts. With most of the world's productive resources already controlled by a minority of private interests, overaccumulated capital has difficulty locating new investment outlets for the development of productive capacity. Ingenious ways of using this capital to generate further profits are thus devised (such as Harvey's spatio-temporal fixes), facilitated by the worldwide interconnection of markets. As Shakespeare points out, the banking system for the most part 'does not allocate money to new productive capacity and, instead, allocates it to derivatives; to the bidding up of existing asset prices (such as house prices); to consumer credit; to putting individuals, companies and whole societies into debt, indeed, to anything but the real productive economy'.[57]

The unprecedented acceleration of this sort of nonproductive investment in fictitious financial instruments really took off after 2000. In that year all the world's countries enacted the New Capital Accord, an international regulatory system for banks devised by the G10 countries through the Bank for International Settlements (BIS) at Basel, Switzerland, with extensive lobbying and input from US and UK financial bodies. The New Capital Accord was implemented in individual countries through their own domestic legislation. For example, in June of that year, the UK passed into law the Financial Services and Markets Act which implemented the New Capital Accord, while the EU's adoption of the accord compelled all its member states to implement it under nationally specific names. The aim of the accord was to establish a new framework for the management of capital by banks and financial institutions, based on the alleged *risks* involved in their investment practices. In effect, the New Capital Accord abolished the system of fractional reserve banking, stipulating that banks could now allocate capital against assets based on the asset's level of risk, *as determined by the bank itself*. Financial institutions were now designated the most capable judges of the levels of risk within a financial instrument, and were therefore permitted to devise their own in-house quantitative risk-assessment models. Yet now, unlike with the fractional reserve system, there was no upper limit placed on the actual quantities of money that banks could lend. *As long as banks calculated that the quantities of money lent involved levels of risk equal to assets held in reserve, they were free to lend as much as they wanted.* Although regulators like the FSA were obligated to 'backtest' the bank's in-house risk-assessment models, the practical requirements for testing were very slim. In effect, the New Capital Accord deregulated the financial system, completely eliminated limits on banks' capacity to create credit, and eroded the possibility of establishing meaningful oversight mechanisms. Banks were now permitted to conduct their own unaccountable forms of risk assessment, in order to generate and invest in new financial instruments. Whereas under the original Capital Accord banks typically were able to leverage their capital up to a maximum ratio of

12.5 to 1 (that is, they could create money as credit in quantities up to 12.5 times what they held in cash as reserves), the New Capital Accord permitted levels of credit creation at any multiples declared safe by the bank's own internal methods of risk assessment. Yet banks routinely exploited the lack of effective regulation and oversight to fudge the levels of risk. This, for instance, permitted JP Morgan to take on a total derivatives nominal exposure of *$93 trillion, against capital holdings of only $45 billion – that is, capital leverage of more than 2000 to 1.*[58]

The vast bulk of these investments, based on the repackaging and sale of debts through innovative structured financial instruments which spread risks throughout the financial system, were rationalized and legitimized by the use of quantitative computer modelling software purportedly designed to measure risk. But the validity of any quantitative model rests on the validity of its assumptions. Given that these models were insulated from reality and inherently biased toward minimizing apparent levels of risk to facilitate lucrative sales of junk financial products, they operated to systematically conceal rather than identify real risks. Worse, the sophistication of information technologies permitted dangerous financial transactions to be completed instantaneously in numbers so large *that their real content was never really known by anyone*. Economist Michael Hudson, who also foresaw the subprime mortgage and global banking crises as early as April 2006, describes this new system of computational finance as utterly divorced from the real economy:

> [T]hanks to the fact that insurance companies are a Milton Friedman paradise – not regulated by the Federal Reserve or any other nation-wide agency, and hence able to get the proverbial free lunch without government oversight – writing such policies was done by computer printouts, and the company collected massive fees and commissions without putting in much capital of its own... It turned out, inevitably, that some of the financial institutions that made billion-dollar gambles – usually in the form of a thousand million-dollar gambles in the course of a few minutes or so, to be precise – couldn't pay up. These gambles all occur in microseconds, at strokes of a keyboard almost without human interference...
>
> The financial machines that placed the trades... were programmed by financial managers to act with the speed of light in conducting electronic trades often lasting only a few seconds each, millions of times a day. Only a machine could calculate mathematical probabilities factored in regarding the squiggles up and down of interest rates, exchange rates and stock and bonds prices – and prices for packaged mortgages. And the latter packages increasingly took the form of junk mortgages, pretending to be payable debts but in reality empty flak.[59]

Hence the primary mechanism of maximizing profits under neoliberal capitalism is no longer even remotely tied to relations of production as such, except in that control over the world's productive resources is a precondition for the overwhelming financial power of the Northern-dominated global banking system. Yet in terms of the actual mechanisms of profit-maximization, the new techniques of computational finance facilitated the spontaneous generation and multiplication of profits by the systematization of debts.

> A hedge fund does not make money by producing goods and services. It does not advance funds to buy real assets or even lend money. It borrows huge sums to leverage its bet with nearly free credit. Its managers are not industrial engineers but mathematicians who program computers to make cross-bets or 'straddles' on which way interest rates, currency exchange rates, stock or bond prices may move – or the prices for packaged bank mortgages. The packaged loans may be sound or they may be junk. It doesn't matter. All that matters is making money in a marketplace where most trades last only a few seconds. What creates the gains is the price fibrillation – volatility.
>
> This kind of transaction may make fortunes, but it is not 'wealth creation' in the form that most people recognize... The pretense, of course, is that all this frenetic trading creates real 'capital.' It certainly does not do so in the classical 19th-century concept of capital. The term has been decoupled from producing goods and services, hiring wage labor or from financing innovation. It is as much 'capital' as the right to conduct a lottery and collect the winnings from the hopes of the losers. But then, casinos from Las Vegas to riverboats have become a major 'growth industry,' muddying the language of capital, growth and wealth itself.[60]

Clarifying that the global financial system nevertheless remains capitalist at its core is the precondition for the proliferation of debt-based structured financial services: the dispossession of the majority of the population from the conditions of production (a dispossession which is the constitutive basis of capitalist social-property relations). In the context of a monetary system derived from interest-laden fractional reserve banking, this social relation made both working and middle classes vulnerable to a growing dependence on credit for the sustenance of housing and other forms of consumption. The massive debts thus accumulated were then fraudulently repackaged and sold in the form of new financial instruments, which were falsely 'insured' by firms lacking sufficient capital to do so.

Key Structural Problem: *Computational finance effectively transformed the banking system's ability to create fiat debt-money through the development of fraudulent quantitative models concealing ballooning levels of risk, making possible the creation of exorbitant profits virtually ex nihilo – but the costs were socialized (spread throughout the financial system and borne by the tax-paying public) and backlogged (inaccurately recorded and forgotten) until the banking system collapsed under the weight of its own unsustainability.*

THE POLITICAL-LEGAL REGULATION OF THE GLOBAL IMPERIAL SYSTEM

The International System of Nation-States

The 'globalization' of capitalism is by no means a smooth or natural process, but rather involves a combined and uneven development which unfolds due to the asymmetrical and fragmented structure of the states-system, and the inherently unequal and exploitative nature of capitalism itself. The political function of nation-states in this context is to establish the regulatory politico-legal framework of private property rights, formal democracy (i.e. elections and voting, even if they are subject to corruption and fraud), and contract law, which allows capitalists to operate freely and relatively unhindered. This does not mean the international system of nation-states is the only structure by which capitalist relations of production can be regulated. But it does mean that nation-states currently provide the simplest politico-territorial form through which appropriate conditions are brought about to allow capitalist forms of exploitation to be implemented around the world. This partly explains the post-Cold War Anglo-American propensity toward 'nation-building' both as a humanitarian discourse legitimizing interventionist foreign policy, and as a process specifically designed to create spaces which by their political-legal structures are susceptible to the penetration of Anglo-American capital.

However, the cohesiveness of national identities within nation-states, and the recognition of the state as legitimate protector of rights and provider of public goods and services, are diminishing as states respond to global crises by militarizing their societies wholesale. These processes illustrate the increasing irrelevance of democratic party-politics for actual policymaking, which continues to be dominated by powerful financial conglomerates via their extensive lobbying power and direct involvement in financing campaigns for the leading political parties – with the result that no political party can win an election without intimate ties to private corporate power.[61]

As such, the nation-state, and with it the increasingly discredited politics of national representative democracy, has ever more provided a vehicle by which the national and transnational capitalist class can oversee the institutionalization of appropriate political-legal regulatory frameworks by which capitalist interests can be maximized locally and regionally – largely at the expense of the needs and interests of the majority. In other words, rather than capitalist globalization erasing the relevance of the nation-state, it is globalizing precisely through the political mechanisms provided by national structures. Yet in doing so, it is both undermining national representative democracy, while accelerating crises that prompt social groups to increasingly construct 'Others' onto whom escalating anxieties, uncertainties and tensions can be projected. This should not come as a surprise, given that capitalist social-property relations played an integral role in emerging grammars of inclusion and exclusion amongst the new classes of male adult property owners, mediated internationally through geopolitical and economic competition, warfare and territorial occupation.[62]

This means that as global crises intensify, elites will increasingly use the state and national identity as principal mechanisms of social control, and attempt to polarize communities to deflect popular anger away from the state and inward toward 'excluded', problematized communities. But, in turn, the state will increasingly be viewed by citizens as an obstacle to genuine popular participation and well-being, and as a major cause of social antagonisms and extremism, while common class interests will unite different ethnic and religious groups across national divides against the transnational elites that dominate them. These opposing forces will react and interact in unpredictable ways, as the unravelling of international order unleashes forgotten but always latent nationalist, cultural and tribal affiliations associated with older forms of social relations.

By the mid twenty-first century, as global crises and hydrocarbon energy scarcity erode the capacity of states to sustain industrial methods of 'total war', those states will find it increasingly difficult to maintain their monopoly of the means of military violence. While this implies a possible proliferation of so-called 'low intensity' conflicts and even de-industrialized methods of warfare, it also suggests an eventual end to mass casualties from interstate war, and greater opportunities for communities to join together in transnational bonds of peace.

Key Structural Problem: *Nation-states and national identity, as the primary loci of 'legitimate' political representation, are increasingly detrimental both in terms of their negative impact on democracy, and the categories by which human social groups ascribe identities to themselves. While current forms of representative democracy are increasingly susceptible to erosion and manipulation by nondemocratic military-financial forces, ethnic and national markers of difference can increasingly be exploited to polarize communities and legitimize political violence against the 'Other'.*

The Structure of the Global Political Economy

In the current global political economy, the international system of states itself must also be stabilized and regulated, which explains the significance of contemporary global governance institutions like the UN, the World Bank, and the IMF. Today this system can only be understood in terms of its domination by American finance capital, backed by overwhelming military power – the very social forces that constructed the system in their own interests after the Second World War.[63]

In this context, the global 'free market' serves the interests of Northern developed economies by extracting wealth from the South. Whereas the core states continue to produce high-value goods, most less developed countries are consigned to supplying raw materials and cheap labour for the North, an inherently unequal division of labour whose consequence is the systematic widening of North-South inequalities. Northern access to 'capital, technology, transportation and large, affluent markets' is unmatched by the South, lending the former a position of structural dominance. The South is compelled to sell labour and land as primary commodities at low prices, such that increasing Southern participation in the global 'free-trade' regime only serves to aggravate poverty.[64]

Notwithstanding decolonization, there are key continuities in the relationships between Southern postcolonial states and Northern former colonial powers. The postcolonial 'world economic order is by far more centralized, concentric, and institutionalized at the top', according to Canadian development economist Professor Jorge Nef:

> Its fundamental components are trade, finance, and the protection of the proprietary rights of international business. Rules, actors, and mechanisms constitute a de facto functional system of global governance where core elite interests in the centre and the periphery are increasingly intertwined.[65]

Nef delineates four central components of the global political economy:

1. The historical and structural context in the form of global macroeconomic restructuring, including the end of the Cold War and the collapse of the Soviet Union, 'construed as a victory of capitalism'; the 'disintegration' and increasing 'marginalization' of the South; and the unprecedented speed and scale of globalization.[66] We may add to these issues the global economic and structural impact on the South of imperialism from the fifteenth to the early twentieth centuries.[67]
2. The global political economy's cultural or ideological underpinning, which is the 'hegemonic and homogenizing' discourse of neoliberalism.[68]
3. The global political economy's formal decision-making structures including the World Bank, the International Monetary Fund (IMF), regional banks such as the European Bank for Reconstruction and Development (EBRD), the Organization for Economic Cooperation and Development (OECD), the World Trade Organization (WTO), the Group of Eight, and the major trading blocs such as ASEAN and NAFTA. These international institutions are correlated with national domestic structures such as ministries of finance, treasury boards and central banks, the two levels formally connected through international agreements and external conditionalities attached to fiscal, monetary, and credit policies.[69]
4. The processes – facilitated by the above mechanisms – by which transnational economic elites negotiate policies that serve their common class interest in the expansion and maximization of private capital accumulation.[70]

Global governance institutions exert a mediating influence within the framework of a 'world order' centred on a single hegemonic power (the United States).[71] These intergovernmental agencies actively reproduce the structure of global trade, production and finance regimes, which are in turn responsible for increasingly transnational class structures which effectively set the 'rules of the game' and systematically create patterns of advantage and disadvantage in the world economy.[72] Countries that industrialized early, establishing a dominant position in the global hierarchy of political and economic power, thereby benefit from structurally entrenched advantages reinforced by a powerful transnational regulatory framework.[73] Thus a 'transnational managerial class,' attempting to construct world order in a manner that conforms to its perceived interests, operates through international institutions on the pretext of development.[74] This is not to underestimate the significance of internal divisions and conflicts between competing factions within this transnational elite, often along regional lines – but the global political economy is regulated in accordance with common class interests. After 9/11, as the US has increasingly sought to assert its interests militarily, this inter-elite rivalry for financial and geopolitical domination of key strategic regions was temporarily ameliorated in Iraq (2009), but it is escalating – particularly in Central Asia and Northwest Africa.[75]

Key Structural Problem: *Global governance institutions are mobilized primarily as mechanisms for the US to regulate the international system in its own interests, rather than in the interests of the majority of the system's members. Calls to reform these institutions (to make the UN more democratic, for example, or to make the World Bank more transparent) have therefore consistently failed.*

CAPITALISM'S PHILOSOPHICAL BASE

Ontological Ideology: The Reduction of Human Nature and the Natural World

Underlying the capitalist system, in particular the neoliberal capitalism system, is an implicit set of ideas about the world. These may not operate as stated facts, but largely as unquestioned axioms. Specifically, neoliberal ideology has its origins in neoclassical economic theory founded after the Second World War, which rests on three primary assumptions:

1. People have rational preferences among outcomes that can be identified and associated with a value.
2. Individuals maximize utility and firms maximize profits.
3. People act independently on the basis of full and relevant information.

These assumptions in turn derive from deeper philosophical and ontological assumptions about human nature, which were meticulously deconstructed by the renowned American economist Thorsten Veblen, 'arguably the most original and penetrating economist and social critic that the United States has produced'.[76] Veblen noted that neoclassical economics attempts to explain everything in terms of rationalizing, egoistical, self-maximizing behaviour, based on a reductionist conception of human nature devoid of serious historical or empirical substantiation. The significance of Veblen's philosophical objections to the theoretical foundations of contemporary neoliberal capitalism is highlighted in the context of the current conjuncture of converging global crises. In Veblen's words, the underlying problem is 'a faulty conception of human nature', which is perceived in purely hedonistic terms:

> The hedonistic conception of man is that of a lightning calculator of pleasures and pains, who oscillates like a homogenous globule of desire of happiness under the impulse of stimuli that shift him about the area, but leave him intact. He has neither antecedent nor consequence. He is an isolated, definite human datum, in stable equilibrium except for the buffets of the impinging forces that displace him in one direction or the other.[77]

Elsewhere, he points out that in neoclassical economic theory:

> [T]he center and circumference of economic life is the production of what a writer on ethics has called 'pleasant feeling.' Pleasant feeling is produced only by tangible, physical objects (including persons), acting somehow upon the sensory. The inflow of pleasant feeling is 'income' – 'psychic income' net and positive. The purpose of capital is to serve this end – the increase of pleasant feeling – and things are capital, in the authentic hedonistic scheme, by as much as they serve this end. Capital, therefore, must be tangible, material goods, since only tangible goods will stimulate the human sensory pleasantly. Intangible assets [such as ethical values], being not physical, do not impinge upon the sensory; therefore they are not capital. Since they unavoidably are thrown prominently on the screen in the show of modern life, they must, consistently with the hedonistic conception, be explained away by construing them in terms of some authentic category of tangible items.[78]

This has the extraordinary implication that all destructive consequences of neoliberal capitalism are considered as perfectly normal, natural and inevitable, in light of the unquestionable axiom that the functioning of capitalism is optimally beneficial to individual and social welfare. Therefore, by definition, capitalism cannot be destructive, and whatever harmful phenomena exist in society have nothing to do with capitalism and must be viewed rather as the consequence of something else entirely (such as the mistakes of governments):

> The hedonistic... economics is a system of taxonomic science – a science of normalities. Its office is the definition and classification of 'normal' phenomena, or, perhaps better, phenomena as they occur in the normal case... In the hedonistically normal scheme of life wasteful, disserviceable, or futile acts have no place. The current competitive, capitalistic business scheme of life is normal, when rightly seen in the hedonistic light. There is not (normally) in it anything of a wasteful, disserviceable, or futile character. Whatever phenomena do not fit into the scheme of normal economic life, as tested by the hedonistic postulate, are to be taken account of by way of exception. If there are discrepancies, in the way of waste, disserviceability, or futility, e.g., they are not inherent in the normal scheme and they do not

call for incorporation in the theory of the situation in which they occur, except for interpretative elimination and correction.[79]

In summary, Veblen recognized that neoclassical economics simplistically assumes that all human activity, in all societies, is invariably utility-maximizing behaviour based on an insatiable drive for material fulfilment. Capitalism only renders these activities more effective, and any failures can without exception be considered as unfortunate exceptions due to human error and the insufficient application of capitalist principles. This ideological framework thus explains the inability of conventional economists to grasp the direct connection between global crises and the structure of the global political economy. But, as Veblen also remarks, if the hedonistic conception was actually accurate – disregarding human values such as trust and cooperation – it would lead to the complete breakdown of society:

> If, in fact, all the conventional relations and principles of pecuniary intercourse were subject to such a perpetual rationalized, calculating revision, so that each article of usage, appreciation, or procedure must approve itself de novo on hedonistic grounds of sensuous expediency to all concerned at every move, it is not conceivable that the institutional fabric [of society] would last over night.[80]

Indeed, neoclassical assumptions about human nature are implicitly embedded in a wider reductionist ontology. The rationalization of human self-interest in the maximized consumption of material goods assumes that individuals are fundamentally separate entities whose needs and desires can be primarily defined in material terms, and for whom social and economic activity can only be competitive and egoistic. Further, the natural world is viewed as a potentially limitless stockpile of material goods which can therefore be subjected to the ceaseless extraction of resources. Nature in general is perceived as a possibly hostile environment whose purely physical units are fundamentally disconnected, atomistic and thus self-interested, and it should therefore be efficiently organized according to the rationalizing imperatives of capitalism. Within this worldview, ethical values such as love, justice and compassion appear to have no real basis, precisely because they cannot be materially located in a natural world seen purely in terms of its potential for maximal exploitation in service of unlimited material consumption.

The global political economy, evolving in the context of these ideological assumptions about human nature and the natural world, is responsible for the generation of multiple global crises which threaten the future not only of industrial civilization, but of all life on earth. These fundamentally destructive consequences of neoliberal capitalism constitute direct empirical confirmation that its underlying ideological assumptions about human nature and the natural world are, in fact, false. American economist Richard Norgaard of the University of California, Berkeley – a former visiting scholar at the World Bank – argues that the bankruptcy of the global political economy and the paucity of obvious solutions evidences a deeper ideological problem in relation to a reductionist and crude overarching materialist philosophy of existence. Global crises are therefore symptoms of a deeper philosophical malaise, and concerted efforts are required to develop an alternative holistic way of understanding human nature and the natural world as inter-embedded:

> We understand environmental crises as material problems. They are 'matters' of too much waste now and insufficient resources in the future, issues of too many species that might be needed by our descendants being driven to extinction, or problems of forcing the global physical parameters that control our climate. And these material imbalances are driven by our material needs and desires. Thus, as we try to resolve our environmental crises, we inevitably emphasize material solutions. Engineers seek ways of getting more from less, economists argue we need to internalize externalities in order to improve economic efficiency, and environmental ethicists encourage us to think of less as more.
>
> Let me suggest an obvious alternate framing. Environmental problems are material problems because materialism is the dominant vision in Western philosophy. Materialism forms both modem understandings of how things work in the world and our visions of the good life. This is not to argue that the material world is irrelevant or reality simply a social construction that can be reconfigured. Nor is it to deny that the vast majority of the people on earth at the end of the twentieth century are materially deprived. Clearly, environmental problems are due to excessive material consumption by the rich. And environmental problems are bound to continue because we have no alternate vision of the future for those who are now poor. But instead of simply thinking less materially, we should be looking for alternatives to materialism in our image of the good life. On a deeper level, however, we need to question our conception of the world as material objects apart from us. We need richer bases for understanding how we interrelate with nature.[81]

The implication is that crude materialism – that is, a reductionist worldview in which reality and even humanity are understood as purely a collection of disconnected material entities – has unwittingly become our defining ethical paradigm, precisely because it has been uncritically adopted as the primary conceptual lens through which we understand the world. Yet the massive destruction wrought by application of this crude materialist ethical paradigm, in the current form and practices of the global political economy, illustrates that the paradigm does not correspond to reality, and that a post-materialist paradigm based on justice and compassion, and which recognizes human life and nature as inherently *inter*connected (rather than *dis*connected), actually corresponds more closely to reality.

Key Structural Problem: *The implicit philosophical and ontological assumptions underlying neoliberal capitalism posit an extreme form of ideological materialism, reducing the world to a collection of physical,* *disconnected, atomistic, self-interested and thus inherently conflictual units. These assumptions are clearly mistaken, as they implicitly shape a global political economy that is simultaneously destroying itself and the natural world.*

Ethical Value System: The Rationalization of Egoism and Consumerism

Based on neoliberal capitalism's ideology of crude materialism, the global political economy operates on the basis of a corresponding value system with equally reductive and deleterious implications for social morality. Implicit within neoliberal capitalism, this value system acts as a subliminal framework of ethical and philosophical assumptions about the way life *ought* to work. Some of the most credible critics of globalization have recognized this. According to Dr David C. Korten, a Stanford University Business School graduate who worked for the US Agency for International Developent (USAID) for more than a decade, underlying capitalism is not merely a specific ethical and ontological outlook on life and nature, but an almost fundamentalist theology in which boundless material profit is the sole criterion of value:

> In the quest for economic growth, the free market ideology has been embraced around the world with the fervour of a fundamentalist religious faith. Money is its sole measure of value, and its practices advance policies that are deepening social and environmental disintegration everywhere. The economic profession serves as its priesthood, it champions values that demean the human spirit. It assumes an imaginary world divorced from reality and it is restructuring our institutions of governance in ways that make our most fundamental problems more difficult to resolve yet to question its doctrine has become virtual heresy.[82]

Implicitly or otherwise, ethical values are always embedded in social systems. Any given social system is linked to its fundamental conception of nature, and a corresponding value system. As noted in greater detail at the beginning of the chapter, energy is the bedrock of society. The way a social system derives and makes use of energy defines its relationship to nature, whose resources are the source of that energy. The way a social system exploits natural resources, the way it produces, consumes and functions, is therefore inseparable from the way that system conceptualizes itself, nature, and the relationship between the two. In other words, a given social system not only consists of a set of particular social, political and economic structures, but rests on a body of (often unconscious) assumptions about human nature, the functioning of the natural world, and the way humankind ought to relate to its natural environment. It is within these assumptions that one finds a set of values, often equally unconscious, regarding what is good and bad for humanity.

In this sense, while moral values may well be human constructs, they are also much more than that – they are *categories that determine the usefulness of different kinds of social behaviour*. Value, in other words, is tied to action, and the essential core of 'value' as a concept is exactly that: *worth*. Action is valuable if it is worth doing, but if it is not worth doing, then it is not valuable. If value is concerned with a measure of *worth*, then an ethical value is a category that implies certain types of action are *intrinsically worth doing*. Moral values therefore designate special kinds of social behaviour as having this sort of *intrinsic worth*.[83] All social systems are tied to values, because they encourage certain types of behaviour while discouraging and prohibiting others. These incentives and disincentives depend on the nature of that specific social system; on the way that social system conceives of human beings and nature; in summary, on a particular conception of life and the natural world – whether or not that conception is unconscious and implicit. There is therefore an objective dimension to values: *whether they work or not, whether they lead to forms of behaviour that generate well-being, or achieve the opposite*. Values are more likely to be objectively useful if they reflect reality – human nature, the nature of the world, and their mutual interrelationship.

This has crucial implications for understanding the value system that underlies neoliberal capitalism. The most penetrating deconstruction of this value system is by the Canadian philosopher John McMurtry, University Professor Emeritus-elect at the University of Guelph in Ontario, and editor of the 'philosophy' volume of the UN *Encyclopedia of Life Support Systems*. McMurtry points out that the global political economy is based on 'an unexamined and absolutist value system'. Capitalist science and technology, transnational trade apparatuses, Anglo-American wars and the intensifying suppression of civil liberties are all symptoms of a 'new totalitarianism cumulatively occupying the world and propelling civil and ecological breakdowns'. Although conventional neoliberal economic theory is supposed to be value-free, objective, scientific:

> To the contrary, the positions of a 'value-free' or positivist economics still presuppose as given and self-evident the value system of private property rights, the pursuit of self-interest and profit, and the monetized production and exchange of needed goods as the foundational, regulating norms of their analyses... The principle of self-serving for money accumulation in all conditions, with no constraining obligation to one's own society or to use-value production, has become the overriding, abstract imperative of market doctrine. The promotion of the public interest, on the other hand, has become a token mantra with no demonstrated connection to money self-maximization.[84]

Like Korten, McMurtry sees in this 'free market' ideology subtle but deeply ingrained fundamentalist strains that elevate materialist market principles of self-interest and profit maximization to an unquestionable God-like status:

> We find that government and their leadership now assume that the value system of the global market is to be the proper order to social organization and that societies must be made to adapt to this order as the needs and demands of the market requires. The market is not now seen as a structure to serve society, rather society is seen as an aggregate of resources to serve

> the global market... No traditional religion had declared more absolutely the universality and necessity of its laws and commandments than the proponents of the global market doctrine.[85]

The direct effect of this crude materialist ethical system can be discerned in relation to the well-being of those populations which live under its jurisdiction. Based on World Health Organization data, British clinical psychologist Oliver James showed in his book *Affluenza* that citizens of English-speaking nations were twice as likely to suffer from mental illness as those of mainland Europe over a twelve-month period.[86] Deeper analysis exposes a direct link between mental illness and the social inequalities generated in the context of neoliberal capitalism, 'which largely explains the greater prevalence among English-speaking nations'. According to James:

> By this I mean a form of political economy that has four core characteristics: judging a business's success almost exclusively by share price; privatisation of public utilities; minimal regulation of business, suppression of unions and very low taxation for the rich, resulting in massive economic inequality; the ideology that consumption and market forces can meet human needs of almost every kind.[87]

James encapsulates this tendency of neoliberal capitalism to generate mental illness using the metaphor of a virus, and suggests that its origins and impact can be explored by treating it as a kind of disease: affluenza.

> Selfish capitalism causes mental illness by spawning materialism, or, as I put it, the affluenza virus – placing a high value on money, possessions, appearances (social and physical) and fame. English-speaking nations are more infected with the virus than mainland western European ones. Studies in many nations prove that people who strongly subscribe to virus values are at significantly greater risk of depression, anxiety, substance abuse and personality disorder. Follow the logic? Selfish capitalism infects populations with affluenza; it fosters mental illness; English-speaking nations are more selfish capitalist – ergo, more prone to illness... Blair's encouragement of free market capitalism has boosted spiralling levels of British mental illness. The net consequence for true Labour voters has been to force us to become more or less severely virus-infected.[88]

James updated these findings in a more detailed assessment of the scientific evidence, *The Selfish Capitalist*, in which he seeks to examine the psychological impacts of the specific form of neoliberal capitalism promoted primarily by the United States and Britain for the last thirty years. Here, James reiterates his three essential arguments: 1) the emergence of industrial capitalism brought with it the proliferation of unprecedented types of mental illness that previously had not existed, across the core industrial states; 2) in particular there has been a rapid rise in the prevalence of mental illness since the 1970s, as compared with the period between 1945 and 1980; 3) comparing the English-speaking nations with other capitalist nations, such as mainland Europe and Japan, evidences a higher prevalence of mental illness in the former, where neoliberal ideology is most influential.

He begins by reviewing the psychological literature on causes of emotional distress, which shows overwhelmingly that 'most cases, perhaps the vast majority of them, are responses to environmental factors'.[89] Citing the findings of cultural psychiatrist Arthur Kleinman, James points out that:

> three-quarters of the hundreds of diseases listed in the DSM [Diagnostic and Statistical Manual of the American Psychiatric Association] are found almost exclusively in the USA and in Westernised elites, whether Asian or European. Problems such as multiple personality disorder, eating disorders and chronic fatigue syndrome are very largely caused by industrialisation and are virtually unknown in pre-industrial communities.[90]

He then compares levels of emotional distress between industrially developed nations. Investigating data from an ongoing World Health Organization (WHO) survey of 15 nations, he shows that in the US 26.4 per cent of the population suffered a period of mental distress in the preceding twelve months, compared with 14.9 per cent for the Netherlands and 4.3 per cent for Shanghai. Reading the WHO survey another way, he shows that, taken together, 23 per cent of the populations of the USA and New Zealand experience emotional distress. This compares with an average of 11.5 per cent for six western European nations and Japan. This difference is due primarily to the shift since the 1970s within the English-speaking nations toward neoliberal capitalism, supplanting the more welfare-driven neo-Keynesian capitalism which had advocated a greater government role in keeping unemployment down and maintaining economic stability. Western Europe and Japan have avoided the extremes of neoliberalism in their domestic spheres, pursuing policies of concerted social spending and safety nets for working people. In particular, James shows that a principal cause of distress in these core industrialized states is simply 'materialism', defined as 'placing a high value on money, possessions, appearances and fame'.[91]

Across the Atlantic, James's work is corroborated by the American psychologist Tim Kasser of Knox College, who concludes:

> People who are highly focused on materialistic values have lower personal well-being and psychological health than those who believe that materialistic pursuits are relatively unimportant. These relationships have been documented in samples of people ranging from the wealthy to the poor, from teenagers to the elderly, and from Australians to South Koreans.[92]

Key Structural Problem: *Neoliberal capitalism is premised on a crude materialist value system which penetrates the entirety of human experience, commodifying everything from human life to the natural world in the service* *of maximizing a 'good' defined largely in terms of material consumption. Yet this ethical system is responsible not only for the escalation of multiple global crises, but simultaneously for high levels of psychological illness and distress among neoliberal capitalist nations. It is therefore a value system divorced from reality, incommensurate with human nature and the natural world: Ideological materialism is unnatural.*

8
Prognosis – The Post-Carbon Revolution and the Renewal of Civilization

Global crises signify the worsening unsustainability of modern industrial civilization, and the inevitable demise of neoliberal industrial capitalism (defined here as a multidimensional social system by which the human species relates to the natural world). By mid century, the non-viability of industrial capitalism will be an undeniable fact. What will the human species develop in its place? Whatever the answer, it will arise in a new landscape where hydrocarbon resources and the social forms associated with their exploitation are ultimately irrelevant. The arrival of the post-carbon age, therefore, cannot be stopped. It can only be prepared for, by building self- and community-resilience, from the ground up. But resilience is not enough. As modern industrial civilization erodes, people will be increasingly attracted to traditional, cultural, and mythical ways of understanding the world that are associated with older social forms and bonds, and which may well be combined haphazardly with aspects of modern scientific thought. There is a clear danger that this will create a space for the emergence of extremist, regressive and exploitative new modes of living.

For civilization to not merely survive, but to make full use of its current predicament as a vehicle of *transition* to a more sustainable and harmonious mode of organization, it is necessary to develop a clear vision of what post-carbon civilization ought to look like, so that this vision can guide efforts to improve resilience and highlight the most fruitful directions that social movements should pursue. In the preceding analysis, we identified the *Key Structural Problems* of the global political economy that require urgent reform. Having interrogated the systemic nature of these problems, we are well equipped to discuss the direction of the reforms that will need to be explored in the passage to a post-carbon world. Hence, these *Key Structural Problems* are each reviewed below, with regard to their implications for meaningful social transformation.

*1. **Key Structural Problem:** The dispossession of labour – the separation of the majority of the population from access to the means of subsistence; the ownership of the means of production – productive resources and the technologies to extract and exploit them – by an elite minority.*

Today about 5 per cent of the world's population owns productive resources. Post-carbon civilization must fundamentally reconfigure the relationship between labour and capital. This will require extensive new thinking on how to increase access to, and ownership of, productive resources for the majority. This does not necessitate the abolition of private property and open markets, nor does it automatically justify the forms of socialist central planning associated with the Soviet Union. Contrary to conventional assumption, there are several viable alternatives. The theory of Binary Economics founded by American economist Louis Kelso with philosopher Mortimer Adler, for example, suggests mechanisms by which *individuals and communities can be apportioned ownership of productive resources in accordance with their labour*. Similarly, Islamic economics advocates that labour itself should be the primary criterion of private ownership of productive resources – ownership is legitimated not through force or tradition, but through work itself. Both perspectives advocate a universal right to individual ownership of productive resources.

There are already several pilot projects which could be models for facilitating the widespread access of individuals to agricultural, industrial and commercial productive enterprises. In the US, over 11,000 companies – together representing 10 million employees – are run on Employee Stock Ownership Plans (ESPOPs). These are majority-owned by their own employees, in some cases with 100 per cent equity participation. Other attempted variants on this scheme which facilitate the spread of capital ownership include Individual Stock Ownership Plans (ISOPs), Consumer Stock Ownership Plans (CSOPs) and Community Investment Corporations (CICs). In any case, all options should be explored in order to develop the widespread distribution of private ownership of productive capital, or in other words, *to facilitate universal access to the means of production by all individuals and communities*.[1]

*2. **Key Structural Problem:** The globalization of capitalist social relations – the total domination of the world's productive resources by a minority, and the corresponding worldwide dispossession of labour.*

Post-carbon civilization must develop mechanisms such as taxes, land reform, and ultimately the reorganization of ownership of the means of production to redistribute productive resources which are now excessively concentrated in the hands of an elite minority. Specific types of resources such as water, energy, minerals, and so on should also be considered part of the Global Commons – that is, not subject to monopoly by private interests, and managed on the basis of participatory organizations. In their theory of Participatory Economics (or 'parecon'), Robin Hahnel, Professor of Economics at the American University in Washington, and US political theorist Michael Albert, propose an economic system based on participatory decision-making through self-managerial producers and consumers councils.[2] This should not necessarily obviate a legitimate role for private enterprise in developing productive resources that are considered to be publicly owned, but ultimately their exploitation (particularly of resources such as water) must aim at meeting the needs of local populations, and this necessarily requires direct community participation in relevant private development enterprises. Once again, this brings to the fore the question of the universal right of capital ownership (that is, use-rights for access to productive resources) for individual labour, with special caveats in place to ensure equality of rights in accessing special types of resources seen as belonging to the Commons. It also necessitates a move to localize and decentralize political power to a degree that permits this sort of grassroots participation.

*3. **Key Structural Problem:** Cyclical generation of socio-economic crises of overaccumulation and underproduction ('boom and bust'); which exacerbates inequalities between historically advantaged core capitalist states and peripheral states relegated to being suppliers of cheap raw materials and labour, and markets for core agricultural (and other) exports.*

An economy that functions without any form of control or oversight designed to regulate private interests is demonstrably crisis-prone. Yet more regulation alone is not the answer. The root of this problem is in the capitalist social relation, the dispossession of labour from productive resources, which generates the pressure to maximize profits through strategies which periodically lead to oversupply and underdemand. With greater capital ownership for labour, this problem will be minimized. However, also relevant here is the role of the fiat monetary system and interest, discussed below. The debt-based monetary system also contributes to periodic inflation, currency devaluation and cyclical crises.

*4. **Key Structural Problem:** The capitalist social relation's pressure to exploit ever larger quantities of the world's natural resources for production for profit, on an ever expanding geographical scale that will soon be beyond the natural world's ability to self-replenish.*

Economic development should be directed not at unlimited growth for its own sake, but at sustainable growth for the specific purpose of catering to the needs of the majority. As economic growth has continued under neoliberal capitalism, profits have been increasingly concentrated among a smaller minority of the world's population, and the benefits of growth accruing to the poor have diminished, creating greater poverty and inequality all round. This brings the very concept of growth into question.[3] As Nobel Prize-winning economist Amartya Sen has argued in his *Development as Freedom* (1999), growth should be redefined as the increase in the capability of all human beings to achieve those things that they most value. While higher income may be part of this, more significant are good health, adequate education, greater longevity, the ability to influence the political decisions that affect one's life, or the freedom to choose alternative lifestyles. Sen in effect calls for the reorientation of economic activity to meet human needs and well-being – as opposed to unlimited material consumption and maximum accumulation of material wealth – which would reverse the tendency for the economy to expand beyond natural limits.[4]

5. Key Structural Problem: *The systemic over-dependence on hydrocarbon resources for industrial production, sustained by an international division of labour designed not to meet the needs of local populations, but purely to maximize profits for primarily Northern banks, corporations and governments.*

The resource base for post-carbon civilization must be renewable. This will include maximum utilization of natural energy sources such as solar, wind, tidal, geothermal and hydroelectric power, as well as research into the potential for new energy technologies including nuclear fusion, zero-point energy, cold fusion, and organic biofuels from human and animal waste (rather than food crops). These alternative methods of energy extraction would be by their very nature far more decentralized and less susceptible to monopolization. This would also provide a new avenue of productive capital investment engendering grassroots participation and opportunities for local ownership, creating new jobs and sustainable economic enterprises.

A cutting-edge model of how this might work in practice was laid out by an interdisciplinary group of experts – including energy consultants, economists, social scientists, and politicians – in a 2008 report by the New Economics Foundation. The report demonstrates that massive investment in renewable energy could make 'every building a power station', creating thousands – potentially hundreds of thousands – of 'green collar jobs'.[5] The Institute of Science in Society (ISIS), an independent London scientific research group, earlier outlined a similar renewable energy blueprint highlighting the necessity for 'a diversity of sustainable renewable energies at medium-, small- and micro-generation scales, according to resources locally available, so that energy is used at the point of generation'. This would save up to 69 per cent of energy currently lost through long-distance transport of electricity from centralized power plants. ISIS also emphasized the revolutionary significance of 'energy-from-waste technologies' largely ignored by governments, for example: 1) producing biogas from organic wastes (agricultural, municipal and industrial) which can be used to generate electricity and even to power cars (as opposed to producing biofuels from food crops); 2) using green algae for capturing carbon dioxide from the exhaust of power plants, coupled with biodiesel production from waste for fuel. New breakthroughs in solar energy mean that available technologies are growing more efficient and more affordable with time, 'and will be an important small- to micro-generation technology especially suited for Third World countries lacking energy infrastructure'. Another little-known technology is the 'production of biodiesel from waste cooking oils and other industrial food wastes, and diesel from waste plastics that cannot be easily recycled into plastics'. Coupled with inputs from wind and hydroelectric power, and with large-scale adoption of organic, locally-produced, low-input sustainable farming, the ISIS's proposals offer serious socio-technological solutions whose implementation could distribute widespread ownership of productive resources and sustainable technologies.[6]

Another notable study in *Scientific American* by Mark Jacobson, Professor of Civil and Environmental Engineering at Stanford University, and Mark

Delucchi, a research scientist at the Institute for Transportation Studies, argued that by 2030 the world could transfer entirely to wind, water and solar energy with an investment of $100 trillion.[7] The study itself suffers from serious problems – such as the omission of costs for constructing millions of new land, sea and air vehicles; costs and practicalities of constructing international electricity transmission lines; failure to account for intermittency; failure to account for global agricultural production; and so on. Apart from retaining centralization in ownership of energy facilities, the model is consequently a technical nightmare. As Gail Tverberg estimates, the real cost of the Jacobson plan is likely double (at least), and a more realistic projected date of completion is 2050 – which would be too late. Catastrophic shortfalls would therefore be unavoidable.[8]

In contrast, an updated 2009 ISIS study of renewable energy potential points to the success of the German model of transition, where feed-in tariffs and other legislation have spurred the rapid growth of a new renewable energy infrastructure combining wind, solar photovoltaic, solar thermal, hydroelectric, geothermal and biomass sources. Germany's renewable energy industry declares that the country can 'reach 100 per cent renewable' by 2050. 'Many politicians and renewable energy experts in Europe see a realistic option of 100 percent renewable energy supply in a commercial market free of any subsidy by 2050', ISIS reports. 'The key is decentralised, distributed generation that provides energy autonomy at the point of use, a model that has proven so successful in Germany.' As Germany is doing, the ISIS blueprint avoids costs and energy losses from electricity transmission over long distances by optimizing small-scale and micro-generation local renewable energy sources.[9]

A more realistic alternative, then, is to drastically reduce industrial overconsumption while transitioning to a localized renewable energy infrastructure – this is already under way, if disparately and unevenly, and could be accelerated through measures like carbon taxing. As of the end of 2008, there were 200,000 households in the US and 40,000 in the UK living entirely off-grid – generating their own electricity, sourcing their own water, and managing their own waste disposal. Richard Perez, a renewable energy researcher at the State University of New York, is an outstanding example. Living entirely off-grid and having paid not a single electricity bill since the 1970s, he nevertheless continues to enjoy modern technology: 'I've got five computers, two laser scanners, two fridge-freezers, a microwave, a convection oven, vacuum cleaners – you name it', he told the *New Scientist*. 'There's an external beam antenna on the roof for the cellphone and a bidirectional satellite for internet connection. I've got 70 kWh stored in batteries that could last me five days. I have too much electricity.' *New Scientist* adds that the costs of going off-grid are increasingly affordable even for individual households, starting at just under $30,000, or £19,000.[10]

Clearly, greater possibilities open up given the prospects of community collaboration and pooling of collective resources. An example from the UK is the borough of Woking, in Surrey. Woking Borough Council has worked on developing local renewable energy systems since the early 1990s, investing money saved through energy-efficiency projects in increasing local energy capacity. The endeavour was a response to the Royal Commission on Environmental Pollution's call for a 60 per cent reduction in carbon dioxide emissions by 2050, making Woking the first local government authority in the UK to adopt a Climate Change Strategy. With centralized methods of power generation wasting two thirds of primary energy, there were potentially vast efficiencies to be gained by generating and transmitting energy close to the communities using it. The results are revolutionary: Woking is currently generating 135 per cent of its energy needs from renewable and sustainable sources. 'Scattered across its borough are mini-power stations, district heating schemes and thousands of electricity-generating cells on roofs', reported the *Guardian*. 'This has made the borough a world leader in providing energy without relying on the national grid. The town centre, including the council offices and Holiday Inn, are entirely energy self-sufficient and surplus electricity is exported.'[11]

6. Key Structural Problem: *The world monetary system is based on fractional reserve banking – that is, the creation of fiat money as credit at interest – which serves to subjugate the population to an enlarging and unrepayable debt that is the basis of self-reproducing profits for banks; and which compounds the imperative for unlimited growth through unconstrained exploitation of natural resources.*

Rather than privately owned banks creating money *ex nihilo* which is then loaned to governments on compound interest, the monetary system requires a principled overhaul. The central bank, on behalf of political bodies representing communities and nations, can create currency as a medium of exchange, without charging interest, on the basis of actually existing depository reserves, with all loans also issued on an interest-free basis for productive capital investments (this would not prevent banks from legitimately charging a nominal administrative fee to depositors). Governments would be able to borrow interest-free from the central bank for investment in public spending to redevelop transport networks, the energy infrastructure, health and education systems, as well as a more localized political and communications infrastructure. The banking system in general would be able to grant interest-free loans to individuals and communities to become directly involved in such innovative development enterprises, promoting more widespread ownership of productive capital throughout the population. Such investment in the development of productive capital would create vast numbers of new jobs as well as permanent incomes from capital ownership. As money supply is created for productive capacity, the private banking system could be restrained from producing new money by gradually increasing the reserves that a bank must deposit with the central bank. With the decrease of debt-money from private banks, interest-free loans from the central bank (administered by the private banking system) will increase, fulfilling the economy's need for credit for productive investment. *The absence of interest would mean that costs socially productive enterprises would be halved, if not quartered.* Private banks would then become depository and investing institutions with two functions: 1) administering the central bank's interest-free loans for productive capital investment; 2) lending depositors' money, with the agreement of those depositors. The effect of this transformation of the banking system would be counter-inflationary, increasing the value and hence purchasing power of money. This effect would therefore create the opportunity for governments to engage in further creation of debt-free money for public spending on productive capital investments, simultaneously maintaining stable prices. Overall, such a restructuring of the monetary and banking system would permit a massive freeing-up of wealth and access to production, feeding into energy (and other) infrastructure transition.[12] A crucial issue which should also be explored is the relationship between currencies and commodities – should mechanisms be introduced to ensure that the money supply is eventually backed by real commodities (such as gold, or energy, or others)?

7. Key Structural Problem: *Computational finance effectively transformed the banking system's ability to create fiat debt-money through the development of fraudulent quantitative models concealing ballooning levels of risk, making possible the creation of exorbitant profits virtually ex nihilo – but the costs were socialized and backlogged until the banking system collapsed under the weight of its own unsustainability.*

The extensive repertoire of quantitative risk-modelling techniques and associated computer-based methods of packaging, reselling and proliferating debts through essentially fictitious structured financial constructs, generating windfall profits, must be prohibited. This of course comes in tandem with the elimination of interest and the gradual reversion of private banking from fractional reserve to a full reserve system, which will abolish the basis of crisis-prone unrestricted profiteering from debt-proliferation. In this case, depositors would be granted a direct stake, proportionate to their deposits, in the investment operations of the private banking system, which would be focused on granting interest-free loans for the development of productive resources, rooting the world of finance in the real economy. Banks would also have a more direct stake in the investment itself, taking the brunt of the risk and sharing the profits, both of these in proportion to the quantity of the investment. Greater regulation is required to ensure the transparency of this process.

8. Key Structural Problem: *Nation-states and national identity, as the primary loci of 'legitimate' political representation, are increasingly detrimental both in terms of their negative impact on democracy, and the categories by which human social groups ascribe identities to themselves. While current forms of representative democracy are increasingly susceptible to erosion and manipulation by nondemocratic military-financial forces, ethnic and national* *markers of difference can increasingly be exploited to polarize communities and legitimize political violence against the 'Other'.*

Post-carbon civilization will need to develop new concepts and mechanisms to facilitate full-scale popular community participation in the political organization of society. These will have to go well beyond our current theories of representative democracy, which have largely failed to guarantee the ability of the people – the *demos* – to participate meaningfully in policymaking, and instead allow powerful vested interests to legitimize their continuing domination of decision-making structures. These new concepts and mechanisms will need to be developed in parallel with more decentralized forms of social organization that allow individuals and communities access to the means of production and custodianship of resources. George Monbiot, in his book *The Age of Consent*, explores a variety of methods and ideas, including new communications technologies, by which grassroots communities can not only form links across national borders, but also develop participatory forums for self-organization.[13] These proposals are not without serious deficiencies,[14] and they require further development in the context of wider social, economic and energy issues, but they are an excellent starting-point for radical new thinking on this subject. The focus of the new thinking, in any case, should be on the decentralization of power and the increasing empowerment of communities in decision-making processes. Communities should mobilize as much as possible through existing political and parliamentary processes, while simultaneously developing their own parallel participatory forums distinct to these, but whose power can potentially increasingly be brought to bear on them.

Post-carbon civilization will also need to be based on recognizing the fundamental unity of the human species, and to see the diversity of ethnic, national and religious groups not as signifiers of irreconcilable difference, but rather as evidence of the depth and richness of the philosophical and cultural resources available for us to draw on in developing civilizational norms and values that are sustainable, in harmony with the Earth, and for the benefit of all peoples. This entails focusing on the deep-commonalities and universal principles that can be used to unite people of different belief-systems, on the basis of respect for all life, and for the Earth.

9. Key Structural Problem: *Global governance institutions are mobilized primarily as mechanisms for the US to regulate the international system in its own interests, rather than the interests of the majority of the system's members. Calls to reform these institutions (such as to make the UN more democratic, or to make the World Bank more transparent) have therefore consistently failed.*

By the mid twenty-first century, the traditional US-dominated global governance architecture will be increasingly irrelevant, impotent and discredited as a means to regulate the international system. As both nation-states and these international structures will lose their relevance, post-carbon civilization will

.o develop new socio-political structures capable of harmonizing the rests of disparate communities and peoples around the world. Yet the most .itical fields of political action in this period will not be the International, but increasingly at local and community levels.

Post-carbon civilization should aim to facilitate and synergize grassroots networks across the world attempting to self-organize on the basis of sustainability, justice and well-being, re-defined in terms of human solidarity with one another and harmonization with nature; as well as to develop the capacity of these groups to coordinate and share ideas, vision and resources through transnational lines of communication that bypass purely hierarchical political structures premised on domination by a few. This does not mean all forms of hierarchy should be automatically excluded, but that they must always be made fully accountable in their entire organizational structure to communities.

As Monbiot notes, some structures already exist in skeleton form, such as the World Social Forum, which brings together grassroots communities from all over the world. These structures could be extended regardless of the lack of political support from mainstream state-institutions, their existence and moral authority acting as a powerful beacon to the rest of the world. The aim would be to 'level' world order away from the existing Western-dominated global governance architecture, by the world's peoples unilaterally erecting their own. Politically, Monbiot suggests we explore the idea of a World Parliament representing peoples of all countries in proportion to population size (he suggests 600 members could represent ten million people each). Economically, he suggests scrapping the World Bank, the IMF and the WTO, in place of an International Clearing Union (ICU) with its own currency and a new International Trade Organisation acting in the interests of communities rather than states. In the ICU, every country is granted an overdraft and credit facility in proportion to the value of its trade. Notwithstanding significant problems with the way Monbiot formulates these ideas,[15] more significant is his unflinching recognition of the possibility of real alternatives. Monbiot's activist imagination illustrates the monumental potential of alternative, decentralized, transnational political, economic and monetary structures that synergize grassroots communities, and demonstrates that communities can collaborate in establishing these structures now. What will give these institutions strength and life is simply the extent to which people actually participate in and benefit from them. By the mid twenty-first century, as global crises render existing national and international politico-economic structures increasingly irrelevant, such alternative structures will become more, rather than less, viable as part of an emerging post-carbon civilization.

10. Key Structural Problem: *The implicit philosophical and ontological assumptions underlying neoliberal capitalism posit an extreme form of materialism, reducing the world to a collection of physical, disconnected, atomistic, self-interested and thus inherently conflictual units. These assumptions are clearly mistaken, as they implicitly shape a global political economy that is simultaneously destroying itself and the natural world.*

Post-carbon civilization will need to be oriented around a fundamentally different understanding of human life as a phenomenon that is inextricably embedded in the natural world. Efforts are required to develop a post-materialist conception of human nature and a holistic philosophy of nature in general, one which sees ethical activity as equivalent to self-fulfilment. This does not require a complete transformation of our knowledge base, but rather a fundamental reinterpretation. This is consistent with the philosophical implications of scientific findings over the last few decades in quantum physics and cell biology. These findings confirm the reality of deep underlying interconnections, not only across the natural world but across the subatomic fabric of physical reality. Such interconnections also find parallels in groundbreaking research demonstrating the role of the environment and consciousness in genetic development. The new science shows that physical spatio-temporal reality is an overarching unity that manifests an imperceptible, all-penetrating deep-structure (the quantum vacuum), all living species are interdependent elements of a single complex system, and that reductionist frameworks of interpretation cannot fully explain the evolution of human consciousness. These revolutionary insights suggest that ethical action is in fact more in tune with the nature of the world than the atomistic and self-centred perspective of neoclassical economics and its ilk.[16]

11. Key Structural Problem: *Neoliberal capitalism is premised on a materialist value system which penetrates the entirety of human life, commodifying everything from human life to the natural world in the service of maximizing a 'good' defined largely in terms of material consumption. Yet this ethical system is responsible not only for the escalation of multiple global crises, but simultaneously for high levels of psychological illness and distress among neoliberal capitalist nations. It is therefore a value system divorced from reality, incommensurate with human nature and the natural world.*

Based on a new, more holistic and scientific understanding of human nature and the natural world, a new ethical system based on human cooperation, grassroots participation and the mutual needs and well-being of all will be increasingly viewed (and justifiably so) as the basis of the rational pursuit of self-maximization. In this alternative post-materialist ethics, the welfare of others will literally be equivalent to the welfare of the individual, and vice versa. The driving force of modern post-carbon societies will not be mass consumerism, but rather the scientifically grounded recognition of the interconnectedness of human beings, all life, and the earth. Self-fulfilment, in other words, will be based on creative communion with others, and with nature, rather than unlimited exploitation of others and of nature for material ends. Such a new ethical system will be capable of uniting different humanist, spiritual and religious traditions in the overarching recognition that moral values objectively harmonize us with nature, and therefore in some way constitute cognitive reflections of humanity's relationship with the natural order. As such, the new ethical system will form the foundation of a new economics, based on what ISIS calls 'a new model of balanced growth based on reciprocity and symbiotic relationships to replace the dominant model of unlimited growth based on rampant competition and the survival of the fittest'.[17]

AFTERWORD

For a viable post-carbon civilization to emerge, scholars, scientists, policymakers, and, above all, peoples, will need to work together in developing alternative social, economic, political, cultural, philosophical and ethical structures. It is not in the scope of this study to explore the possibilities for civilizational renewal, and the various alternative ideas, strategies and solutions currently being proposed and developed around the world in response to global crises. What has been discussed here should by no means be interpreted as a final answer, but rather as a tentative template for the full array of issues, concepts, theories and values that we will need to explore to make viable an ecologically harmonious, politically popular, socially just and spiritually fulfilling post-carbon civilization. It is also worth noting that business as usual is likely to continue to define official state policies over the coming years and decades. While this may partially guarantee horrendous forms of ecological, economic and socio-political upheaval, it does not obviate the necessity for *peoples* to continue working to develop post-carbon, post-materialist civilizational forms that can emerge from the ashes of such upheaval. This period of civilizational transition will, indeed, be one of intense struggle between people's movements for global social justice and outmoded regressive socio-political forces desperate to maintain their power against the odds, by polarizing communities to monopolize the world's resources. The outcome of this struggle is not set in stone: There are more of 'us' than there are of 'them'. This is grounds for genuine, pragmatic optimism about the long-term future of human civilization.

The task here has been to identify the *key concepts*, illustrating the direction that we should take, and the most critical issues that must be evaluated to find meaningful solutions to these crises of our own making. Policies that ignore these Key Structural Problems and their associated implications for future strategic thinking will be ultimately irrelevant. *The post-carbon age is coming*. Regardless of the chaos and destruction that may or may not arrive in the meantime as global crises unfold, the future nevertheless represents a realm of unprecedented hope and the possibility to build a world based on compassion, peace and justice for all. And it will be grassroots communities that will lead the way to the new world.

Notes and References

(Internet sources last checked December 2009.)

Introduction

1. Although contemporary Western societies have often been described as 'post-industrial', in fact they have in large part delegated industrial production to selected Southern peripheries. Hence, our contemporary civilizational conjuncture retains a social form that is fundamentally dependent on modern industry, even in the 'post-industrial' centres where service sectors seem to predominate.
2. David Jackson, 'Bush: Vigilance is key in war on terrorism', *USA Today* (11 September 2006) <http://www.usatoday.com/news/washington/2006-09-11-bush-sept-11-address_x.htm>
3. Tony Blair, 'A Battle for Global Values', *Foreign Affairs* (January/February 2007) <http://www.foreignaffairs.org/20070101faessay86106/tony-blair/a-battle-for-global-values.html>
4. Samuel Huntington, 'The Clash of Civilizations?', *Foreign Affairs* (Vol. 72, No. 3, Summer 1993) pp. 22–49
5. Francis Fukuyama, 'After the end of history', openDemocracy (2 May 2006) <http://www.opendemocracy.net/democracy-fukuyama/revisited_3496.jsp>
6. Piki Ish-Shalom, 'For a Democratic Peace of Mind', *Harvard International Review* (2008) <http://hir.harvard.edu/index.php?page=article&id=1503>
7. See Nafeez Ahmed, *The Violence of Empire: The Logic and Dynamic of Strategies of Violence and Genocide in Historical and Contemporary Imperial Systems* (Brighton: Department of International Relations, University of Sussex, 2009)
8. Thanks to Dr Richard Levins from Harvard University for his definition of a system.

Chapter 1

1. This is the primary preoccupation of most studies of climate change by scholars of international relations and political science. One of the better and more accessible examples of this is Gwynne Dyer, *Climate Wars* (London: Random House Canada, 2008).
2. John Cook, 'Human CO2 is a tiny % of CO2 emissions', *Skeptical Science* <http://www.skepticalscience.com/human-co2-smaller-than-natural-emissions.htm>. Cook documents these points using scientific papers published in *Nature* and *Science*. Also see Niels Bohr Institute Press Release, 'Ice cores reveal fluctuations in the Earth's greenhouse gases' (14 May 2008) <http://www.nbi.ku.dk/english/news/greenhouse_gases>
3. See for example James Hansen and Andrew Lacis, et al, 'How sensitive is the world's climate?', *Research & Exploration* (1993, Vol. 9, No 2) <http://pubs.giss.nasa.gov/docs/1993/1993_Hansen_etal_1.pdf>; J. M. Gregory and R. J. Stouffer, et. al, 'An Observationally Based Estimate of the Climate Sensitivity', *Journal of Climate* (15 November 2002, Vol. 15, No. 22) <http://www.gfdl.noaa.gov/reference/bibliography/2002/jmgregory0201.pdf> For more references see John Cook, 'Climate sensitivity is low' <http://www.skepticalscience.com/climate-sensitivity.htm>
4. The draft of Pierrehumbert's book is available online here <http://geosci.uchicago.edu/~rtp1/ClimateBook/ClimateVol1.pdf >
5. IPCC Report, *Climate Change 2007: The Physical Science Basis – Summary for Policymakers*, Contribution of Working Group I to the Fourth Assessment Report of the Intergovernmental Panel on Climate Change (Geneva: United Nations Environment Programme, February 2007) <http://ipcc-wg1.ucar.edu/wg1/docs/WG1AR4_SPM_

PlenaryApproved.pdf>; for a basic summary see UN News Centre, 'Evidence is now "unequivocal" that humans are causing global warming' (2 February 2007) <http://www.un.org/apps/news/story.asp?NewsID=21429&Cr=climate&Cr1=change>

6. Seth Shulman, Kate Abend and Alden Meyer, *Smoke, Mirrors & Hot Air: How ExxonMobil Uses Big Tobacco's Tactics to Manufacture Uncertainty on Climate Science* (Cambridge, MA: Union of Concerned Scientists, January 2007) <http://www.ucsusa.org/assets/documents/global_warming/exxon_report.pdf>
7. Naomi Oreskes, 'Beyond the Ivory Tower: The Scientific Consensus on Climate Change', *Science* (3 December 2004, vol. 306, no. 5702) pp. 1688 <http://www.sciencemag.org/cgi/content/full/306/5702/1686>
8. Tim Lambert's discussion of the 34 abstracts and Peiser's retraction can be read here: <http://scienceblogs.com/deltoid/2006/10/peiser_admits_he_was_97_wrong.php>
9. ABC Media Watch, 'Bolt's Minority View' (30 October 2006) <http://www.abc.net.au/mediawatch/transcripts/s1777013.htm>
10. Ibid. A more recent attempt at refutation has come from an endocrinologist, Klaus-Martin Schulte, who claimed that 6 per cent of 539 abstracts reviewed from January 2004 to February 2007 reject the consensus, and that 'fewer than half now endorse it'. The paper was eventually published by *Energy & Environment*, which is a social science, not a physical science, journal, and is not recognized as peer-reviewed. See Klaus-Martin Schulte, 'Scientific Consensus on Climate Change?' *Energy & Environment* (March 2008, Vol. 19, No 2) pp. 281–286. Further, a detailed examination of the paper shows that the 'fewer than half' are simply papers that say nothing either way about anthropogenic climate change because of their focus on specific issues; while the alleged 6 per cent that purportedly reject climate change either do not do so, or are unconvincing and not properly peer-reviewed. Highlighting seven of the best papers flagged up as rejecting human-induced global warming by Schulte, Tim Lambert of the University of New South Wales, for instance, demonstrates how bad or irrelevant they really are. See Lambert, 'Classifying abstracts on global climate change' (30 August 2007) <http://scienceblogs.com/deltoid/2007/08/classifying_abstracts_on_globa.php>. Worse still, a close analysis by John Mashey, former chief scientist at Silicon, concludes that Schulte 'commits clear (and incredibly silly and incompetent) plagiarism. He claims competence at a task where his own words show him to lack such competence. He cooperates closely with well-known contrarians/denialists who lack relevant scientific expertise, but generate masses of... disinformation. Even after being exposed as a plagiarist, he persists in writing threatening letters to an expert in the field.' See Mashey, 'Another Attack on Global Warming's Scientific Consensus' (23 March 2008) <http://www.desmogblog.com/sites/beta.desmogblog.com/files/monckton%20schulte%20oreskes%207%200%20%282%29.pdf>
11. US Senate Environment and Public Works Committee, 'Over 400 Prominent Scientists Disputed Man-Made Global Warming Claims in 2007' (20 December 2007) <http://epw.senate.gov/public/index.cfm?FuseAction=Minority.SenateReport>
12. CSR Summary of Federal Election Commission Data, 'Top Industries – Senator James Inhofe 2005–2010' (Washington: Center for Responsive Politics, 9 August 2009) <http://www.opensecrets.org/politicians/industries.php?type=C&cid=N00005582&newMem=N&cycle=2010>
13. Mark V. Johnson, 'Inhofe's 400 Global Warming Deniers Debunked', *The Daily Green* (11 January 2008) <http://www.thedailygreen.com/environmental-news/latest/inhofe-global-warming-deniers-47011101>. A detailed breakdown of the list is provided by Johnson here <http://www.thedailygreen.com/environmental-news/latest/inhofe-global-warming-deniers-scientists-46011008>
14. Cited in Joseph Romm, "More on the 'scientific' attacks on global warming", *Gristmill* (27 December 2007) <http://gristmill.grist.org/story/2007/12/21/16436/710>
15. Ibid.
16. Dessler, 'The "Inhofe 400": Busting the "consensus busters"', (27 December 2007) <http://gristmill.grist.org/story/2007/12/26/1971/6517>; 'The "Inhofe 400" Skeptic of the Day' (15 January 2008) <http://gristmill.grist.org/story/2008/1/14/231236/019>; <http://

gristmill.grist.org/story/2008/1/1/182558/9615> (2 January 2008). Scientist Dessler continues to expose individuals on the Inhofe list at this column.

17. US Senate Minority Report, 'More Than 650 International Scientists Dissent Over Man-Made Global Warming Claims' (11 December 2008) <http://epw.senate.gov/public/index.cfm?FuseAction=Minority.Blogs&ContentRecord_id=2674e64f-802a-23ad-490b-bd9faf4dcdb7>
18. Bradford Plumer, 'Inhofe's 650 "Dissenters" (Make That 649... 648...)', *The New Republic* (15 December 2008) <http://blogs.tnr.com/tnr/blogs/environmentandenergy/archive/2008/12/15/inhofe-s-650-quot-dissenters-quot-make-that-649-648.aspx>
19. An excellent and exhaustive step-by-step review of each individual on the list, their qualifications, and their arguments, is being provided and updated here: <http://650list.blogspot.com>
20. AGU Position Statement, 'Human Impacts on Climate' (Washington DC: American Geophysical Union, December 2007) <http://www.agu.org/sci_soc/policy/positions/climate_change2008.shtml>
21. See <http://www.eastangliaemails.com/emails.php?eid=154&filename=942777075.txt>
22. David Lungren, 'EPW Policy Beat: "Hide the Decline"' (14 December 2009) <http://epw.senate.gov/public/index.cfm?FuseAction=Minority.Blogs&ContentRecord_id=8f16552a-802a-23ad-465f-8858beb85ac2&Issue_id=>
23. Union of Concerned Scientists, 'Debunking Misinformation About Stolen Climate Emails in the "Climategate" Manufactured Controversy' (11 December 2009) <http://www.ucsusa.org/global_warming/science_and_impacts/global_warming_contrarians/debunking-misinformation-stolen-emails-climategate.html>
24. Jeffrey Ball and Keith Johnson, 'Push to Oversimplify at Climate Panel', *Wall Street Journal* (26 February 2010) <http://online.wsj.com/article/SB100014240527487041881045750836813198349 78.html?mod=rss_Today's_Most_Popular>
25. Michael E. Mann, 'Michael Mann responds to "false and misleading claims" in the error-ridden, defamatory WSJ piece', Climate Progress (26 February 2010) <http://climateprogress.org/2010/02/26/michael-mann-false-and-misleading-claims-wall-street-journal-oversimplify-piece-by-jeffrey-ball-and-keith-johnson>
26. Michael E. Mann, Zhihua Zhang, Malcolm K. Hughes et al, 'Proxy-based reconstructions of hemispheric and global surface temperature variations over the past two millennia', *Proceedings of the National Academy of Sciences* (September 2008) doi:10.1073/pnas.0805721105 <http://www.pnas.org/content/early/2008/09/02/0805721105.abstract>
27. Mason Inman, 'Earth Hotter Now Than in Past 2,000 Years, Study Says', *National Geographic* (2 September 2008) <http://news.nationalgeographic.com/news/2008/09/080902-hottest-earth.html>
28. P D Jones and M.E. Mann, 'Climate Over Past Millennia', *Reviews of Geophysics*, (May 2004, Vol. 42) RG2002, doi: 10.1029/2003RG000143, 2004 <http://iri.columbia.edu/~goddard/EESC_W4400/CC/jones_mann_2004.pdf> Also see Mann et al, 'On Past Temperatures and Anomalous Late-20th Century Warmth', *Eos* (July 2003, Vol. 84, No. 27) <http://holocene.meteo.psu.edu/shared/articles/eos03.pdf>
29. Damian Carrington, 'IPCC officials admit mistake over melting Himalayan glaciers', *Guardian* (20 January 2010)
30. Jonathan Leake, 'UN climate chief Rajendra Pachauri "got grants through bogus claims"', *Sunday Times* (24 January 2010) <http://www.timesonline.co.uk/tol/news/environment/article6999975.ece>
31. Christopher Booker, 'A perfect storm is brewing for the IPCC', *Telegraph* (27 February 2010) <http://www.telegraph.co.uk/comment/columnists/christopherbooker/7332803/A-perfect-storm-is-brewing-for-the-IPCC.html>
32. Jonathan Leake, 'The UN climate panel and the rainforest claim', *Sunday Times* (31 January 2010) <http://www.timesonline.co.uk/tol/news/environment/article7009705.ece>
33. Robin McKie, 'What planet are these people on?' *Observer* (9 December 2007) <http://www.guardian.co.uk/books/2007/dec/09/scienceandnature.features>

34. Real Climate, 'IPCC errors: facts and spin', (14 February 2010) <http://www.realclimate.org/index.php/archives/2010/02/ipcc-errors-facts-and-spin>
35. Martin Parry, 'The Implications of Climate Change for Crop Yields, Global Food Supply and Risk of Hunger,' *Journal of Semi Arid-Tropics (SAT) Agricultural Research* (December 2008, Vol. 6) <http://www.icrisat.cgiar.org/Journal/SpecialProject/sp14.pdf>
36. Risk Management Solutions FAQ, 'Research on Climate Change and Disaster Loss Costs and the IPCC' (2010) <http://www.rms.com/Publications/2010_FAQ_IPCC.pdf>
37. See for instance surveys of the peer-reviewed literature by Jason Anderson and Camilla Bausch, 'Climate Change and Natural Disasters: Scientific evidence of a possible relation between recent natural disasters and climate change', (Brussels: European Parliament Policy Department for Economic and Scientific Policy, January 2006) <http://www.europarl.europa.eu/comparl/envi/pdf/externalexpertise/ieep_6leg/naturaldisasters.pdf>; and Maarten K. Van Aalst, 'The impact of climate change on the risk of natural disasters', *Disasters* (March 2006, Vol. 3, No. 1) pp. 5–18
38. Woods Hole Research Center, 'Senior Scientist Daniel Nepstad endorses the correctness of the IPCC's (AR4) statement on Amazon forest susceptibility to rainfall reduction', (February 2010) <http://www.whrc.org/resources/online_publications/essays/2010-02-Nepstad_Amazon.htm>
39. I. G. Usokken and M. Schussler et al, 'Solar Activity Over the Last 1150 Years: Does it Correlate with Climate?'(Klatenburg-Lindau: Max Planck Institute for Solar System Research, July 2004) <http://www.mps.mpg.de/dokumente/publikationen/solanki/c153.pdf>, p. 22
40. S. K. Solanki and I. G. Usokken, et al, 'Unusual activity of the Sun during the recent decades compared to the previous 11,000 years', *Nature* (28 October 2004, vol. 431), p. 1087 <http://cc.oulu.fi/~usoskin/personal/nature02995.pdf>
41. Mike Lockwood and Claus Frohlich, 'Recently oppositely directed trends in solar climate forcings and the global mean surface air temperature', *Proceedings of the Royal Society* (25 May 2007) <http://publishing.royalsociety.org/media/proceedings_a/rspa20071880.pdf>
42. For a documented summary see John Cook, 'CO_2 lags temperature – what does it mean?' *Skeptical Science* (October 2007) <http://www.skepticalscience.com/co2-lags-temperature.htm>
43. A. Berger, 'Eight glacial cycles from an Antarctic ice core', *Nature* (10 June 2004, Vol. 429) pp. 623–628 <http://www.nature.com/nature/journal/v429/n6992/abs/nature02599.html>; A. Berger and M. F. Loutre, 'An Exceptionally Long Interglacial Ahead?' *Science* (August 2002, Vol. 297, No. 5585) pp. 1287–1288 <http://www.sciencemag.org/cgi/content/summary/297/5585/1287>
44. Don J. Easterbrook, 'The Next 25 Years: Global Warming of Global Cooling? – Geological and Oceanographic Evidence for Cyclical Climate Oscillations' (Boulder, CO: Geological Society of America Annual Meeting, 5–8 November 2001) Also see the more recent revamping of this argument in Easterbrook, 'Global cooling is here' (Montreal: Centre for Research on Globalization, 2 November 2008) <http://www.globalresearch.ca/index.php?context=va&aid=10783>
45. John Cross and John Cook, 'Is Pacific Decadel Oscillation the Smoking Gun?', *Skeptical Science* (3 May 2008) <http://www.skepticalscience.com/Is-Pacific-Decadal-Oscillation-the-Smoking-Gun.html>
46. N. S. Keenlyside and M. Latif, et al, 'Advancing decadal-scale climate prediction in the North Atlantic sector', *Nature* (1 May 2008, Vol. 453) <http://www.usclivar.org/Pubs/2May08Keenlyside.pdf>
47. See 'Global Cooling – Wanna Bet?' and 'The Global Cooling Bet – Part 2' at Real Climate (May 2008) <http://www.realclimate.org/index.php/archives/2008/05/global-cooling-wanna-bet> and <http://www.realclimate.org/index.php/archives/2008/05/the-global-cooling-bet-part-2>
48. Joseph Romm, '*Nature* article on "cooling" confuses media, deniers: Next decade may see rapid warming', *Climate Progress* (2 May 2008) <http://climateprogress.org/2008/05/02/

nature-article-on-cooling-confuses-revkin-media-deniers-next-decade-may-see-rapid-warming>

49. Bob Carter, 'There IS a problem with global warming... it stopped in 1998', *Telegraph* (9 April 2006) <http://www.telegraph.co.uk/opinion/main.jhtml?xml=/opinion/2006/04/09/do0907.xml>
50. David Whitehouse, 'Has global warming stopped?', *New Statesman* (19 December 2007) <http://www.newstatesman.com/scitech/2007/12/global-warming-temperature>
51. Met Office, 'Climate Change – Fact 2: temperatures continue to rise' (2008) <http://www.metoffice.gov.uk/corporate/pressoffice/myths/2.html>
52. John Cook, 'Did global warming stop in 1998?', *Skeptical Science* <http://www.skepticalscience.com/global-warming-stopped-in-1998.htm>
53. Roger Harrabin, 'Global temperatures to "decrease"' BBC News (4 April 2008) <http://news.bbc.co.uk/2/hi/science/nature/7329799.stm>
54. Anthony Watts, '4 sources say "globally cooler" in the past 12 months', *Watts Up With That?* (19 February 2008) <http://wattsupwiththat.com/2008/02/19/january-2008-4-sources-say-globally-cooler-in-the-past-12-months>
55. Michael Asher, 'Temperature Monitors Report Widescale Global Cooling', *DailyTech* (26 February 2008) <http://www.dailytech.com/Temperature+Monitors+Report+Widescale+Global+Cooling/article10866.htm>
56. Michael Asher, 'Solar Activity Diminishes: Researchers Predict Another Ice Age', *DailyTech* (9 February 2008) <http://www.dailytech.com/Solar+Activity+Diminishes+Researchers+Predict+Another+Ice+Age/article10630.htm>
57. Charles D. Camp and Ka Kit Tung, 'Surface warming by the solar cycle as revealed by the composite mean difference projection', *Geophysical Research Letters* (18 July 2008, Vol. 34) <http://www.amath.washington.edu/~cdcamp/Pub/Camp_Tung_GRL_2007b.pdf> John Cook, 'Global cooling: the new kid on the block', *Skeptical Science* (4 March 2008) <http://www.skepticalscience.com/Global-cooling-the-new-kid-on-the-block.html>
58. NCAR News Release, 'Scientists Issue Unprecedented Forecast of Next Sunspot Cycle' (Boulder, Co: University Corporation for Atmospheric Research, 6 March 2006) <http://www.ucar.edu:80/news/releases/2006/sunspot.shtml>
59. Peter Schwartz and Doug Randall, 'An Abrupt Climate Change Scenario and its Implications for United States National Security' (Washington: Department of Defense, October 2003) pp. 1-3, <http://www.climate.org/PDF/clim_change_scenario.pdf>; Mark Townsend and Paul Harris, 'Now the Pentagon tells Bush: climate change will destroy us', *Observer* (22 February 2004) <http://observer.guardian.co.uk/international/story/0,6903,1153513,00.html> For a detailed analysis see Dave Webb, 'Thinking the Worst: The Pentagon Report' in David Cromwell and Mark Levene (ed.) *Surviving Climate Change: The Struggle to Avert Global Catastrophe* (London: Pluto Press, 2007)
60. Ian Traynor, 'Climate change may spark conflict with Russia, EU told', *Guardian* (10 March 2008) <http://www.guardian.co.uk/world/2008/mar/10/eu.climatechange?gusrc=rss&feed=networkfront>
61. Schwartz and Randall, 'An Abrupt Climate Change'; Townsend and Harris, 'Now the Pentagon tells Bush'.
62. IPCC Fourth Assessment Report, Climate Change 2007 [Working Group 1] (Cambridge: Cambridge University Press, February 2007). Available online, <http://ipcc-wg1.ucar.edu/wg1/wg1_home.html>
63. Mark Lynas, *Six Degrees: Our Future in a Hotter Planet* (London: HarperCollins Fourth Estate, 2007)
64. See Richard Girling, 'To the ends of the Earth', *Sunday Times Magazine* (15 March 2007) <http://www.timesonline.co.uk/tol/news/uk/science/article1480669.ece>
65. Catherine Brahic, 'Carbon emissions rising faster than ever", New Scientist News Service <http://www.newscientist.com/article/dn10507-carbon-emissions-rising-faster-than-ever.html>
66. Michael R. Raupach, Gregg Marland, Philippe Ciais, et al, 'Global and regional drivers of accelerating CO2 emissions', *Proceedings of the National Academy of Sciences* (12 June

2007, Vol. 104, No. 24, pp. 10288-10293) <http://www.pnas.org/content/104/24/10288.abstract>

67. Ibid.
68. Juliet Eilperin, 'Carbon is Building Up in Atmosphere Faster than Predicted', *Washington Post* (26 September 2008) <http://www.washingtonpost.com/wp-dyn/content/article/2008/09/25/AR2008092503989.html?hpid=moreheadlines>
69. Peter M. Cox, Richard A. Betts, Chris D. Jones, Stephen A. Spall, Ian J. Totterdell, 'Acceleration of global warming due to carbon-cycle feedbacks in a coupled climate model', *Nature* (9 November 2000, Vol. 408, No. 6809) pp. 184–7; David Wasdell, 'Feedback Dynamics and the Acceleration of Climate Change: An Update of the Scientific Analysis', Meridian Programme (London: March 2007) <http://www.meridian.org.uk/Resources/Global%20Dynamics/Feedback%20Dynamics/index.htm>
70. Richard Girling, "To the ends of the Earth", *Sunday Times Magazine* (15 March 2007) <http://www.timesonline.co.uk/tol/news/uk/science/article1480669.ece>
71. Fred Pearce, 'Climate report was "watered down"', *New Scientist* (8 March 2007); David Wasdell, 'Political Corruption of the IPCC Report: Changes in the Final Text of the "Summary for Policy Makers"', (London: Meridian Programme, March 2007) <http://www.meridian.org.uk/Resources/Global%20Dynamics/IPCC/index.htm>
72. George Monbiot, 'The Real Climate Censorship', *Guardian* (10 April 2007)
73. Pearce, 'Climate report'
74. Union of Concerned Scientists and Government Accountability Project, "Atmosphere of Pressure: Political Interference in Federal Climate Science" (February 2007) <http://www.ucsusa.org/assets/documents/scientific_integrity/Atmosphere-of-Pressure.pdf> The number of respondents to the survey was 279.
75. Fred Pearce, 'Global meltdown', *Guardian* (30 August 2006)
76. National Research Council, *Abrupt Climate Change: Inevitable Surprises* (Washington: National Academy Press, National Academy of Sciences, 2002)
77. Pearce, 'Global meltdown'.
78. Geoffrey Lean, 'Apocalypse Now: How Mankind is Sleepwalking to the End of the Earth' *Independent* (6 February 2005) <http://news.independent.co.uk/world/environment/story.jsp?story=608209>; Michael McCarthy, 'Global warming: passing the "tipping point"', *Independent* (11 February 2006) <http://news.independent.co.uk/environment/article344690.ece>; OST Report, "Rapid Climate Change" (London: Parliamentary Office of Science and Technology, July 2005, No. 245) <http://www.parliament.uk/parliamentary_offices/post/pubs2005.cfm>
79. Reuters, 'Greenhouse gas emissions hit danger mark – scientist' (9 October 2007); IPCC Fourth Assessment Report, *Synthesis Report* (November 2007) <http://www.ipcc.ch/pdf/assessment-report/ar4/syr/ar4_syr.pdf>
80. He was Special Adviser to three Secretaries of State for the Environment from 1991–97 and statutory adviser on biodiversity to the British Government from 1999–2005. In 2002 he served as an adviser to the Central Policy Group in the Deputy Prime Minister's Office.
81. McCarthy, 'Global warming: passing the "tipping point"'
82. Paul Brown, *Global Warning: The Last Chance for Change* (London: *Guardian* and A&C Black, 206).
83. Stephen Byers and Olympia J. Snowe, *Meeting the Climate Challenge: Recommendations of the International Climate Change Taskforce* (London, Washington, Canberra: International Climate Change Taskforce, January 2005) <http://www.americanprogress.org/kf/climatechallenge.pdf> [emphasis added]
84. Ed Pilkington, 'Climate target is not radical enough – study', *Guardian* (7 April 2008) <http://www.guardian.co.uk/environment/2008/apr/07/climatechange.carbonemissions>; See J. Hanson, et al. 'Target atmospheric CO2: Where should humanity aim?' arXiv.org (7 April 2008, revised 18 June 2008) <http://arxiv.org/abs/0804.1126> Also see 'Biochar potential overlooked', *Weekly Times* (26 January 2009) <http://www.weeklytimesnow.com.au/article/2009/01/26/45315_national-news.html>

85. Philip Sutton, comment on 350 ppm limit <http://noimpactman.typepad.com/blog/2008/06/the-planets-mos.html>
86. Interview with David Wasdell (6 August 2009); David Adam, 'Roll back time to safeguard climate, expert warns', *Guardian* (15 September 2008) <http://www.guardian.co.uk/environment/2008/sep/15/climatechange.carbonemissions>
87. Jonathan Leake, 'Britain faces big chill as ocean current slows', *Sunday Times* (8 May 2005) <http://www.timesonline.co.uk/tol/news/uk/article520013.ece>
88. Ian Sample, 'Alarm over dramatic weakening of the Gulf Stream', *Guardian* (1 December 2005)
89. Quirin Schiermeier, 'Ocean circulation noisy, not stalling', *Nature* (2007, Vol. 448, No. 7156), pp. 844–845.
90. Michael Vellinga and Richard A. Wood, 'Global climatic impacts of a collapse of the Atlantic thermohaline circulation', (London: Met Office, Hadley Centre for Climate Prediction and Research, February 2001) <http://www.metoffice.gov.uk/research/hadleycentre/pubs/HCTN/HCTN_26.pdf>
91. James E Kloeppel, 'Global warming could halt ocean circulation with harmful results', University of Illinois Press Release (6 December 2005) [emphasis added]
92. Steve Conner, 'World Nearing Point of No Return on Global Warming', *Independent* (17 September 2005). David Adam, 'Meltdown fear as Arctic ice cover falls to record winter low', *Guardian* (15 May 2006). Corroborated by other studies, such as Marika M. Holland, Cecilia M. Bitz, Bruno Temblay, 'Future abrupt reductions in the summer Arctic sea ice', *Geophysical Research Letters* (Vol. 33, 12 December 2006) <http://www.agu.org/pubs/crossref/2006/2006GL028024.shtml>
93. Steve Connor, 'Collapse of Arctic sea ice "has reached tipping point"', *Independent* (16 March 2007)
94. Indymedia, 'NASA climatologist predicts disastrous sea level rise' (16 March 2007) <http://www.indybay.org/newsitems/2007/03/16/18377388.php>; Michael Byrnes, 'Scientist says sea level rise could accelerate', Reuters (12 March 2007) <http://www.alertnet.org/thenews/newsdesk/SYD298897.htm>
95. Press agencies, 'Arctic ice "could be gone in five years"', *Telegraph* (12 December 2007) [emphasis added]. Also see Lieutenant Commander John Whelan, *Understanding Recent Variability in the Arctic Sea Ice Cover – Synthesis of Model Results and Observations*, Submitted in partial fulfillment of the requirements for the degree of Master of Science in Meteorology and Physical Oceanography (Naval Postgraduate School, September 2007)
96. Steve Connor, 'Has the Arctic passed the point of no return?', *Independent* (16 December 2008) <http://www.independent.co.uk:80/environment/climate-change/arctic-melt-passes-the-point-of--no-return-1128197.html>
97. R. Kwok, G. F. Cunningham, M. Wensnahan et al. 'Thinning and volume loss of the Arctic Ocean sea ice cover: 2003–2008', *Journal of Geophysical Research* (July 2009, Vol. 114) C07005, doi:10.1029/2009JC005312
98. BBC News, 'Gulf Stream', *Guide to Climate Change*, <http://news.bbc.co.uk/1/shared/spl/hi/sci_nat/04/climate_change/html/gulf.stm>
99. NCAR Press Release, 'Drought's Growing Reach' (Boulder, CO: National Center for Atmospheric Research, 10 January 2005) <http://www.ucar.edu/news/releases/2005/drought_research.shtml>
100. Thomas Fuller, 'For Asia, a vicious cycle of flood and drought', *International Herald Tribune* (1 November 2006) <http://www.iht.com/articles/2006/11/01/news/water.php>; Associated Press, 'Warming report to warn of coming drought', (10 March 2007). Richard Black, 'Climate water threat to millions', BBC News (20 October 2006) <http://news.bbc.co.uk/2/hi/science/nature/6068348.stm>
101. Alan Dupont, 'The Strategic Implications of Climate Change', *Survival: Global Politics and Strategy* (Vol. 50, No. 3, June 2008), p. 33
102. 'Antarctic sea ice increases despite warming', *New Scientist* (12 September 2008) <http://www.newscientist.com/article/dn14724-antarctic-sea-ice-increases-despite-warming.html>

103. Mark A. J. Curran, Tas D. van Ommen, Vin I. Morgan, et al. 'Ice Core Evidence for Antarctic Sea Decline Since the 1950s', *Science* (14 November 2003, Vol. 302, No. 5648), pp. 1203–1206, <http://www.sciencemag.org/cgi/content/full/302/5648/1203?maxtoshow=&HITS=&hits=&RESULTFORMAT=&fulltext=antarctic+curran&searchid=1&FIRSTINDEX=0&resourcetype=HWCIT>
104. NASA Press Release, 'NASA Mission Detects Significant Antarctic Ice Mass Loss' (2 March 2006) <http://www.nasa.gov/home/hqnews/2006/mar/HQ_06085_arctic_ice.txt>
105. American Geophysical Union and NASA Press Release, 'Warmer air may cause increased Antarctic sea ice cover' (29 June 2005) <http://www.agu.org/sci_soc/prrl/prrl0522.html>
106. Drew T. Shindell, David Rind, and Patrick Lonergan, 'Increased Polar Stratospheric Ozone Losses and Delayed Eventual Recovery Owing to Increasing Greenhouse-gas Concentrations', *Nature* (9April 1998, Vol. 392) pp. 589–592
107. Alister Doyle, 'Man-made climate change seen in Antarctica, Arctic', Reuters (30 October 2008); Steve Conner, 'Climate change at the Poles IS man-made', *Independent* (31 October 2008) <http://www.independent.co.uk/environment/climate-change/climate-change-at-the-poles-is-manmade-980256.html>
108. Alex Kirby, 'Earth warned on "tipping points"', BBC News (26 August 2004) <http://newsvote.bbc.co.uk/mpapps/pagetools/print/news.bbc.co.uk/1/hi/sci/tech/3597584.stm>
109. James Hansen et al. 'Dangerous human-made interference with climate: a CISS modelE study', *Atmospheric Chemistry and Physics* (7 May 2007) <http://www.atmos-chem-phys.net/7/2287/2007/acp-7-2287-2007.pdf> [emphasis added]
110. Sergey A. Zimov, Edward A. G. Schuur, F. Stuart Chapin III, 'Permafrost and the Global Carbon Budget', *Science* (Vol. 312, No. 5780, 16 June 2006) <http://www.sciencemag.org/cgi/content/summary/312/5780/1612>
111. K. M. Walter, S. A. Zimov et al. 'Methane bubbling from Siberian thaw lakes as a positive feedback to climate warming', *Nature* (No. 443, 7 September 2006) <http://www.nature.com/nature/journal/v443/n7107/abs/nature05040.html>
112. Steve Conner, 'The Methane Time Bomb', *Independent* (23 September 2008) <http://www.independent.co.uk/environment/climate-change/exclusive-the-methane-time-bomb-938932.html>
113. N. Shakhova, I. Semiletov, A. Salyuk, D. Kosmach, 'Anomalies of methane in the atmosphere over the East Siberian shelf: Is there any sign of methane leakage from shallow shelf hydrates?', EGU General Assembly 2008, *Geophysical Research Abstracts* (2008, Vol. 10, EGU2008-A-01526)
114. John Atcheson, 'Ticking Time Bomb', *Baltimore Sun* (15 December 2004) <http://www.baltimoresun.com/news/opinion/oped/bal-op.warming15dec15,1,290031.story?coll=bal-oped-headlines> [emphasis added]
115. W. A. Kurz, C. C. Dymond, G. Stinson, et al. 'Mountain pine beetle and forest carbon feedback to climate change', *Nature* (Vol. 452, No. 7190, 24 April 2008) <http://www.nature.com/nature/journal/v452/n7190/full/nature06777.html> [emphasis added]
116. R. Warren, 'Impacts of Global Climate Change at Different Annual Mean Global Temperature Increases', in H. J. Schellnhuber (ed.), *Avoiding Dangerous Climate Change* (Cambridge: Cambridge University Press, 2006) pp. 93–131
117. Geoffrey Lean and Fred Pearce, 'Amazon rainforest "'could become a desert"', *Independent* (23 July 2006) <http://www.independent.co.uk/environment/amazon-rainforest-could-become-a-desert-408977.html>
118. Wenhong Li, Rong Fu et al., 'Observed change of the standardized precipitation index, its potential cause and implications to future climate change in the Amazon region', *Philosophical Transactions of the Royal Society B* (Vol. 363, 2008) pp. 1767–1772 <http://journals.royalsociety.org/content/0p45m32442323027/fulltext.pdf>; Peter Good, Jason A. Lowe et. al., 'An objective tropical Atlantic sea surface temperature gradient index for studies of south Amazon dry-season climate variability and change', *Philosophical Transactions of the Royal Society B* (Vol. 363, 2008) pp. 1761–1766 <http://journals.royalsociety.org/content/a170k12rk5340013/fulltext.pdf>

119. Daniel C. Nepstad, Claudia M. Stickler et al., 'Interactions among Amazon land use, forests and climate: prospects for a near-term forest tipping point' *Philosophical Transactions of the Royal Society B* (Vol. 363, 2008) pp. 1737–1746
120. Guy J. D. Kirk, Pat H. Bellamy, Peter J. Bradley et al., 'Carbon losses from all soils across England and Wales 1978–2003', *Nature* (Vol. 437, No. 7056, 8 September 2005) pp. 245–248 <http://www.nature.com/nature/journal/v437/n7056/abs/nature04038.html>
121. Chris D. Jones, Peter M. Cox, Richard Essery et al., 'Strong carbon cycle feedbacks in a climate model with interactive CO_2 and sulphate aerosols', *Geophysical Research Letters* (Vol. 30, No. 9, May 2003) p. 1479 <http://www.agu.org/pubs/crossref/2003/2003GL016867.shtml>.
122. Isaac M. Held and Brian J. Soden, 'Water Vapor Feedback and Global Warming', *Annual Review of Energy and the Environment* (November 2000, Vol. 25) pp. 441–475
123. A.P. Sokolov et al., 'Probabilistic Forecast for Twenty-First Century Climate Based on Uncertainties in Emissions (Without Policy) and Climate Parameters', *Journal of Climate* (October 2009, Vol. 22, No. 19), pp. 5175–5204
124. Vicky Pope et al., (eds.), *Avoiding Dangerous Climate Change* (London: Met Office Hadley Center, 2008) p. 13. Also see Vicky Pope, 'Met office warns of "catastrophic" rise in temperature' *Times* (19 December 2008) <http://www.timesonline.co.uk/tol/news/environment/article5371682.ece>
125. David Adam, 'Met Office warns of catastrophic global warming in our lifetimes', *Guardian* (28 September 2009) <http://www.guardian.co.uk/environment/2009/sep/28/met-office-study-global-warming>
126. Margaret Torn and John Harte, 'Missing feedbacks, asymmetric uncertainties, and the underestimation of future warming', *Geophysical Research Letters* (Vol., 33, May 2006) <http://www.agu.org/pubs/crossref/2006/2005GL025540.shtml>; 'Feedback Loops In Global Climate Change Point To A Very Hot 21st Century' *Science Daily* (22 May 2006) <http://www.sciencedaily.com/releases/2006/05/060522151248.htm>
127. Union of Concerned Scientists, 'Warning to Humanity' (Cambridge, MA: 1992) [emphasis added]
128. United Nations Environmental Programme (UNEP) Press Release, 'Our present course is unsustainable – postponing action is no longer an option' (Nairobi: 15 September 1999) [emphasis added]
129. CBS News, 'Two-thirds of Earth's ecosystems at risk: UN' (31 March 2005) <http://www.cbc.ca/story/world/national/2005/03/30/UNEnvironment0330.html>
130. Report of the United Nations Millennium Ecosystem Assessment, *Ecosystems and Human Well-Being: Synthesis* (Washington DC: Island Press, 30 March 2005) <http://www.maweb.org/proxy/document.aspx?source=database&TableName=Documents&IdField=DocumentID&Id=356&ContentField=Document&ContentTypeField=ContentType&TitleField=Title&FileName=MA+General+Synthesis+-+Final+Draft.pdf&Log=True>
131. David Edwards, *Free To Be Human* (Devon: Green Books, 1995) p. 144
132. Cited in John Rowley, 'Forecasting the Future', *People & the Planet*, (Vol. 8/4, 1999)
133. Mihajilo D. Mesarovic and Eduard Pestel, *Mankind at the Turning Point* (New York: E. P. Dutton, 1974)
134. Gwyn Prins, 'Time to ditch Kyoto: the sequel' (Colorado: University of Colorado Center for Science and Technology Policy, October 2008) <http://sciencepolicy.colorado.edu/prometheus/wp-content/uploads/2008/11/time-to-ditch-kyoto_the_sequelfinal-aspubl251008.pdf>
135. Patrick McCully, 'Discredited Strategy', *Guardian* (21 May 2008) <http://www.guardian.co.uk/environment/2008/may/21/environment.carbontrading>
136. Patrick Wintour, 'Blair signs carbon pact with Schwarzenegger', *Guardian* (1 August 2006) <http://www.guardian.co.uk/climatechange/story/0,,1834662,00.html>
137. Roy Wilkes, 'Is "Contraction" and "Convergence" the Answer to Climate Change', *Climate and Capitalism* (31 March 2007) <http://climateandcapitalism.com/?p=54>
138. BBC News, 'Copenhagen deal: Key points' (19 December 2009) <http://news.bbc.co.uk/1/hi/sci/tech/8422307.stm>

139. HM Government, *The UK Low Carbon Transition Plan: National strategy for climate and energy* (London: Department of Energy & Climate Change, 15 July 2009) <http://www.decc.gov.uk/Media/viewfile.ashx?FilePath=White Papers\UK Low Carbon Transition Plan WP09\1_20090724153238_e_@@_lowcarbontransitionplan.pdf&filetype=4>
140. Christian Aid, 'Government "eviscerates" climate bill' (10 June 2008) <http://www.alertnet.org/thenews/fromthefield/218275/121309157467.htm>
141. James Kanter, 'Carbon trading: where greed is green', *New York Times* (20 June 2007) <http://www.nytimes.com/2007/06/20/business/worldbusiness/20iht-money.4.6234700.html>
142. Terry Macalister, 'Carbon trading could be worth twice that of oil in next decade', *Guardian* (29 November 2009) <http://www.guardian.co.uk/environment/2009/nov/29/carbon-trading-market-copenhagen-summit>
143. Aubrey Meyer, 'The Case for Contraction and Convergence', in David Cromwell and Mark Levene (eds.), *Surviving Climate Change: The Struggle to Avert Global Catastrophe* (London: Pluto Press, 2007) pp. 29–58
144. DB Climate Change Advisors, *Global Climate Change Policy Tracker: An Investor's Assessment* [Executive Summary] (Deutsche Bank Group, October 2009) <http://www.dbcca.com/dbcca/EN/_media/Global_Climate_Change_Policy_Tracker_Exec_Summary.pdf>. See summary at 'Carbon markets not working, says Deutsche Bank', *Ecologist* (2 November 2009) <http://www.theecologist.org/News/news_round_up/348754/carbon_markets_not_working_says_deutsche_bank.html>
145. Sir Nicolas Stern, *The Economics of Climate Change: The Stern Review* (Cambridge: Cambridge University Press, 2006) <http://www.hm-treasury.gov.uk/independent_reviews/stern_review_economics_climate_change/ sternreview_index.cfm>; Juliette Jowit and Patrick Wintour, 'Cost of tackling global climate change has doubled, warns Stern', *Guardian* (26 June 2008) <http://www.guardian.co.uk/environment/2008/jun/26/climatechange.scienceofclimatechange>
146. James Randerson, 'Climate change: Prepare for global temperature rise of 4C, warns top scientist', *Guardian* (7 August 2008) <http://www.guardian.co.uk/environment/2008/aug/06/climatechange.scienceofclimatechange>
147. Mark Lynas, 'Six steps to hell', *Guardian* (23 April 2007) <http://www.guardian.co.uk/books/2007/apr/23/scienceandnature.climatechange>
148. David Adam, 'Too late? Why scientists say we should expect the worst', *Guardian* (9 December 2008)
149. See for instance Statement for the Record of Thomas Fingar, Deputy Director of National Intelligence for Analysis and Chairman of National Intelligence Council, *National Intelligence Assessment on the National Security Implications of Global Climate Change to 2030* (Washington DC: House Permanent Select Committee on Intelligence and House Select Committee on Energy Independence and Global Warming, 25 June 2008). Also see Noah Shachtman, 'Nation's Spies: Climate Change Could Spark War', *Wired* (23 June 2008) <http://blog.wired.com/defense/2008/06/environmental-g.html>
150. George Monbiot, *Heat: How to Stop the Planet Burning* (London: Allen Lane, 2006)
151. Clive Hamilton, 'Building on Kyoto', *New Left Review* (May–June 2007, Vol. 45)

Chapter 2

1. M. Shahid Alam, 'A Short History of the Global Economy Since 1800', University Library of Munich, MPRA Paper (June 2003, No. 1263) <http://mpra.ub.uni-muenchen.de/1263/01/MPRA_paper_1263.pdf>
2. Joshua Karliner, *The Corporate Planet: Ecology and Politics in the Age of Globalization* (San Francisco, CA: Sierra Club Books, 1997)
3. Ibid.
4. Ibid.
5. Mike Tooke, 'Oil Peak – A Summary', *Powerswitch* (October 2004, No. 1.7)

6. International Energy Agency, 'World Energy Outlook 2007: Executive Summary', <http://www.worldenergyoutlook.org/2007.asp>, p. 4
7. Ron Swenson, 'The Hubbert Peak for World Oil: Summary', *The Coming Global Oil Crisis: The Hubbert Peak for Oil Production* (22 December 2003) <http://www.hubbertpeak.com/summary.htm>
8. Ibid.
9. Ugo Bardi, 'Abiotic Oil: Science or Politics?', *From The Wilderness* (4 October 2004) <http://www.fromthewilderness.com/free/ww3/100404_abiotic_oil.shtml>
10. See for example Kenneth F. Deffeyes, *Hubbert's Peak: The Impending World Oil Shortage* (Princeton: Princeton University Press, 2001)
11. Richard Heinberg, 'Bush Administration Suppresses Peak Oil Study' *Counterpunch* (30/31 July 2005) <http://www.counterpunch.org/heinberg07302005.html>; Robert L. Hirsch et al., 'Peaking of World Oil Production: Impacts, Mitigation and Risk-Management', Science Applications International Corporation (February 2005) <http://www.projectcensored.org/newsflash/The_Hirsch_Report_Proj_Cens.pdf>
12. Hirsch et al., 'Peaking of World Oil Production', p. 8, 4–5
13. Donald F. Fournier and Eileen T. Westervelt, *Energy Trends and their Implications for US Army Installations* (Washington DC: US Army Corps of Engineers, September 2005) p. 7, <http://stinet.dtic.mil/cgi-bin/GetTRDoc?AD=A440265&Location=U2&doc=GetTRDoc.pdf>
14. Ibid., pp. vii, xi
15. Colin J. Campbell and J. H. Laherrere *The World Oil Supply: 1930–2050* (Geneva: Petroconsultants, 1995), pp. 19, 27. Cited in B. J. Fleay, *Climaxing Oil: How Will Transport Adapt?* (Perth: Institute for Sustainability and Technology Policy, Murdoch University, 1999) <http://wwwistp.murdoch.edu.au/publications/projects/oilfleay/04discoverandprodn.html>
16. Ibid. G. Cope, 'Will improved oil recovery avert an oil crisis?' *Petroleum Review* (June 1998).
17. Richard Heinberg, *The Party's Over: Oil, War and the Fate of Industrial Societies* (London: Clairview, 2003) p. 117–9
18. IHS Press Release, 'CERA: Oil & Liquids Capacity to Outstrip Demand Until At Least 2010' (24 June 2005) <http://petrochemical.ihs.com/news-05Q2/cera-global-oil-production-capacity.jsp>; CERA Press Release, 'Peak Oil Theory – "World Running Out of Oil Soon" – Is Faulty; Could Distort Policy & Energy Debate' (14 November 2006) <http://www.cera.com/aspx/cda/public1/news/pressReleases/pressReleaseDetails.aspx?CID=8444>; CERA Press Release, 'No Evidence of Precipitous Fall on Horizon for World Oil Production', (17 January 2008) <http://www.cera.com/aspx/cda/public1/news/pressReleases/pressReleaseDetails.aspx?CID=9203>
19. David Ebner, 'Cera v Peak Oil', *Globe and Mail* Inside Energy (17 January 2008) <http://www.theglobeandmail.com/servlet/story/RTGAM.20080117.WBwenergyblog061320080117113730/WBStory/WBwenergyblog0613>
20. Jim Puplava, Interview with Matthew R. Simmons, 'Critique of the CERA Report', Financial Sense Newshour (25 November 2006) <http://www.financialsense.com/transcriptions/2006/1125simmons.html>
21. Terry Macalister, 'Key oil figures were distorted by US pressure, says whistleblower', *Guardian* (9 November 2009) <http://www.guardian.co.uk/environment/2009/nov/09/peak-oil-international-energy-agency>
22. Randy Udall and Steve Andrews, 'CERA's Depletion Study: The "Good News" About Running Up the Down Escalator' (Denver: Association for the Study of Peak Oil & Gas, 21 January 2008) <http://www.aspo-usa.com/index.php?option=com_content&task=view&id=301&Itemid=93>
23. Robert L. Hirsch, 'The WSJ Article on a CERA Oil Decline Study' (Denver: Association for the Study of Peak Oil & Gas, 28 January 2008) <http://www.aspo-usa.com/index.php?option=com_content&task=view&id=305&Itemid=93>

24. David Galland, 'What the Export Land Model Means for Energy Prices', *Casey Energy Speculator* (19 May 2008) <http://www.investorsinsight.com/blogs/john_mauldins_outside_the_box/archive/2008/05/19/what-the-export-land-model-means-for-energy-prices.aspx>
25. Ibid.
26. *Oilwatch Monthly* (Amsterdam: Association for the Study of Peak Oil, December 2008) <http://www.peakoil.nl/wp-content/uploads/2008/12/2008_december_oilwatch_monthly.pdf>
27. Matthew R. Simmons, 'Kicking the oil habit', *New Scientist* (25 June 2008)
28. Tony 'Ace' Eriksen, 'World Oil Production Forecast – Update May 2009', *The Oil Drum* (19 May 2009) <http://www.theoildrum.com/node/5395>
29. Paul Roberts, 'Over a Barrel', *Mother Jones* (November/December 2004) <http://www.motherjones.com/news/feature/2004/11/10_401.html>
30. Cited in John Vidal, 'The end of oil is closer than you think', *The Guardian* (21 April 2005) <http://www.guardian.co.uk/life/feature/story/0,13026,1464050,00.html>
31. Ibid.
32. Ibid.
33. Cited in Alfred J. Cavello, 'Oil: the caveat', *Bulletin of the Atomic Scientists* (May–June 2005, Vol. 61, No. 3) p. 16–18 < http://www.thebulletin.org/article.php?art_ofn=mj05cavallo>
34. Cited in Michael C. Ruppert and Michael Kane, 'The Markets React to Peak Oil: Industrial Society Rides an Unstable Plateau Before the Cliff', *From The Wilderness* (4 October 2006) <http://www.fromthewilderness.com/members/100406_markets_react.shtml>
35. Werner Zittel and Jorg Schindler, *Crude Oil: The Supply Outlook* (Berlin: Energy Watch Group, October 2007, EWG-Series No. 3/2007) pp. 5–17 <http://www.energywatchgroup.de/fileadmin/global/pdf/EWG_Oilreport_10-2007.pdf>;Uppsala University, 'World Oil Production Close to Peak', *ScienceDaily* (1 April 2007) <http://www.sciencedaily.com/releases/2007/03/070330100802.htm>. See Fredrik Robelius, *The Highway to Oil: Giant Oil Fields and their Importance for Future Oil Production* (Uppsala: Department of Nuclear and Particle Physics, Uppsala University, 2007) <http://publications.uu.se/abstract.xsql?dbid=7625>
36. Lord Meghnad Desai, 'High Oil Price Bubble – Driven by Speculation?' *ATCA Briefing* (London: Asymmetric Threats Contingency Alliance, 6 June 2008) <http://www.mi2g.com/cgi/mi2g/press/060608.php>
37. Malcolm Wicks cited in Jon Hughes and Graham Ennis, 'From Our Own Correspondence', *Fourth World Review* (No. 17, October 2008)
38. Richard Heinberg, 'Peak coal: sooner than you think', *Energy Bulletin* (21 May 2007) <http://www.energybulletin.net/node/29919>
39. Werner Zittel and Jörg Schindler, *Coal: Resources and Future Production* (Berlin: Energy Watch Group, March 2007, EWG-Paper No. 1/07) pp. 4–8. Also see David Strahan, 'Lump sums: Oil production may soon peak, but what about coal?', *Guardian* (5 March 2008) <http://www.guardian.co.uk/environment/2008/mar/05/fossilfuels.energy>
40. Jean Laherrere, *Oil and Gas: What future?* Paper prepared for Groningen Annual Energy Convention (21 November 2006) <http://www.oilcrisis.com/laherrere/groningen.pdf>. Also see review and summary of this paper in Luis de Sousa, 'Natural gas: how big the problem?', *The Oil Drum* (5 December 2006) <http://www.theoildrum.com/story/2006/11/27/61031/618>
41. Kieth Kohl, 'The Decline of Canadian Natural Gas', *Energy & Capital* (9 October 2007) <http://www.energyandcapital.com/articles/peak-natural-gas/529>
42. Gail Tverberg, 'US natural gas: the role of unconventional gas', *Energy Bulletin* (18 May 2008) <http://www.energybulletin.net/node/44389>
43. Green Car Congress, 'Study: Unconventional Natural Gas Resources Boost US Reserves to 118 Years Worth at Current Production Levels' (11 August 2008) <http://www.greencarcongress.com/2008/08/study-unconvent.html>
44. Pierre Noel, *Beyond Dependence: How to Deal with Russian Gas* (London: European Council on Foreign Relations, November 2008) <http://ecfr.3cdn.net/c2ab0bed62962b5479_ggm6banc4.pdf>

45. Philippa Runner, 'EU Still Waiting for Russian Gas', *Business Week* (12 January 2009) <http://www.businessweek.com/globalbiz/content/jan2009/gb20090112_679659.htm?campaign_id=rss_daily>. Agence France Presse, 'US, Ukraine risk irking Russia with strategic accord' (19 December 2008) <http://www.google.com/hostednews/afp/article/ALeqM5hJTaVBZdnCpMevQ5gF1M0rEnHhsg>
46. John Busby, 'Oil and Gas Net Exports', *Real Resources Review* (London: Sanders Research Associates, 2 July 2008)
47. Editorial comment, 'Unconventional reserves: think of the volumes, not the quality', *Financial Times* (21 Feruary 2007) <http://www.ft.com/cms/s/77a4de16-c1fa-11db-ae23-000b5df10621.html>
48. Cavello, 'Oil: the caveat'
49. Ibid.
50. Ibid.
51. Mary O'Driscoll, 'Tar sand companies try balancing oil gains, environmental pains', *Environment & Energy* (17 August 2005) <http://www.eenews.net/specialreports/tarsands/sr_tarsands2.htm>; Stephen Leahy, 'Burning Energy to Produce it', Inter Press Service (25 July 2006) <http://www.ipsnews.net:80/news.asp?idnews=34076>
52. Carl Mortished, 'Costs explode at Shell Canadian venture', *Times* (7 July 2006) <http://business.timesonline.co.uk/tol/business/industry_sectors/natural_resources/article684327.ece>
53. Bengt Söderbergh, Fredrik Robelius and Kjell Aleklett, 'A Crash Program Scenario for the Canadian Oil Sands Industry' (Uppsala University, 8 June 2006) <http://www.peakoil.net/uhdsg/20060608EPOSArticlePdf.pdf>
54. Robin McKie, 'Nuclear waste could power Britain', *Observer* (23 December 2007) <http://www.guardian.co.uk/science/2007/dec/23/scienceofclimatechange.climatechange>
55. Edwin Lyman, 'Factsheet - Nuclear Processing: Dangerous, Dirty and Expensive', (Washington DC: Union of Concerned Scientists, 17 June 2008) <http://www.ucsusa.org/nuclear_power/nuclear_power_risk/nuclear_proliferation_and_terrorism/nuclear-reprocessing.html>
56. Ibid.
57. Ibid. [emphasis added]
58. David A. Lochbaum, *Nuclear Waste Disposal Crisis* (Tulsa, Oklahoma: PennWell Publishing, 1996) pp. 69–72
59. Magnus Linklater, 'Who says nuclear power is clean?' *Times* (23 November 2005) <http://www.timesonline.co.uk/tol/comment/columnists/magnus_linklater/article593023.ece>; Jan Willem Storm van Leeuwen and Philip Smith, *Nuclear Power: The Energy Balance* (February 2008) <http://www.stormsmith.nl/> John Busby, 'Why nuclear power is not a sustainable source of low carbon energy', in *The Busby Report: A National Plan for UK Survival in the 21st Century* (2002) <http://after-oil.co.uk/nuclear/>. Criticisms of Leeuwen and Smith's work are unconvincing. For a review see Mark Diesendorf, *Greenhouse Solutions with Sustainable Energy* (Sydney: University of New South Wales Press, 2007) chapter twelve.
60. Linklater, 'Who says nuclear power is clean?'
61. Alan E. Walter, *America the Powerless: Facing Our Nuclear Energy Dilemma* (Cogito, 1995), p. 56; Nando.net, 'Japan puts reactor program on back burner' (1 October 1997) <http://dieoff.com/page155.htm>. For further discussion of the overwhelming technical, financial and other practical difficulties with much vaunted 'fast breeder' nuclear reactors, see especially Sheila Newman (ed.), *The Final Energy Crisis* (London: Pluto Press, 2008).
62. John Willem Storm van Leeuwan, 'Breeders' (London: Institute of Physics, April 2006) p. 10 <http://journals-of-physics.org/activity/groups/subject/Energy/Group_events/file_6891.doc>
63. Background Paper, 'Uranium Resources and Nuclear Energy' (Energy Watch Group, December 2006) pp. 1–5, <http://www.lbst.de/publications/studies__e/2006/EWG-paper_1-06_Uranium-Resources-Nuclear-Energy_03DEC2006.pdf>

64. IAEA Report, *Analysis of Uranium Supply to 2050* (Virginia: International Atomic Energy Agency, 2001) pp. 5, 32
65. Gina Teel, 'The Rush for Alberta's Uranium', *Calgary Herald* (27 March 2007) <http://www.canada.com/calgaryherald/news/calgarybusiness/story.html?id=e1da89f1-0f1e-4aa0-ac04-a34fb02cabb8>
66. John Vidal, 'Labour's plan to abandon renewable energy targets', *Guardian* (23 October 2007) <http://www.guardian.co.uk/environment/2007/oct/23/renewableenergy.energy>
67. Dennis Anderson in debate with Peter Hodgson, 'Do we need nuclear power?' *Physics World* (5 June 2001)
68. David Biello, 'Super-efficient, cost-effective solar cell breaks conversion records', *Scientific American* (8 December 2006)
69. Cristina L. Archer and Mark Z. Jacobsen, 'Evaluation of Global Wind Power', *Journal of Geophysical Research* (30 June 2005, Vol. 110) <http://www.stanford.edu/group/efmh/winds/2004jd005462.pdf>
70. BWEA, 'Marine Renewable Energy' (London: British Wind Energy Association, 2007) <http://www.bwea.com/marine/resource.html>
71. For a comprehensive overview see Hans-Holger Rogner, *World Energy Assessment 2001*, chapter five: Energy Resources, available at <http://www.undp.org/energy/activities/wea/drafts-frame.html>
72. Renewable Energy Campaign Germany, 'Background Paper: The Combined Power Plant' (Berlin: Renewable Energy Campaign, October 2007) <http://www.kombikraftwerk.de/fileadmin/downloads/Background_Information_Combined_power_plant.pdf>. Also see 'Technical summary of the Combined Power Plant' (Berlin: Renewable Energy Campaign, October 2007) <http://www.kombikraftwerk.de/fileadmin/downloads/Technik_Kombikraftwerk_EN.pdf>
73. Stefan Peter and Harry Lehmann, *Renewable Energy Outlook 2030: Global Renewable Energy Scenarios* (Berlin: Energy Watch Group, November 2008) <http://www.energywatchgroup.org/fileadmin/global/pdf/2008-11-07_EWG_REO_2030_E.pdf>
74. Cited in Heinberg, *The Party's Over*, p. 200
75. Richard M. Duncan, 'The Peak of World Oil Production and the Road to Olduvai Gorge', Pardee Keynote Symposia, Geological Society of America, Summit 2000, Reno, Nevada (Seattle, WA: Institute on Energy and Man, November 2000) <http://dieoff.org/page224.htm>. Duncan's major mistake is to state that post-carbon existence will of necessity be equivalent to that of the Stone Age – this, of course, is not correct, since there were myriad forms of civilization prior to the era of industrialization and electricity. Also see Duncan, 'The Olduvai Theory – Energy, Population and Industrial Civilization', *The Social Contract* (Winter 2005–2006).
76. Richard C. Duncan, 'The Olduvai Theory'
77. John Michael Greer, *The Long Descent: A Users Guide to the End of the Industrial Age* (Gabriola Island: New Society, 2008)

Chapter 3

1. Brian Halweil, 'Where have all the farmers gone?' *World Watch Magazine* (September/October 2000) <http://www.worldwatch.org/node/490>
2. A.V. Krebs, *Agribusiness Examiner*, (No. 57, 23 November 1999); Krebs, *Agribusiness Examiner*, (No. 67, 20 March 2000); 'Should corporations be farmers or should farmers be farmers', Turning Project (Washington DC: 1999); *RAFI and Agrow* (No. 335, 27 August 1999)
3. Steven Gorelick, 'Facing the Farm Crisis – poor economic health of farmers', *Ecologist* (June 2000)
4. Halweil, 'Where have all the farmers gone?'
5. Stephen Lendman, 'Agribusiness Giants Seek to Gain Worldwide Control over our Food Supply' (Montreal: Centre for Research on Globalization, 7 January 2008) <http://www.globalresearch.ca/index.php?context=va&aid=7735>

6. OECD Background Note, 'Agricultural policy and trade reform: potential effects at global, national and household levels' (Paris: OECD, 2006) <http://www.oecd.org/dataoecd/52/23/36896656.pdf>
7. UN Food and Agriculture Organization, *Key Statistics Of Food And Agriculture External Trade* <http://www.fao.org/es/ess/toptrade/trade.asp?lang=EN&dir=exp&country=100>
8. Halweil, 'Where have all the farmers gone?'
9. Dale Allen Pfeiffer, 'Eating Fossil Fuels', *From The Wilderness* (3 October 2003) <http://www.fromthewilderness.com/free/ww3/100303_eating_oil.html>; David Pimentel and Mario Giampietro, 'Executive Summary', *Food, Land, Population and the US Economy*, (Washington DC: Carrying Capacity Network, 21 November 1994) <http://www.dieoff.com/page40.htm>; David Pimentel and Marcia Pimental, *Land, Energy and Water: the constraints governing Ideal US Population Size* (New Jersey: Negative Population Growth Forum Paper, 1990. Focus, Spring 1991) <http://www.dieoff.com/page136.htm>; Mario Giampietro and David Pimentel, *The Tightening Conflict: Population, Energy Use, and the Ecology of Agriculture*, (1994) <http://www.dieoff.com/page69.htm>
10. Christopher B. Field and David B. Lobell, 'Global scale climate–crop yield relationships and the impacts of recent warming', *Environmental Research Letters* (Vol. 2, 16 March 2007) <http://www.iop.org/EJ/abstract/1748-9326/2/1/014002/>
11. Steve Connor, 'World's most important crops hit by global warming effects', *Independent* (19 March 2007) <http://news.independent.co.uk/environment/climate_change/article2371569.ece>
12. Jonathan Amos, 'Climate food crisis "to deepen"', BBC News (5 September 2005) <http://newsvote.bbc.co.uk/mpapps/pagetools/print/news.bbc.co.uk/1/hi/sci/tech/4217480.stm>
13. Michael McCarthy, 'The century of drought', *Independent* (4 October 2006)
14. Rob Taylor, 'Millions to Go Hungry by 2080: Report', Reuters (20 January 2007) <http://www.truthout.org/article/millions-go-hungry-2080>
15. Sandi Doughton, 'Food crisis is global warming's biggest threat, say UW, Stanford scientists', *Seattle Times* (8 January 2009) <http://seattletimes.nwsource.com/html/localnews/2008604722_webwarming09m.html>
16. SAGE Press Release, 'New Map Reveals True Extent of Human Footprint on Earth' (San Francisco: Center for Sustainability and the Global Environment, University of Wisconsin-Madison, 5 December 2005) <http://www.news.wisc.edu/releases/11907.html>; Kate Ravilous, 'Food crisis feared as fertile land runs out', *Guardian* (6 December 2005) <http://www.guardian.co.uk/food/Story/0,,1659112,00.html>
17. Ole Hendrickson, 'The coming global food crisis', *Watershed Ways* (Ontario: Ottawa River Institute, 5 March 2004) <http://www.ottawariverinstitute.ca/WatershedWays04/wwfood-crisis.htm>; Agence France Presse, 'China's rising grain prices could signal global food crisis' (19 November 2003) <http://www.terradaily.com/2003/031119092535.t70a5roc.html>
18. Lester R. Brown, 'World facing huge new challenge on food front: Business-as-usual not a viable option', *Plan B Update* 72 (Washington DC: Earth Policy Institute, 16 April 2008) <http://www.earth-policy.org/Updates/2008/Update72.htm>
19. Alok Jha, 'Biofuel farms make CO2 emissions worse', *Guardian* (8 February 2008) <http://www.guardian.co.uk/science/2008/feb/08/scienceofclimatechange.biofuels>
20. Dale Allen Pfeiffer, 'Eating Fossil Fuels', *From The Wilderness* (3 October 2003) <http://www.fromthewilderness.com/free/ww3/100303_eating_oil.html>; Henry H. Kindell and David Pimentel, 'Constraints on the Global Food Supply', *Ambio: Journal of the Human Environment* (Vol. 23, No. 3, May 1994)
21. Richard Heinberg, 'Threats of Peak to the Global Food Supply', *Museletter* (July 2005, No. 159) <http://www.richardheinberg.com/museletter/159>
22. David Gutierrez, 'Food riots have already begun as global grain prices skyrocket, supplies dwindle', *Natural News* (4 October 2008) <http://www.naturalnews.com/024372.html>
23. Correspondence with Jeff Vail, former US Department of the Interior energy infrastructure counterterrorism specialist.
24. Correspondence with Richard Levins, Professor of Population Sciences, Harvard University.

25. Frederic Mousseau, *Food Aid or Food Sovereignty? Ending World Hunger in Our Time* (Oakland, CA: Oakland Institute, October 2005) <http://www.oaklandinstitute.org/pdfs/fasr.pdf>
26. M. Jahi Chappell, 'Shattering Myths: Can sustainable agriculture feed the world?' (Oakland, CA: Institute for Food & Development Policy, 4 October 2007) <http://www.foodfirst.org/node/1778>
27. William Easterling, 'Feeding the World', *Research/PennState* (Pennsylvania: Pennsylvania State University, May 2002, Vol. 23, No. 2) <http://www.rps.psu.edu/0205/feeding.html>. Finally, it is worth noting that drastic attempts to engineer population reduction would be impossible without considerable violence involving acts of genocide against specific groups, and it would take decades to accomplish the massive million-, if not billionfold, scale reductions that overpopulation advocates claim is necessary for sustainability.

Chapter 4

1. See Nafeez Ahmed, 'US Army Contemplates Redrawing Middle East Map to Stave-Off Looming Global Meltdown', *OpEd News* (31 August 2006) <http://www.opednews.com/articles/opedne_nafeez_m_060831_us_army_contemplates.htm>
2. Nafeez Ahmed, 'The Hidden Holocaust: Our Civilizational Crisis', Lecture at Imperial College London (21 November 2007). Material on which this lecture was based was published on my blog as the third of a series of articles, 'hidden holocaust—civilizational crisis, Part 3: The End of the World as we Know it?' *The Cutting Edge* (1 January 2008) <http://nafeez.blogspot.com/2008/01/hidden-holocaust-civilizational-crisis.html>
3. UCS Report, 'Warning to Humanity', (Cambridge, MA: Union of Concerned Scientists, 1992)
4. UNEP Press Release, 'Our present course is unsustainable – postponing action is no longer an option', (Nairobi: United Nations Environment Programme, 15 September 1999)
5. Wolfgang Sachs, Reinhard Loske and Manfield Linz et al., *Greening the North: A Post-Industrial Blueprint for Ecology and Equity. A study by the Wuppertal Institute for Climate, Environment and Energy* (London: Zed, 1998)
6. Sarah Anderson and John Cavanagh, *The Rise of Global Corporate Power* (Washington DC: Institute for Policy Studies, 1996)
7. David C. Korten, 'The Post-Corporate', *Yes! A Journal of Positive Futures* (Spring 1999)
8. Lester R. Brown, 'The Future of Growth', in Worldwatch Institute Annual Report, *State of the World 1998* (Washington DC: Worldwatch Insitute, January 1998) p. 4. The cancer 'analogy' can be taken literally – the best and most comprehensive treatment is John McMurtry, *The Cancer Stage of Capitalism* (London: Pluto, 1999)
9. World Bank Policy Research Report, *Globalization, Growth, and Poverty: Building an Inclusive World Economy* (New York: Oxford University Press, 2002) p. 2
10. Ibid. How seriously the statistics cited in this report can be taken is a matter of considerable uncertainty, since the Bank, peculiarly, absolves itself of responsibility for the accuracy of the data published therein.
11. Michel Chossudovsky, 'Global Poverty in the Late 20th Century', *Journal of International Affairs* (Vol. 52, No. 1, Fall 1998) pp. 145–163 <http://www.mtholyoke.edu/acad/intrel/chossu.htm>
12. Shaohua Chen and Martin Ravillion, 'The Developing World is Poorer Than We Thought, But No Less Successful in the Fight Against Poverty', Policy Research Working Paper 4703 (Washington DC: World Bank Development Research Group, August 2008) http://www-<wds.worldbank.org/external/default/WDSContentServer/IW3P/IB/2008/08/26/000158349_20080826113239/Rendered/PDF/WPS4703.pdf>
13. Sanjay G. Reddy, 'The World Bank's New Poverty Estimates: Digging Deeper into a Hole', (New York: Columbia University, 2008) <http://www.columbia.edu/~sr793/response.pdf>; Sanjay G. Reddy and Thomas Pogge, 'How Not to Count the Poor', in Joseph Stiglitz et al. (eds), *Debates on the Measurement of Global Poverty* (Oxford: Oxford University Press, 2010)

14. Mark Weisbrot, Dean Baker, Egor Kraev, and Judy Chen, 'The Scorecard on Globalization: Twenty Years of Diminished Progress' (Washington DC: Center for Economic and Policy Research, 11 July 2001) <http://www.cepr.net/documents/publications/globalization_2001_07_11.pdf>
15. Ibid.
16. SAPRI Report, *Structural Adjustment: The Policy Roots of Economic Crisis, Poverty and Inequality. A Report on a Joint Participatory Investigation by Civil Society and the World Bank on the Impact of Structural Adjustment Policies* (London: Zed, 2004) p. 203–217.
17. Michael Hudson, *Super Imperialism: The Origin and Fundamentals of US World Dominance* (London: Pluto, 2003) pp. 137, 144. Frederick F. Clairmont, *The Rise and Fall of Economic Liberalism* (Penang: Third World Network 1996)
18. Hudson, *Super Imperialism*, p. 139
19. Ibid., p. 140
20. Ibid., p. 143
21. Ibid., pp. 11–14
22. Joseph R. Stromberg, 'The Role of Monopoly Capitalism in the Age of Empire', *Journal of Libertarian Studies* (Vol. 14, No. 3, Summer 2001) p. 69
23. Hudson, *Super Imperialism*, pp. 14–15
24. Ibid., p. 17
25. Ibid., pp. 18–19
26. Ibid., p. 35
27. Cheryl Payer, *The Debt Trap: The International Monetary Fund and the Third World* (New York: Monthly Review, 1974) pp. 217–218.
28. Benjamin J. Cohen, 'Bretton Woods System', in R. J. Barry Jones (ed.), *Routledge Encyclopedia of International Political Economy* (London: Routledge, 2001); Michael P. Dooley, 'The Revived Bretton Woods System: The Effects of Periphery Intervention and Reserve Management on Interest Rates & Exchange Rates in Center Countries', *NBER Working Paper No. w10332* (Cambridge, MA: National Bureau of Economic Research, August 2004)
29. Hans W. Singer, 'Bretton Woods and the UN system - relationship of the International Monetary Fund and the World Bank to the UN', *Ecumenical Review* (July 1995)
30. Greg Palast, 'IMF's Four Steps to Damnation: How crises, failures, and suffering finally drove a Presidential adviser to the wrong side of the barricades', *The Observer* (29 April 2001)
31. Martin H. Wolfson, 'Crony capitalism comes to America' (Madison, WI: Progressive Media Project, October 1998) <http://www.progressive.org/mpwolfson1098.htm> [emphasis added]
32. Robert Naiman and Neil Watkins, *A Survey of the Impacts of IMF Structural Adjustment in Africa: Growth, Social Spending, and Debt Relief* (Washington DC: Center for Economic and Policy Research, April 1999)
33. FAO, 'Poverty and Food Security in Africa' *Food for All* (Rome: World Food Summit, 13–17 November 1996) <http://www.fao.org/DOCREP/x0262e/x0262e21.htm>
34. John Serioux, 'Debt and Poverty in Africa', *Submission to the Sub-Committee on Human Rights and International Development of the Standing Committee of Foreign Affairs and International Trade, House of Commons* (Ottawa: North-South Institute, 1 December 1999) <http://www.nsi-ins.ca/ensi/news_views/oped09.html>
35. WHO Report, *World Health Report 2003: Shaping the Future* (Geneva: World Health Organization, 2004) available online at <http://www.who.int/whr/2003/overview/en> (viewed 17 August 2004)
36. WHO/UNICEF Report, *Africa Malaria Report 2003* (Geneva: World Health Organization/ United Nations International Children's Emergency Fund, 2003)
37. Johan Galtung, 'Violence, Peace and Peace Research', *Journal of Peace Research* (1969, Vol. 6, No. 3) pp. 167–191
38. Mark Weisbrot, Dean Baker, Egor Kraev and Judy Chen, *The Scorecard on Globalization 1980-2000: 20 Years of Diminished Progress* (Washington DC: Center for Economic

Policy and Research, July 2001) <http://www.cepr.net/documents/publications/globalization_2001_07_11.pdf>

39. Mark Weisbrot, Dean Baker and David Rosnick, *The Scorecard on Development: 25 Years of Diminished Progress* (New York: United Nations Department of Economic and Social Affairs, September 2006) p. 17 <http://www.un.org/esa/desa/papers/2006/wp31_2006.pdf>
40. FAO Report, *The State of Food and Agriculture 1990* (Rome: Food and Agricultural Organization, 1991) p. 30
41. Cited in David Edwards, *Free To Be Human: Intellectual Self-Defence in an Age of Illusions* (Devon: Green Books, 1995) p. 140
42. United Nations Development Programme, *Human Development Report* (New York: Oxford University Press, 1999) p. 3.
43. Ibid., p. 3
44. United Nations Development Programme, *Human Development Report* (New York: Oxford University Press, 2003) pp. 2–3
45. Ignacio Ramonet, 'The Politics of Hunger', *Le Monde diplomatique* (November 1998) <http://mondediplo.com/1998/11/01leader>; United Nations Development Programme (2003), p. 9; Ramonet, 'The Year 2000', *Le Monde diplomatique* (December 1999) <http://mondediplo.com/1999/12/01leader>; UNMP, 'Fast Facts: The Faces of Poverty' (2006) <http://www.unmillenniumproject.org/documents/UNMP-FastFacts-E.pdf>
46. PRB Report, 2005 World Population Data Sheet (Washington DC: Population Reference Bureau, 2005) <http://www.prb.org/pdf05/05WorldDataSheet_Eng.pdf>; for summary see PRB Press Release, 'More than half the world lives on less than $2 a day' <http://www.prb.org/Journalists/PressReleases/2005/MoreThanHalftheWorldLivesonLessThan2aDayAugust2005.aspx>
47. UNDP (1999) p. 2
48. Ibid., pp. 2–3
49. Ibid., p. 3. Also see UNDP data cited in Ramonet, 'The Politics of Hunger'.
50. World Bank data, 'Total debts of the developing world in 2006: $2.7 trillion' <http://web.worldbank.org/WBSITE/EXTERNAL/DATASTATISTICS/0,,contentMDK:20394689~menuPK:1192714~pagePK:64133150~piPK:64133175~theSitePK:239419,00.html>; 'Total official development assistance in 2006: $106 billion' <http://web.worldbank.org/WBSITE/EXTERNAL/DATASTATISTICS/0,,contentMDK:20394658~menuPK:1192714~pagePK:64133150~piPK:64133175~theSitePK:239419,00.html>
51. *Economist* (10 July 1993). See William Easterly, *The White Man's Burden: Why the West's Efforts to Aid the Rest Have Done So Much Ill and So Little Good* (New York: Penguin, 2006)
52. Hudson, *Super Imperialism*, pp. 28–35 [emphasis added]
53. Faisal Islam and Will Hutton, 'Global crash fears as German bank sinks', *Observer* (6 October 2002)
54. Kenneth Rogoff, 'Can the IMF Avert a Global Meltdown?' *Japan Times* (10 September 2006)
55. Garry J. Schinasi, *Safeguarding Financial Stability: Theory and Practice* (New York: IMF, 2006). Also see Kern Alexander, Rahul Dhumale and John Eatwell, *Global Governance of Financial Systems: The International Regulation of Systemic Risk* (Oxford: Oxford University Press, 2005). Cited in Gabriel Kolko, 'Weapons of mass financial destruction: An economy of buccaneers and fantasists', *Le Monde diplomatique* [English edition] (October 2006)
56. Kolko, 'Weapons'
57. Associated Press, 'Experts warn that heavy debt threatens American economy', *USA Today* (27 August 2005) <http://www.usatoday.com/money/economy/2005-08-27-growing-debt_x.htm>; BBC News, 'US deficit "hits record $1.4tn"' (8 October 2009) <http://news.bbc.co.uk/1/hi/8296079.stm>
58. AP, 'Experts warn'.

59. Bruce Stannard, 'Dumping of US dollar could trigger "economic September 11"', *The Australian* (29 August 2005) <http://www.theaustralian.news.com.au/common/story_page/0,5744,16416680%255E28737,00.html>.
60. Ibid.
61. Douglas Casey, 'The Monetary Crisis', *Financial Web* (5 September 2006) <http://financialweb.org/publications.asp?Page=ViewPub&PubID=2&Action=ReadExcerpt&ArticleID=30> [emphasis added]
62. Anthony Faiola, 'The dollar's fall is felt overseas', *Washington Post* (29 October 2009) <http://www.washingtonpost.com/wp-dyn/content/article/2009/10/28/AR2009102802347.html>
63. Stephen Roach, *Global Economic Forum* (New York: Morgan Stanley, 16 June 2006 and 24 April 2006). Cited in Kolko, 'Weapons'.
64. *Financial Times*, 6 September 2006. Cited in Kolko, 'Weapons'.
65. Brad DeLong, 'The odds of economic meltdown', UC Berkeley News Center (7 August 2006)
66. David Martin, 'Assymetrical Collateral Damage: Basel II, the Mortgage House of Cards and the Coming Economic Crisis', (Virginia: Arlington Institute, 12 July 2006) <http://archive.arlingtoninstitute.org/library/ArlingtonInstituteAddressTranscript_eng.pdf>
67. Martin, 'Assymetrical Collateral Damage'
68. For a technical discussion of the link see Nourel Roubini and Brad Setser, 'The effects of the recent oil price shock on the US and the global economy', Working Paper, Stern School of Business (New York: New York University, August 2004) <http://pages.stern.nyu.edu/~nroubini/papers/OilShockRoubiniSetser.pdf>
69. Gail Tverberg, "Where we are headed: Peak oil and the financial crisis", *The Oil Drum* (25 March 2009) <http://www.theoildrum.com/node/5230>
70. Korten, 'The Post-Corporate'
71. Richard C. Cook, 'Grand Larceny on a Monumental Scale: Does the Bailout Bill Mark the End of America as we Know it?' (Montreal: Center for Research on Globalization, 2 October 2008) <http://www.globalresearch.ca/index.php?context=va&aid=10413>
72. Cook, 'Grand Larceny'
73. Michael Hudson, 'America's Own Kleptocracy: The largest transformation of America's financial system since the Great Depression' (Institute for the Study of Long-Term Economic Trends, 20 September 2008) <http://www.michael-hudson.com/articles/debt/080920BailoutKleptocracy.html>
74. Dan Amoss, 'What mortgage-backed securities mean for the US housing market', *Money Week* (5 September 2007) <http://www.moneyweek.com/personal-finance/what-mortgage-backed-securities-mean-for-the-us-housing-market.aspx>
75. Ellen H Brown, 'It's the Derivatives, Stupid! Why Fannie, Freddie, AIG, had to be Bailed Out' (Montreal: Centre for Research on Globalization, 18 September 2008) <http://www.globalresearch.ca/index.php?context=va&aid=10265>
76. Ibid.
77. Nassim Nicholas Taleb interviewed on 'Newsnight' (BBC, 10 October 2008)
78. Kolko, 'Weapons'
79. Dean Baker, 'Why Bail? The Banks have a gun pointed to their head and are threatening to pull the trigger' (Washington DC: Center for Economic and Policy Research, 29 September 2008) <http://www.cepr.net/index.php/op-eds-&-columns/op-eds-&-columns/why-bail-the-banks-have-a-gun-pointed-at-their-head-and-are-threatening-to-pull-the-trigger>
80. Rich Miller, 'US Heading for Slump, With or Without Bailout', Bloomberg (30 September 2008) <http://www.bloomberg.com/apps/news?pid=20601068&sid=ayyjXs25Y7oE&refer=home>; Ye Xie and Daniel Kruger, 'Yen Declines as Bailout Plan Spurs Demand for Higher Yields', Bloomberg (25 September 2008) <http://www.bloomberg.com/apps/news?pid=20601101&sid=aVt4pZqvHnlU&refer=japan>
81. Michel Chossudovksy, 'The Great Depression of the 21st Century: Collapse of the Real Economy' (Montreal: Center for Research on Globalization, 15 November 2008) <http://www.globalresearch.ca/index.php?context=va&aid=10977>

82. Rich Miller, 'Recession's grip forces US to flood world with more dollars' Bloomberg (24 November 2008) <http://www.bloomberg.com/apps/news?pid=20601109&sid=aCqvVS7Zk7ZQ&refer=home>
83. Nourel Roubini, 'The Deadly Dirty D-Words', *Global EconoMonitor* (21 November 2008) <http://www.rgemonitor.com/blog/roubini/254515/the_deadly_dirty_d-words__deflation_debt_deflation_and_defaults__and_how_central_banks_will_have_to_resort_to_crazy_policies_as_we_have_reached_such_bermuda_triangle_of_a_liquidity_trap>
84. Gonzala Vina, 'Brown takes the UK "back to the 70s" with debt plans', Bloomberg (25 November 2008) <http://www.bloomberg.com/apps/news?pid=20601087&sid=acjY8lNUQVp4&refer=home>
85. 'The Greatest Depression', *Trends Journal* (New York: Trends Research Institute, Winter 2008, Vol. XVII, No. 1)
86. Linda Moulton Howe, Interview with Gerald Celente, 'After Wall Street Bailout, Is Main Street Heading for Depression', *Earth Files* (17 October 2008) <http://www.earthfiles.com/news.php?ID=1485&category=Environment>; Celente, Interview on Fox News (10 November 2008), available online at <http://www.youtube.com/watch?v=46MEqEgdLTg>
87. Walter 'John' Williams, 'Hyperinflation Special Report' *Shadow Government Statistics* (2 December 2009) <http://www.shadowstats.com/article/hyperinflation-2010.pdf> [emphasis added]

Chapter 5

1. Cited in Lawrence H. Shoup and William Minter, *Imperial Brain Trust: The Council on Foreign Relations and US Foreign Policy* (New York: Monthly Review Press, 1977) pp. 163–4.
2. These documents are as follows: *Defense Planning Guidance for the 1994–1999 Fiscal Years* (Draft), Office of the Secretary of Defense, 1992; *Defense Planning Guidance for the 1994–1999 Fiscal Years* (Revised Draft), Office of the Secretary of Defense, 1992; *Defense Strategy for the 1990s*, Office of the Secretary of Defense, 1993; *Defense Planning Guidance for the 2004–2009 Fiscal Years*, Office of the Secretary of Defense, 2002.
3. Draft 'Defense Planning Guidance' document leaked to *The New York Times* (8 March 1992); *International Herald Tribune* (9 March 1992); *Washington Post* (22 March 1992); *The Times* (25 May 1992).
4. Daniel Yergin, *The Prize: The Epic Quest for Oil, Money and Power* (New York: Simon & Schuster, 1993)
5. Memorandum by the Acting Chief of the Petroleum Division, 1 June 1945, *Foreign Relations of the United States*, 1945, Vol. VIII, p. 54
6. Cited in Mark Curtis, *Web of Deceit: Britain's Real Role in the World* (London: Vintage, 2003)
7. NSC 5401. Cited in Mohammed Heikal, *Cutting the Lion's Tail; Suez through Egyption Eyes* (London: Andre Deutsch, 1986), p. 38
8. Cited in Curtis, *The Great Deception: Anglo-American Power and World Order* (London: Pluto, 1998)
9. John Zogby, 'America as seen through Arab eyes: Polling the Arab world after September 11th', Zogby International (March 2004)
10. Bruce Lawrence (ed.), *Messages to the World: The Statements of Osama bin Laden* (London: Verso, 2005)
11. Nafeez Mosaddeq Ahmed, *The War on Truth: 9/11, Disinformation and the Anatomy of Terrorism* (New York: Olive Branch, 2005); *The London Bombings: An Independent Inquiry* (London: Duckworth, 2006)
12. Thomas D. Kraemer, 'Addicted to Oil: Strategic Implications of American Oil Policy', *Carlisle Papers in Security Strategy*(Pennsylvania: US Army War College, Strategic Studies Institute, May 2006) <http://www.strategicstudiesinstitute.army.mil/pdffiles/PUB705.pdf>
13. Prologue to Richard Labévière, *Dollars for Terror: The United States and Islam* (New York: Algora Publishing, 2000)

14. Sibel Edmonds, State Secret Privilege Gallery <http://www.justacitizen.com/images/Gallery%20Draft2%20for%20Web.htm>
15. Muriel Kane, 'Whistleblower: Bin Laden was US proxy until 9/11', *Raw Story* (31 July 2009) <http://rawstory.com/08/news/2009/07/31/whistleblower-bin-laden-was-us-proxy-until-911>
16. Ibid.
17. Sibel Edmonds and Philip Giraldi, 'Who's Afraid of Sibel Edmonds', *American Conservative* (1 November 2009) <http://www.amconmag.com/pdfissue.html?Id=AmConservative-2009nov01&page=06>
18. William Arkin, 'The Secret War', *Los Angeles Times* (27 October 2002) <http://www.commondreams.org/views02/1028-11.htm> [emphasis added]
19. Seymour M. Hersh, 'The Coming Wars: What the Pentagon can now do in secret', *New Yorker* (24 January 2005) <http://www.newyorker.com/archive/2005/01/24/050124fa_fact>
20. US Department of the Army Field Manual FM 3-05.130, *Army Special Operations Forces: Unconventional Warfare* (Washington DC: Department of the Army, 30 September 2008) <http://www.wikileaks.org/leak/us-fm3-05-130.pdf> pp. 1–7 [emphasis added]. Also see Tom Burghardt, *Unconventional Warfare in the 21st Century: US Surrogates, Terrorists and Narcotraffickers* (London: Institute for Policy Research & Development, January 2009) <http://www.iprd.org.uk/images/stories/pdf/CISS/A%20Hidden%20History%20Covert%20Operations%20and%20Military%20Intelligence%20Policy%20Post-World%20War%20II/Unconventional%20Warfare%20in%20the%2021st%20Century.pdf>, and especially Michael McClintock, *Instruments of Statecraft: US Guerrilla Warfare, Counterinsurgency and Counterterrorism, 1940–1990* (New York: Pantheon, 1992) chapter nine. Available at <http://www.statecraft.org/chapter9.html>
21. Ola Tunander, 'Humanitarian Intervention', Confidential Report to the Norwegian Ministry of Foreign Affairs (2009), pp. 20–27, 39–43 [not in public domain – received through correspondence with Tunander]
22. US Department of the Army Field Manual FM 3-05.130, p. 2
23. Gerald Posner, *Why America Slept: The Failure to Prevent 9/11* (New York: Random House, 2003) pp. 105–6
24. Cited in Ahmed Rashid, *Taliban: Militant Islam, Oil and Fundamentalism in Central Asia* (New Haven, Conn: Yale University Press, 2000) p. 166
25. Agence France-Presse, 'US gave silent blessing to Taliban rise to power: analysts' (7 October 2001)
26. Statement of Congressman Dana Rohrabacher, 'US Policy Toward Afghanistan', Senate Foreign Relations Subcommittee on South Asia (14 April 1999)
27. Nafeez Mosaddeq Ahmed, 'Afghanistan, the Taliban and the United States: The Role of Human Rights in Western Foreign Policy', Institute for Afghan Studies (January 2001) <http://www.institute-for-afghan-studies.org/AFGHAN%20CONFLICT/TALIBAN/afghanistan%20taliban%20and%20us.htm>. For extensive references also see Ahmed, *The War on Truth: 9/11, Disinformation and the Anatomy of Terrorism* (New York: Olive Branch, 2005); *The War on Freedom: How and Why America was Attacked, September 11, 2001* (Joshua Tree, California: Progressive Press, 2002)
28. 'US Companies Eye Trans-Afghan Pipeline', *Forbes* (19 January 2005) <http://www.forbes.com/markets/feeds/ap/2005/01/18/ap1764703.html>
29. For a thorough and concise appraisal of the Trans-Afghan pipeline project see Adnan Vatansever, 'Prospects for Building the Trans-Afghan Pipeline and its Implications' (Washington: Advanced International Studies Unit, Pacific Northwest National Laboratory, 31 August 2003, PNNL-14550) <http://www.pnl.gov/aisu/pubs/tapvatan.pdf>
30. Colin Freeman, 'US "disappointed" at British failure to stem opium trade', *Telegraph* (17 July 2005) <http://www.telegraph.co.uk/news/main.jhtml?xml=/news/2005/07/17/wafg17.xml&sSheet=/news/2005/07/17/ixworld.html>
31. BBC News, 'Afghanistan retakes heroin crown' (3 March 2003) <http://news.bbc.co.uk/2/hi/business/2814861.stm>

32. Robert Fox, 'CIA is undermining British war effort, say military chiefs', *The Independent* (10 December 2006) <http://www.independent.co.uk/news/world/middle-east/cia-is-undermining-british-war-effort-say-military-chiefs-427848.html>
33. Thomas Schweich, 'Is Afghanistan a Narco-State?' *New York Times* (27 July 2008) <http://www.nytimes.com/2008/07/27/magazine/27AFGHAN-t.html?_r=1&hp&oref=slogin>; Tom Lasseter, 'West looked the other way as Afghan drug trade exploded', *McClatchy* (10 May 2009) <http://www.mcclatchydc.com/homepage/story/67722.html>; Ahmed Rashid, *Descent into Chaos – How the War against Islamic Extremism is being Lost in Pakistan, Afghanistan and Central Asia* (London: Allen Lane, 2008) pp. 320–325
34. Philip Smucker, *Al Qaeda's Great Escape: The Military and the Media on Terror's Trail* (Washington: Brassey's, 2004) p. 9
35. B. Raman, 'Assassination of Haji Abdul Qadeer in Kabul' (South Asia Analysis Group, 7 August 2002, Paper No. 489) <http://www.saag.org/papers5/paper489.html>
36. Peter Dale Scott and Jonathan Marshall, *Cocaine Politics: The CIA, Drugs and Armies in Central Asia* (Berkeley, CA: University of California Press, 1998) pp. x-xi
37. Paul Zabriskie, 'Afghanistan's drug czar: world's toughest job', *Fortune* (30 September 2009) <http://money.cnn.com/2009/09/29/news/international/afghanistan_taliban_drugs.fortune/index.htm>
38. Michel Chossudovsky, 'Heroin is "Good for your health": Occupation forces support Afghan narcotics trade' (Montreal: Centre for Research on Globalization, 29 April 2007) <http://www.globalresearch.ca/index.php?context=va&aid=5514>. Colum Lynch, 'Afghanistan drug trade hits $4 billion a year', *The Age* (28 June 2008) <http://www.theage.com.au/world/afghanistan-drug-trade-hits-4-billion-a-year-20080627-2y43.html?page=-1>
39. Loretta Napoleoni, *Terror Incorporated: Tracing the Dollars Behind the Terror Networks* (New York: Seven Stories Press, 2005) pp. 90–97
40. Sibel Edmonds, 'Hijacking of a Nation, Part II', *Dissident Voice* (1 December 2006) <http://www.dissidentvoice.org/Dec06/Edmonds01.htm>
41. Edmonds and Giraldi, 'Who's Afraid of Sibel Edmonds?'
42. Vladimir Radyuhin, 'Russia: victim of narco-aggression', *The Hindu* (4 February 20008) <http://www.hindu.com/2008/02/04/stories/2008020453271000.htm>
43. Hugh O'Shaughnessy, 'US waves white flag in disastrous "war on drugs"', *Independent* (17 January 2010) <http://www.independent.co.uk/news/world/americas/us-waves-white-flag-in-disastrous-war-on-drugs-1870218.html>
44. George Friedman, 'Tora Bora and Nuclear Nightmares', in *America's Secret War: Inside the Hidden Worldwide Struggle Between America and its Enemies* (New York: Doubleday, 2004) p. 223
45. Gareth Porter, 'US shrugs off Pakistan-Taliban links' *Asia Times* (6 August 2009) <http://www.atimes.com/atimes/South_Asia/KH06Df01.html>
46. Akiva Eldar, 'Perles of wisdom for the Feithful', *Ha'aretz*, (1 October 2002) <http://www.haaretz.com/hasen/pages/ShArt.jhtml?itemNo=214635>
47. Stratfor, 'Uniting Jordan and Iraq Might Be Prime Post-War Strategy', (26 September 2002); Stratfor Press Release, 'US plan to merge Iraq, Jordan after war', (26 September 2002), available at <http://www.globalresearch.ca/articles/KHA209A.html>; Gary D. Hallbert, 'US Considers Dividing Iraq into Three Separate States after Saddam is Gone', *Forecasts & Trends* (1 October 2002) <http://www.profutures.com/article.php/91/%20>
48. Thomas H. Henriken, 'The War: Divide et Impera', *Hoover Digest* (No. 1, 2006) <http://www.hoover.org/publications/digest/2904886.html>
49. Correspondents in Dubai, 'Iraq extremists threaten attacks', *The Australian* (24 November 2004) <http://www.theaustralian.news.com.au/common/story_page/0,5744,11488568%255E1702,00.html>
50. Marie Colvin, 'Al-Qaeda directs Iraqi hit squad', *Sunday Times* (10 August 2003). Excerpts available online at <http://watch.windsofchange.net/03_0804_0810.htm#directs>
51. See Nafeez Ahmed, *The War on Truth*
52. Syed Saleem Shahzad, 'US fights back against "rule by clerics"',*Asia Times* (15 February 2005) <http://www.atimes.com/atimes/Middle_East/GB15Ak02.html>

53. CNN, 'Hersh: Bush Funnelling Money to al-Qaeda Related Groups' (25 February 2007), transcript available at <http://www.truthout.org/docs_2006/022607A.shtml> [emphasis added]
54. Seymour M Hersh, 'The Redirection', *New Yorker* (5 March 2007) <http://www.newyorker.com/reporting/2007/03/05/070305fa_fact_hersh>
55. Robert Baer, *Sleeping with the Devil: How Washington Sold Our Soul for Saudi Crude* (London: Crown, 2003)
56. Alexander Cockburn, 'Exclusive: Secret Bush "Finding" Widens Covert War on Iran', *Counterpunch* (2 May 2008) <http://www.counterpunch.org/andrew05022008.html>. Also see Brian Ross, 'Bush Authorizes New Covert Action Against Iran', ABC News (22 May 2007) <http://blogs.abcnews.com/theblotter/2007/05/bush_authorizes.html>
57. Ron Synovitz, 'Reports suggest Obama faces early choice on Iran covert ops', Radio Free Europe/Radio Liberty (13 January 2009) <http://www.rferl.org/content/Reports_Suggest_Obama_Faces_Early_Choice_On_Iran_Covert_Ops/1369640.html>
58. Also see Peter Dale Scott, *The Road to 9/11: Wealth, Empire, and the Future of America* (Berkeley, CA: University of California Press, 2007)
59. Loretta Napoleoni, *Rogue Economics* (New York: Seven Stories Press, 2008); Eric Wilson (ed.) *Government of the Shadows: Parapolitics and Criminal Sovereignty* (London: Pluto Press, 2009)
60. See Nafeez Ahmed, 'Terrorism and Western Statecraft: Al-Qaeda and Western Covert Operations After the Cold War', in Paul Zarembka (ed.), *Research in Political Economy* (Elsevier, 2006, Vol. 23) pp. 149–188
61. See Burghardt, *Unconventional Warfare in the 21st Century*

Chapter 6

1. Ole Wæver, 'Securitization and Desecuritization', in Ronnie D. Lipschutz (ed.) *On Security* (New York: Columbia University Press, 1995); Tunander, 'Democratic State vs Deep State: Approaching the Dual State of the West', in Eric Wilson and Tim Lindsey (eds.), *Government of the Shadows: Parapolitics and Criminal Sovereignty* (London: Pluto Press, 2009) pp. 56–72
2. Paul Rogers, *Losing Control: Global Security in the Early Twenty-first Century* (London: Pluto, 2002)
3. Lawrence H. Shoup and William Minter, *Imperial Brain Trust: The Council on Foreign Relations and US Foreign Policy* (New York: Monthly Review Press, 1977) p. 164
4. P. J. Cain and A. G. Hopkins, *British Imperialism: Crisis and Deconstruction 1914–1990* (Longman, 1993), pp. 39–41, 308; Denis Judd, *Empire: The British Imperial Experience from 1765 to the Present* (Harper Collins, 1996), chapter 19
5. Alejandro Colas, *Empire* (Cambridge: Polity, 2007)
6. Mark Curtis, *The Ambiguities of Power: British Foreign Policy Since 1945* (London: Zed, 1995), p. 17
7. Colas, *Empire*, pp. 178, 183
8. *Marine Corps Gazette* (May 1990). Cited in Noam Chomsky, *Deterring Democracy* (London: Vintage, 1992)
9. Colas, pp. 176–7
10. Ralph Peters, 'Constant Conflict', *Parameters* (Carlisle: US Army War College, Summer 1997) pp. 4–14 <http://www.carlisle.army.mil/USAWC/parameters/97summer/peters.htm>
11. Edward L. Morse, *Strategic Energy Policy Challenges for the 21st Century* (Washington DC: Council on Foreign Relations and James A. Baker III Institute for Public Policy, April 2001) <http://www.cfr.org/publication/3942/strategic_energy_policy_challenges_for_the_21st_century.html>; William Clark, *Petrodollar Warfare: Oil, Iraq and the Future of the Dollar* (Gabriola Island: New Society, 2005)
12. Kees van der Pijl, *Transnational Classes and International Relations* (London: Routledge, 1998)

13. Michael T. Klare, *Resource Wars: The New Landscape of Global Conflict* (New York: Owl Books, 2002); *Blood and Oil: The Dangers and Consequences of America's Growing Dependency on Imported Petroleum* (New York: Owl Books, 2005). Stephen Blank, *US Military Engagement with Transcaucasia and Central Asia.* Strategic Studies Institute (Carlisle: US Army War College, June 2000) <http://www.strategicstudiesinstitute.army.mil/pdffiles/PUB113.pdf>; Stephen Blank, *US Interests in Central Asia and the Challenge to Them,* Strategic Studies Institute (Carlisle: US Army War College, March 2007) <http://www.strategicstudiesinstitute.army.mil/pdffiles/PUB758.pdf>
14. Department of Trade and Industry, *Our Energy future: Creating a Low Carbon Economy* (London: February 2003). Also see Curtis, *Web of Deceit*, p. 68–69
15. Rob Evans and David Hencke, 'UK and US in joint effort to secure African oil', *Guardian* (14 November 2003)
16. Foreign and Commonwealth Office, *UK international priorities: A strategy for the FCO*, Cm 6052 (December 2003)
17. Ministry of Defence, *Delivering security in a changing world* (December 2003). Also see Curtis, *Web of Deceit*, p. 76–77
18. Hibernia is a petroleum field in the North Atlantic Ocean, approximately 315 kilometres east-southeast of St. John's, Newfoundland, Canada. It contains about three billion barrels of oil in place and recoverable reserves are estimated to be at around 615 million barrels.
19. Cited in Kjell Aleklett, 'Dick Cheney, Peak Oil and the Final Countdown', Association for the Study of Peak Oil (London: 12 May 2004) p. 1 <http://www.peakoil.net/Publications/Cheney_PeakOil_FCD.pdf> [emphasis added]. See full text of Cheney's speech at *Energy Bulletin* (8 June 2004) <http://www.energybulletin.net/node/559>
20. Report of an Independent Task Force, *Strategic Energy Policy Challenges For The 21st Century* (Houston: James A. Baker III Institute for Public Policy/Council on Foreign Relations, April 2001) <http://www.rice.edu/projects/baker/Pubs/workingpapers/cfrbipp_energy/energytf.htm>
21. Seumas Milne, 'Britain leaves Iraq in shame. The US won't go so quietly', *Guardian* (11 December 2008); Jeremy R. Hammond, 'US Would Control Profits from Iraqi Oil Exports Under Agreement', *Foreign Policy Journal* (23 November 2008) <http://www.foreignpolicyjournal.com/articles/2008/11/23/hammond_status_of_forces.htm>
22. Hammond, 'US Would Control Profits'.
23. Nick Turse, 'Pentagon Hands Iraq Oil Deal to Shell', AlterNet (2 October 2008) <http://www.alternet.org/waroniraq/101012/pentagon_hands_iraq_oil_deal_to_shell>; Stephen C. Webster, 'In secret agreement, Shell nets 25-year monopoly on S. Iraq's gas', *Raw Story* (8 November 2008) <http://rawstory.com/news/2008/In_secret_agreement_Shell_nets_25year_1108.html>;
24. Patrick Martin, 'US firms lose out in bidding for Iraq oil fields', WSWS (14 December 2009) <http://www.wsws.org/articles/2009/dec2009/iraq-d14.shtml>
25. Pepe Escobar, 'Iraq's oil auction hits the jackpot', *Asia Times* (16 December 2009) <http://www.atimes.com/atimes/Middle_East/KL16Ak02.html>
26. Greg Palast, 'Baghdad Coup d'Etat for Peak Oil', *Harpers Magazine* (10 April 2005) <http://www.gregpalast.com/harpers-baghdad-coup-detat-for-big-oil>
27. Deutsche Welle, 'Iraq oil auctions cause concerns over stability in Gulf hierarchy' (23 December 2009) <http://www.dw-world.de/dw/article/0,,5033275,00.html>
28. Nafeez Ahmed, 'Ex-British Army Chief in Iraq Confirms Peak Oil Motive for War', *Digital Journal* (17 July 2008) <http://www.digitaljournal.com/article/256227>
29. Klare, *Resource Wars*, p. 72
30. General James Henry Binford Peay, Address at Asia Society, New York, printed in *US Department of State Dispatch* (July 1998) p. 8. Cited in ibid.
31. US Government Energy Information Administration, 'Iran Country Analysis Brief', (March 2005) <http://www.eia.doe.gov/emeu/cabs/iran.html>
32. Matthew Simmons, *Twilight in the Desert: The Coming Saudi Oil Shock and the World Economy* (London: Wiley & Sons, 2005). See also for instance Jim Landers, 'Skeptics doubt Saudi Arabia can boost oil supply', *Dallas Morning News* (24 June 2008) <http://www.

dallasnews.com/sharedcontent/dws/bus/columnists/jlanders/stories/DN-Landers_24bus.ART.State.Edition1.4d8234d.html>

33. Michael Klare, 'Oil, Geopolitics and the Coming War with Iran', Globalpolicy.org (11 April 2005) <http://www.globalpolicy.org/empire/intervention/iran/economy/2005/0411bloodoiliran.htm>
34. 'China, Iran sign biggest oil and gas deal', *China Business Weekly* (8 May 2004) <http://www2.chinadaily.com.cn/english/doc/2004-10/31/content_387140.htm> Cited in ibid.
35. Fars News Agency, 'Iran Sees Gas Deal with India, Pakistan by mid-year' (25 June 2008) <http://english.farsnews.com/newstext.php?nn=8703090479>
36. Praful Bidwai, 'India should reciprocate Zardari's overture', *Rediff India Abroad* (10 October 2008) <http://www.rediff.com/news/2008/oct/10guest2.htm>
37. 'India, Pak made "good progress" on Iran pipeline', *Outlook* (12 July 2008) <http://www.outlookindia.com/pti_news.asp?id=310098>; Swaminathan S Anklesaria Aiyar, 'IPI gas pipeline, RIP', *Times of India* (28 December 2008) <http://timesofindia.indiatimes.com/Opinion/Columnists/IPI_gas_pipeline_RIP/articleshow/3902150.cms>
38. Klare, 'Oil, Geopolitics and the Coming War with Iran'
39. BBC News, 'Japan signs huge Iranian oil deal' (19 February 2004) <http://news.bbc.co.uk/1/hi/world/asia-pacific/3499155.stm>
40. Reuters, 'Swiss foreign minister to sign Iran gas deal' (16 March 2008) <http://uk.reuters.com/article/oilRpt/idUKL1613888820080316>
41. Daniel Domby et al., 'Iran-Europe gas deals anger Washington' *Financial Times* (30 April 2008) <http://www.ft.com/cms/s/0/a473f7de-16d5-11dd-bbfc-0000779fd2ac.html>
42. 'US to create bases in Kazakhstan, Uzbekistan – Russian General', *Uzbekistan Daily* (17 December 2008) <http://www.uzdaily.com/articles-id-4411.htm>; Agence France Presse, 'US, Ukraine risk irking Russia with strategic accord' (19 December 2008); Agence France-Presse, 'Georgia, US to sign strategic accord Jan 4: Tbilisi' (25 December 2009)
43. Ralph Peters, 'Blood borders: How a better Middle East would look', *Armed Forces Journal* (June 2006) <http://www.armedforcesjournal.com/2006/06/1833899> [emphasis added]
44. Ibid.
45. Ibid.
46. Ibid.
47. Ibid.
48. Ibid. [emphasis added]
49. Oded Yinon, 'A Strategy for Israel in the Nineteen Eighties', *Kivunim* (Jerusalem: World Zionist Organization, No. 14, Winter, February 1982) <http://student.cs.ucc.ie/cs1064/jabowen/IPSC/articles/article0005345.html>
50. David Wurmser, Douglas Feith, Richard Perle, et al., 'A Clean Break: A New Strategy for Securing the Realm' (Jerusalem: Institute for Advanced Strategic and Political Studies, 1996) <http://www.iasps.org/strat1.htm>
51. CIA sources cited in Richard Sale and Nicholas M. Morris, 'War talk sweeps city', United Press International (11 February 2003). Reprinted in *Washington Times* (12 February 2003) <http://www.washtimes.com/upi-breaking/20030211-065953-3776r.htm>
52. Kathleen and Bill Christison, 'The Bush Administration's Dual Loyalties: A Rose By Any Other Name' *Counterpunch* (13 December 2003) <http://www.counterpunch.org/christison1213.html>
53. PNAC Report, 'Rebuilding America's Defenses: Strategy, Forces and Resources for a New Century' (Washington DC: Project for the New American Century, September 2000) <http://www.newamericancentury.org/RebuildingAmericasDefenses.pdf>
54. Oxfam Press Release, 'Gaza siege puts public health at risk as water and sanitation services deteriorate, warns Oxfam' (London: Oxfam International, 21 November 2008) <http://www.oxfam.org/node/228>
55. PCHR Report, 'Israeli forces kill Gaza civilians in botched execution attempt' (Palestine Center for Human Rights, 16 January 2008)

56. Richard Falk, 'Slouching toward a Palestinian holocaust', Transnational Foundation for Peace and Future Research (29 June 2007) <http://www.transnational.org/Area_MiddleEast/2007/Falk_PalestineGenocide.html>
57. Barak Ravid, 'Disinformation, secrecy and lies: How the Gaza offensive came about', *Haaretz* (28 December 2007) <http://www.haaretz.com/hasen/spages/1050426.html>
58. 'Israeli minister warns of Palestinian "holocaust"', *Guardian* (29 February 2008) <http://www.guardian.co.uk/world/2008/feb/29/israelandthepalestinians1>. The term used was 'shoah', 'the Hebrew word normally reserved to refer to the Jewish Holocaust. It is rarely used in Israel outside discussions of the Nazi extermination of Jews during the second world war, and many Israelis are loath to countenance its use to describe other events.'
59. For sources see Nafeez Ahmed, 'Is the Gaza Catastrophe Really About Natural Resources?'*AlterNet* (8 January 2009) <http://www.alternet.org/environment/118039/is_the_gaza_catastrophe_really_about_natural_resources_/?page=4>
60. Gil Feiler and Don Peskin, 'Lebanon may claim gas deposit found off Israel's coast', *Ynet News* (21 January 2009) < http://www.ynet.co.il/english/articles/0,7340,L-3659319,00.html>
61. *Ghana Web* (23 February 2006); *Houston Chronicle* (25 December 2008); *Agence France-Presse* (30 November 2008); *Stars And Stripes* (9 March 2006); *Reuters* (27 November 2006). Cited in Rick Rozoff, 'Global Energy War: Washington's New Kissinger', *OpEd News* (22 January 2009) <http://www.opednews.com/articles/Global-Energy-War-Washing-by-Rick-Rozoff-090120-873.html>
62. United Press International (13 October 2005). Cited in Rozoff, 'Global Energy War'
63. Associated Press (24 April 2006); Associated Press (2 May 2006). Cited in Rozoff, 'Global Energy War'
64. Jeffrey J. Brown, 'A Simple Explanation for Oil Prices', *Financial Sense* (23 December 2008) <http://www.financialsense.com/fsu/editorials/brown/2008/1223.html>
65. US Department of the Army, *Army Modernization Strategy 2008* (Washington DC: Office of the Deputy Chief of Staff, July 2008) <http://downloads.army.mil/docs/08modplan/Army_Mod_Strat_2008.pdf> pp. 5–8. See summary by Tom Clonan, 'US generals planning for resource wars', *Irish Times* (22 September 2008) <http://www.irishtimes.com/newspaper/opinion/2008/0922/1221998220381.html>
66. MoD Report, *DCDC Global Strategic Trends Programme 2007–2036* (Swindon: Development, Concepts and Doctrine Centre, Ministry of Defence, December 2006, updated and revised March 2007) < http://www.mod.uk/NR/rdonlyres/94A1F45E-A830-49DB-B319-DF68C28D561D/0/strat_trends_17mar07.pdf> For summary see Richard Norton-Taylor, 'Revolution, flash mobs and brain chips. A grim vision of the future', *Guardian* (9 April 2007) <http://www.guardian.co.uk/science/2007/apr/09/frontpagenews.news>. On the issues surrounding Western conflict with Iran, see Ahmed, *The Iran Threat: An Assessment of the Middle East Nuclear Stalemate* (London: Institute for Policy Research & Development, September 2008) <http://www.iprd.org.uk/images/stories/pdf/the%20iran%20threat.pdf>
67. Chris Hedges, *War is a Force that Gives Us Meaning* (New York: Anchor, 2003)
68. Cited in Sami Zemni and Christopher Parker, 'European Union, Islam and the Challenges of Multiculturism', in Shireen T. Hunter (ed.), *Islam in Europe: The New Social, Cultural and Political Landscape* (Washington: Praeger, 2002) pp. 231–244
69. Leon T. Hadar, 'The "Green Peril": Creating the Islamic Fundamentalist Threat', *Cato Policy Analysis* (No. 17, August 1992) <http://www.cato.org/pubs/pas/pa-177.html>
70. Samuel P. Huntington, *The Clash of Civilizations and the Remaking of World Order* (New York: Simon & Schuster, 1998)
71. Paul Sperry, 'The Pentagon Breaks the Islam Taboo', *FrontPage Magazine* (14 December 2005) <http://www.frontpagemag.com/Articles/ReadArticle.asp?ID=20539>
72. Alan Travis, 'MI5 report challenges views on terrorism in Britain', *Guardian* (20 August 2008) <http://www.guardian.co.uk/uk/2008/aug/20/uksecurity.terrorism1>
73. Laurent Bonelli, 'On Suspicion of Being One of US', *Le Monde diplomatique* (April 2005)

74. Timothy M. Savage, 'Europe and Islam: Crescent Waxing, Cultures Clashing', *Washington Quarterly* (2004, Vol. 27, No. 3) pp. 28, 31–3
75. Joby Warrick, 'CIA Chief Sees Unrest Rising With Population', *Washington Post* (1 May 2008) <http://www.washingtonpost.com/wp-dyn/content/article/2008/04/30/AR2008043003258.html>
76. News Agencies, 'Sarkozy – Too many Muslims in Europe' (14–20 November 2007) <http://www.euro-islam.info/spip/article.php3?id_article=174>. Cites French and Swedish broadsheets *Libération* and *Aftonbladet*.
77. Daniel Pipes, 'Swiss Minarets and European Islam', *Jerusalem Post* (9 December 2009) <http://www.danielpipes.org/7808/swiss-minarets-european-islam>
78. Curtis, *Ambiguities*. Edward Herman and Noam Chomsky, *Manufacturing Consent: The Political Economy of the Mass Media* (New York: Pantheon, 1988)
79. Diana Ralph, 'Islamophobia and the "War on Terror": The Continuing Pretext for US Imperial Conquest', *Research in Political Economy* (Vol. 23, 2006) pp. 261–298
80. Jack Shaheen, *Reel Bad Arabs: How Hollywood Vilifies a People* (New York: Olive Branch Press, 2001)
81. Greater London Authority, *The search for common ground: Muslims, non-Muslims and the UK media* (London: Greater London Authority, November 2007)
82. Kerry Moore, Paul Mason and Justin Lewis, *Images of Islam in the UK: The Representation of British Muslims in the National Print News Media 2000–2008* (Cardiff: Cardiff University, July 2008) p. 3 <http://www.irr.org.uk/pdf/media_muslims.pdf>
83. Hussein Ibish, *Islamophobia: A Crisis of Hate Speech* (Washington DC: Muslim Public Affairs Council, 2006) <http://www.mpac.org/truthoverfear/special-report/index.php>
84. Sacha Evans et al., 'Media Coverage' in Nancy Tranchet and Dianna Rienstra (eds.), *Islam and the West: Annual Report on the State of Dialogue* (Geneva: World Economic Forum, January 2008) pp. 102–106
85. Ralph, 'Islamophobia'
86. Ibid.
87. Ibid.
88. Lee Harris, *The Suicide of Reason: Radical Islam's Threat to the West* (New York: Basic Books, 2007)
89. Williamson Murray, 'Professional Military Education and the 21st Century', in Murray (ed.), *Strategic Challenges for Counterinsurgency and the Global War on Terrorism*. Strategic Studies Institute (Carlisle: US Army War College, September 2006)
90. Ralph, 'Islamophobia'
91. ACLU Press Release, 'ACLU of Maryland Lawsuit Uncovers Maryland State Police Spying Against Peace and Anti-Death Penalty Groups' (New York: American Civil Liberties Union, 17 July 2008) <http://www.aclu.org/police/spying/36025prs20080717.html>
92. Lisa Rein, 'Marlyand Police Put Activists' Names On Terror Lists', *Washington Post* (8 October 2008) <http://www.washingtonpost.com/wp-dyn/content/article/2008/10/07/AR2008100703245.html>
93. CAMPACC, Submission to Joint Parliamentary Committee on Human Rights. 'Policing and Protest' (London: Campaign Against Criminalising Communities, June 2008) <http://www.campacc.org.uk/Library/policing_220608.doc>
94. Kitty Donaldson and Gonzala Vina, 'UK Used Anti-Terrorism Law to Seize Icelandic Bank Assets', Bloomberg (9 October 2008) <http://www.bloomberg.com/apps/news?pid=20601102&sid=aXjIA5NzyM5c&refer=uk>
95. OneWorld US, 'Worldwide, "War on Terror" Used as Excuse to Oppress, Charge Activists' (25 January 2007) <http://us.oneworld.net/node/145396> See also Beth Whitaker, 'Exporting the Patriot Act: US Pressure and Anti-Terror Laws in the Third World', Paper presented at the annual meeting of the International Studies Association, Hilton Hawaiian Village, Honolulu, Hawaii (5 March 2005) <http://www.allacademic.com/meta/p71143_index.html>

96. Lewis Seiler and Dan Hamburg, 'Rule by fear or rule by law?', *San Francisco Chronicle* (4 February 2008) <http://www.sfgate.com/cgi-bin/article.cgi?f=/c/a/2008/02/04/ED5OUPQJ7.DTL>
97. Michel Chossudovsky, 'Pre-election Militarization of the North American Homeland. US Combat Troops in Iraq repatriated to "help with civil unrest"', (Montreal: Centre for Research on Globalization, 26 September 2008) <http://www.globalresearch.ca/index.php?context=va&aid=10341>
98. Ambrose Evans-Pritchard, 'Citigroup says gold could rise above $2,000 next year as world unravels', *Telegraph* (27 November 2008)
99. Jim Meyers, 'US Military Preparing for Domestic Disturbances', *Newsmax* (23 December 2008) <http://www.newsmax.com/headlines/military_domestic_use/2008/12/23/164765.html>
100. Statewatch Briefing, 'Civil Contingencies Bill: Britain's Patriot Act - revised, and just as dangerous as before' (January 2004) <http://www.statewatch.org/news/2004/jan/12uk-civil-contingencies-bill-revised.htm>; Richard Tyler, 'Britain prepares its own version of US patriot act', WSWS (21 January 2004) <http://www.wsws.org/articles/2004/jan2004/patri-j21.shtml>
101. Graham Chick, 'Civil Contingencies Act: safeguarding Britain or simply hot air?', *BAPCO Journal: information management for civil contingencies responders* (London: British Association of Public Safety Communications Officers, 5 October 2007) <http://www.bapcojournal.com/news/fullstory.php/aid/886/Civil_Contingencies_Act:_safeguarding_Britain_or_simply_hot_air_.html>
102. EU Court of Justice Press Release, 'The European Community has the Power to Require the Member States to Lay Down Criminal Penalties for the Purpose of Protecting the Environment' (13 September 2005). Available online at <http://www.statewatch.org/news/2005/sep/ecj-environment-dec.pdf>
103. PI Press Release, 'What is Wrong with Europe?' (London: Privacy International, 14 December 2005) <http://www.privacyinternational.org/article.shtml?cmd%5B347%5D=x-347-494877> For regular monitoring of these processes at EU level see <http://www.statewatch.org>
104. Robert Kagan, 'Obama the Interventionist', *Washington Post* (29 April 2007) <http://www.washingtonpost.com/wp-dyn/content/article/2007/04/27/AR2007042702027.html>; Philip Giraldi, 'Obama's Neocon in Residence', Antiwar.com (4 November 20008) <http://www.antiwar.com/orig/giraldi.php?articleid=13712>; Doug Bandow, 'Hillary's nuclear umbrella', *National Interest* (15 December 2008) <http://www.nationalinterest.org/Article.aspx?id=20382>
105. Editorial, 'Mr. Obama's Economic Advisors', *New York Times* (24 November 2008) <http://www.nytimes.com/2008/11/25/opinion/25tue1.html>
106. Centre for Responsive Politics, 'Top contributors: Barack Obama' (17 November 2008) <http://www.opensecrets.org/pres08/contrib.php?id=N00009638&cycle2=2008&goButt2.x=7&goButt2.y=9>
107. Thomas Blanton and Peter Kornbluh (eds.), *Prisoner Abuse: Patterns from the Past*, National Security Archive Electronic Briefing Book No. 122 (Washington DC: George Washington University, 2004) < http://www.gwu.edu/~nsarchiv/NSAEBB/NSAEBB122/>
108. Spencer Ackerman, 'What to look for as the Obama detention/interrogation review process proceeds', *Washington Independent* (23 January 2009) <http://washingtonindependent.com/26990/what-to-look-for-as-the-obama-detentioninterrogation-review-process-proceeds>; Tom Eley, 'Obama's orders leave framework of torture, indefinite detention intact', WSWS (23 January 2009) <http://www.wsws.org/articles/2009/jan2009/guan-j23.shtml>
109. Jacques Semelin, *Purify and Destroy: The Political Uses of Massacre and Genocide* (London: C. Holt and Company, 2007) p. 14
110. Ibid., p. 49
111. Ibid., p. 91
112. Ibid., pp. 258–267

Chapter 7

1. British Petroleum, *Strategic Energy Review* (2007)
2. John McMurtry, *The Cancer Stage of Capitalism* (London: Pluto, 1999); Michael P. Byron, *Infinity's Rainbow: The Politics of Energy, Climate and Globalization* (2005)
3. Jim Yong Kim et al. (eds.), *Dying for Growth: Global Inequality and the Health of the Poor* (Monroe, Maine: Common Courage Press, 2000). SAPRI Report, *Structural Adjustment: The Policy Roots of Economic Crisis, Poverty and Inequality. A Report on a Joint Participatory Investigation by Civil Society and the World Bank on the Impact of Structural Adjustment Policies* (London: Zed, 2004)
4. Mark J. Lacy, *Security and Climate Change: International Relations and the Limits of Realism* (London: Routledge, 2005)
5. Carolyn Pumphrey (ed.), *Global Climate Change: National Security Implications* (Carlisle: US Army Strategic Studies Institute, May 2008) p. 8; EC Report S113/08, 'Climate Change and International Security', Paper from the High Representative and the European Commission to the European Council (14 March 2008) p. 2
6. Chukwumerije Okereke, *Global Justice and Neoliberal Environmental Governance: Ethics, sustainable development and international co-operation* (London: Routledge, 2008)
7. Joseph Tainter, *The Collapse of Complex Societies* (Cambridge: Cambridge University Press, 1988) p. 4
8. John Michael Greer, *The Long Descent: A User's Guide to the End of the Industrial Age* (Gabriola Island: New Society, 2008); Greer, 'How Civilizations Fall: A Theory of Catabolic Collapse' (2005) pp. 2–8 <http://www.xs4all.nl/~wtv/powerdown/greer.htm>
9. Thomas Homer-Dixon, *The Upside of Down: Catastrophe, Creativity and the Renewal of Societies* (London: Souvenir Press, 2007)
10. Ibid.
11. Michael Perelman, *The Invention of Capitalism: Classical Political Economy and the Secret History of Primitive Accumulation* (Durham: Duke University Press, 2000); Ellen Meiksins Wood, *Democracy Against Capitalism* (Cambridge: Cambridge University Press, 1995); Wood, *The Origin of Capitalism: A Longer View* (London: Verso, 2002)
12. Fred Halliday, 'The pertinence of imperialism' in Mark Rupert and Hazel Smith (eds.), *Historical Materialism and Globalization* (London: Routledge, 2001)
13. Hugo Radice, 'Globalization, State Failure and Underdevelopment: Imperialism?' (13 March 2001) <http://www.sussex.ac.uk/Units/CGPE/Failed%20States/radice.pdf>. Paper prepared for presentation at Centre for Global Political Economy conference, *The Global Constitution of Failed States: the Consequence of a New Imperialism* (Brighton: University of Sussex, 18–20 April 2001)
14. Richard Levins and Yrjo Haila, *Humanity and Nature: Ecology, Science and Society* (London: Pluto Press, 1992)
15. Eric Wolf, *Europe and the People Without History* (Berkeley: University of California Press, 1997) p. 72
16. Adam Tooze, *The Wages of Destruction: The Making and Breaking of the Nazi Economy* (London: Penguin, 2007)
17. Correspondence with Dr Richard Levins, Harvard.
18. Ellen Meiksins Wood, 'The Separation of the Economic and the Political in Capitalism', *New Left Review* (Vol. 127, 1981) pp. 77–79
19. Further specification and elaboration, including critical reviews of the relevant literature in social theory, can be found in my doctoral thesis, *The Violence of Empire* (Brighton: Department of International Relations, University of Sussex, 2009)
20. Kees van der Pijl, *Nomads, Empires, States: Modes of Foreign Relations and Political Economy, Vol. 1* (London: Pluto, 2007) p. 24
21. Benno Teschke, *The Myth of 1648* (London: Verso, 2005) p. 11
22. Karl Marx, *Grundrisse: Foundations of the Critique of Political Economy* (Harmondsworth: Penguin, 1973) pp. 407–10

23. Teschke, *Myth of 1648*. Also see Justin Rosenberg, *The Empire of Civil Society: A Critique of the Realist Theory of International Relations* (London: Verso, 1994)
24. Marx, *Capital, Vol. 1* (London: Penguin, 1976) pp. 240–253
25. Marx, *Theories of Surplus Value* (Progress Publishers, Moscow, 1966) Chapter II, 4d; Marx, *Capital, Vol. 1*, chapter 25, section 2
26. Judy Cox, 'Can capitalism be sustained?', *International Socialism* (Spring 2000, No. 86) <http://pubs.socialistreviewindex.org.uk/isj86/cox.htm>
27. Chris Harman, 'The rate of profit and the world today', *International Socialism* (July 2007, No. 115) <http://www.isj.org.uk/?id=340>
28. Ibid.; Gerard Duminel, 'Brenner on competition', *Capital & Class* (Summer 2001)
29. Immanuel Wallerstein, *Alternatives: The United States Confronts the World* (Boulder, CO: Paradigm Publishers, 2004); Wallerstein, *The Decline of American Power* (New York: New Press, 2003)
30. Robert Brenner, *Economics of Global Turbulence* (London: Verso, 1998). Also see Brenner, 'The Capitalist Economy, 1945–2000: A Reply to Konings and to Panitch and Gindin', in D. Coates (ed.), *Varieties of Capitalism, Varieties of Approaches* (Basingstoke: Palgrave Macmillan, 2005)
31. David Harvey, *The New Imperialism* (Oxford: Oxford University Press, 2003) p. 184
32. Leo Panitch and Sam Gindin, 'Global Capitalism and American Empire' (London: Merlin Press, 2004)
33. Marx, *Grundrisse* (1973), p. 407
34. Marx, 'The Future Results of British Rule', *New-York Daily Tribune* (8 August 1853), reprinted in the *New-York Semi-Weekly Tribune*, No. 856 (9 August 1853); Marx, 1867, cited in 'Progressive and Negative Perspectives of Capitalism and Imperialism', in Ronald Chilcote (ed.), *Imperialism: Theoretical Directions* (New York: Humanity Books, 2000) p. 138
35. See for example Kunibert Raffer, *Unequal Exchange and the Evolution of the World System* (London: Macmillan, 1987)
36. Alan Freeman, 'The modernity of backwardness', Paper prepared for International Confederation for Pluralism in Economics (July 2007) pp. 23–34 <http://mpra.ub.uni-muenchen.de/6831>
37. Teresa Brennan, 'Economy for the Earth: The labour theory of value without the subject/object orientation', *Ecological Economics* (February 1997, Vol 12, No 2) pp. 175–85
38. Ibid., p. 268
39. Ibid., p. 182
40. Ibid., p. 182
41. Ibid., p. 183
42. Ibid., pp. 184–5
43. Ibid., p. 184
44. Kevin O'Rourke and Jeffrey Williamson, *Globalisation and History: The Evolution of a Nineteenth-Century Atlantic Economy* (Cambridge, MA: MIT Press, 1999)
45. Eric Hobsbawm, *The Age of Empire, 1875–1914* (London: Abacus, 1987) p. 41
46. Ibid., p. 50
47. Eric R. Wolf, *Europe and the People Without History* (London: University of California Press, 1982), pp. 310–13
48. Ibid., pp. 310–314. See the whole of chapter eleven for Wolf's detailed examination of commodity production in different world regions in the service of capitalist accumulation, and some of the impact and implications of this process for indigenous social relations.
49. Hobsbawm, pp. 64–5
50. Ellen Brown, 'Dollar Deception: How Banks Secretly Create Money' (3 July 2007) <http://www.webofdebt.com/articles/dollar-deception.php>
51. Ibid.
52. John Gilbody, *The UK Monetary and Financial System: An Introduction* (London: Routledge, 1998) p. 41

53. Arleen and John Hoag, *Introductory Economics* (London: Prentice-Hall, 1990) pp. 379–80 [emphasis added]
54. US Federal Reserve, *Modern Money Mechanics: A Workbook on Bank Reserves and Deposit Expansion* (Chicago: Federal Reserve Bank of Chicago, 1992)
55. Anne Marie L. Gonczy, Timothy P. Schilling and Dorothy M. Nichols, *Two Faces of Debt* (Chicago: Federal Reserve Bank of Chicago, September 1992) p. 19 <http://www.fdrs.org/two_faces_of_debt.pdf>
56. Rodney Shakespeare, *The Modern Universal Paradigm* (Jakarta: University of Trisakti, 2007) p. 60
57. Ibid., p. 85
58. Comments by a British adviser to the Bank of International Settlements, directly involved in the official consultation process for establishment of the New Capital Accord, 'Instant Crisis Experts – Dilute to Taste', a discussion paper prepared for Globalvision2000 symposium on global banking crisis and emailed to this author (Dubai: Latticework Management Consultancy, 2008) pp. 3–4
59. Michael Hudson, 'America's Own Kleptocracy'
60. Ibid.
61. See Ted Nace, *Gangs of America: The Rise of Corporate Power and the Disabling of Democracy* (San Francisco: Berrett-Koehler, 2005), and George Monbiot, *Captive State: The Corporate Takeover of Britain* (London: Pan, 2001)
62. Frédérick Guillaume Dufour, 'Social-property Regimes and the Uneven and Combined Development of Nationalist Practices', *European Journal of International Relations* (2007, Vol. 13, No. 4) pp. 583–604
63. Aswini K. Ray, 'The international political system and the developing world: A view from the periphery', in Caroline Thomas and Paikiasothy Saravanamuttu (eds.), *Conflict and Consensus in South/North Security* (Cambridge: Cambridge University Press, 1989) p. 16
64. Vincent Ferraro, Ana Cristina R. Santos and Julie Ginocchio, 'The Global Trading System and International Politics', in Michael T. Klare and Yogesh Chandrani (eds.), *World Security: Challenges for a New Century* (New York: St. Martin's Press, 1998) pp. 310–311
65. Jorge Nef, *Human Security and Mutual Vulnerability: An Exploration into the Global Political Economy of Development and Underdevelopment* (Ottawa: International Development Research Centre, 1997) chapter three <http://www.idrc.ca/library/document/103249/chap3.html>
66. Ibid.
67. Robert Biel, *The New Imperialism: Crisis and Contradictions in North/South Relations* (London: Zed, 2000) pp. 24–50
68. See Stephen Gill's critique of 'disciplinary neo-liberalism' in his 'The Constitution of Global Capitalism'. Paper presented to 'Panel: The Capitalist World Past and Present' at the International Studies Association Annual Convention, Los Angeles (Brighton: First Press, 2000) <http://www.theglobalsite.ac.uk/ press/ 010gill.pdf>.
69. Nef, *Human Security and Mutual Vulnerability*
70. Ibid. See also Kees van der Pijl, *The Making of an Atlantic Ruling Class* (London: Verso, 1984)
71. Robert W. Cox, *Production, Power, and World Order: Social Forces in the Making of History* (New York: Columbia University Press, 1987) pp. 29, 7
72. Mark Beeson and Stephen Bell, 'Capitalism and the Fate of Nations: Structure, Agency and Institutions in the International Political Economy', in Nicola Phillips (ed.), *Globalising IPE* (London: Palgrave, 2003). Available online at Cybrary, University of Queensland, <http://eprint.uq.edu.au/archive/00000341/01/beesonbell.pdf> p. 18. (viewed 17 August 2004)
73. Ibid., p. 22
74. Cox, *Production, Power, and World Order*, p. 359
75. See Kees van der Pijl, *Global Rivalries: From the Cold War to Iraq* (London: Pluto Press, 2006)

76. Rick Tilman, *Thorstein Veblen and His Critics, 1891–1963* (Princeton: Princeton University Press, 1992) p. ix
77. Thorstein Veblen, 'Why is Economics Not an Evolutionary Science', *Quarterly Journal of Economics* (1898, Vol. 12)
78. Veblen, 'Fisher's Capital and Income', *Political Science Quarterly* (1908, Vol. 23)
79. Ibid.
80. Veblen, *The Place of Science in Modern Civilization* (New York: Cosimo, 1919) p. 251
81. Richard B. Norgaard, 'Beyond materialism: a coevolutionary reinterpretation of the environmental crisis', *Review of Social Economy* (Vol. 54, 1995)
82. Korten, *When Corporations Rule the World*, p. 69
83. For a discussion of relevant philosophical debates see Saul Smilansky, 'The Paradoxical Relationship between Morality and Moral Worth', *Metaphilosophy* (July 2005, Vol. 36, No. 4) pp. 490–500
84. John McMurtry, *Unequal Freedoms: The Global Market as an Ethical System*, p. 20
85. Ibid., p. 67
86. Oliver James, *Affluenza* (London: Vermilion, 2007)
87. Oliver James, 'Infected by affluenza', *Guardian* (24 January 2007) <http://www.guardian.co.uk/commentisfree/2007/jan/24/comment.politics>
88. Ibid.
89. Oliver James, *The Selfish Capitalist: Origins of Affluenza* (London: Vermilion, 2008) p. 17
90. Ibid., p. 39
91. Ibid., p. 43
92. Tim Kasser, *The High Price Of Materialism* (Massachusetts: MIT Press, 2002) p. 22

Chapter 8

1. See for example Rodney Shakespeare and Peter Challen, *Seven Steps to Justice* (London: New European Publications, 2002)
2. See for instance Michael Albert, *Parecon: Life After Capitalism* (London: Verso, 2003), and for a more academic treatment Michael Albert and Robin Hahnel, *The Political Economy of Participatory Economics* (Princeton University Press, 1991). I have serious points of disagreement with many of Albert and Hahnel's specific ideas about 'participatory planning' as they are in danger of reducing legitimate scope for individuals to engage in private enterprise according to their preferred skill-set and available resources, but theirs represents at least a fruitful starting point for debate, particularly in relation to other alternatives such as binary and Islamic economics.
3. Tim Jackson, *Prosperity Without Growth? – The transition to a sustainable economy* (London: Sustainable Development Commission, March 2009)
4. See for example Herman E. Daly, *Beyond Growth: The Economics of Sustainable Development* (Boston: Beacon, 1997). Daly is a former World Bank economist.
5. Andrew Simms, Ann Pettifor, Caroline Lucas et al., *A Green New Deal: Joined up policies to solve the triple crunch of the credit crisis, climate change and high oil prices* (London: New Economics Foundation, July 2008)
6. Mae-Wan Ho, Peter Bunyard, Peter Saunders et al., *Which Energy?* (London: Institute of Science in Society, 2006)
7. Mark Jacobson and Mark Delucchi, 'A path to sustainable energy by 2030', *Scientific American* (November 2009) pp. 58–65
8. Gail Tverberg, 'Scientific American's Path to Sustainability', *The Oil Drum* (9 November 2009) <http://www.theoildrum.com/node/5939>
9. Mae-Wan Ho, Brett Cherry, Sam Burcher and Peter Saunders, *Green Energies: 100% Renewables by 2050* (London: Institute of Science in Society, September 2009)
10. Gaia Vince, 'How to unplug from the grid', *New Scientist* (3 December 2008)
11. Alan Simpson MP, 'Food and Energy Security' (London: Institute of Science in Society, September 2005) < http://www.i-sis.org.uk/FAES.php>; Paul Brown, 'Woking shines in providing renewable energy', *Guardian* (26 January 2004) <http://www.guardian.co.uk/

environment/2004/jan/26/energy.renewableenergy>. For a more detailed assessment of the Woking case study, see Dave Elliot, 'Energy, Efficiency and Renewables', *Energy and Environment* (2004, Vol. 15, No. 6) pp. 1099–1105 <http://oro.open.ac.uk/3113/1/ElliottE&Ehandover.pdf>

12. Shakespeare and Challen, *Seven Steps to Justice*
13. Monbiot, *The Age of Consent: A Manifesto for a New World Order* (London: New Press, 2004)
14. See for instance Sam Ashram, 'The Columnist's Manifesto', *Socialist Review* (July 2003) <http://www.socialistreview.org.uk/article.php?articlenumber=8530>
15. For instance, Monbiot draws on John Maynard Keynes, who argued that poor countries that use up their overdraft in the International Clearing Union should be charged interest so as to reduce their currency value and prevent capital export; while rich countries that build up credit should be charged interest so as to increase the value of their currency and encourage capital export. Thus, rich countries' exports seem less attractive, while capital flight from poor to rich nations is blocked. But as Ashram points out, Monbiot's proposal is based on the mistaken belief that 'trade is the key relationship' between North and South, when actually unequal relations of production, skewed property ownership structures, domination of world productive resources by a minority in the North, and a financial system based on debt, interest and money creation out of nothing are key issues that his proposal has absolutely no bearing on whatsoever. His continued advocacy of an interest-based international lending system reveals a serious misunderstanding of the structure of the global political economy and compound interest as a structural cause of debt-proliferation. As Ashram concludes: 'Production and investment are not located as the source of the problem, when it is precisely capitalism's "accumulation for accumulation's sake" that creates massive concentrations of investment (and the systematic tendency to crisis), and which is further reinforced by imperialism both past and present. Trade is a symptom of this problem, not its source.' (Ibid.)
16. Fritjof Capra, *The Tao of Physics: An Exploration of the Parallels Between Modern Physics and Eastern Mysticism* (London: Flamingo, 1992); Bruce Lipton, *The Biology of Belief* (London: Hay House, 2008)
17. Ho et al., *Which Energy?*

Index